Physics

Phase 3

NOTION PRESS

NOTION PRESS

India. Singapore. Malaysia.

ISBN xxx-x-xxxxx-xx-x

Dedicated to our Beloved Prime Minister Shree Narendra Modi ji, whose life gives the Inspiration to every Individual

*If a Man **Decides** to achieve something in his life nothing is Impossible*

CHAPTERS

1. GRAVITATION

1. INTRODUCTION

Have you ever wondered whether we would still be studying about with Gravitation if a stone had fallen on Newton's head instead of an apple? Anyways, the real question is, why does an apple fall down rather than go upward?

2. NEWTON'S LAW OF UNIVERSAL GRAVITATION

"Every particle of matter in the universe attracts every other particle with a force equal to the product of masses of particles and inversely proportional to the square of the distance between them"

If m_1 and m_2 are two point masses separated by a distance r, the gravitational force of attraction F is given by

Figure 10.1

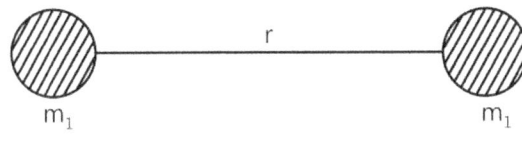

$$F \propto \frac{m_1 m_2}{r^2}$$

$$F = \frac{G m_1 m_2}{r^2}$$

Figure 10.2

Where G is a constant and is called the Universal gravitational constant.

Magnitude (and unit) of G   : 6.67×10^{-11} Newton. m^2 / kg^2

Dimension of G   : $M^{-1}L^3T^{-2}$

NOMORECLASS CONCEPTS

The direction of force F is independent of the medium, not affected by the presence of the other bodies and acts along the line joining the two particles.

If two persons come very close to each other such that the distance between them is almost 0, the two persons should experience a high force of attraction. Observe keenly the value of G. It's of order -11.

The Universal gravitational constant G is an experimental value calculated by Cavendish 71 years after the law was formulated.

Always remember Gravitational Force is conservative in nature i.e. work done doesn't depend on the path taken and depends only on the end points.

Illustration 1: Two particles of masses 1.0 kg and 2.0 kg are placed at a separation of 50 cm. Assuming that the only forces acting on the particles are their mutual gravitation, find the initial accelerations of the two particles.

(JEE MAIN)

Sol: The force of mutual gravitation acting on particles is $F = \dfrac{Gm_1m_2}{r^2}$. As the particle are accelerating under the force of gravitation, the acceleration is obtained using Newton's laws of motion.

The force of gravitation exerted by one particle on the other is

$$F = \frac{Gm_1m_2}{r^2} = \frac{6.67 \times 10^{-11} \dfrac{N-m^2}{kg^2} \times (1.0kg) \times (2.0kg)}{(0.5m)^2} = 5.3 \times 10^{-10} N.$$

The acceleration of 1.0 kg particle is $a_1 = \dfrac{F}{m_1} = \dfrac{5.3 \times 10^{-10} N}{1.0kg} = 5.3 \times 10^{-10} ms^{-2}$

This acceleration is towards the 2.0 kg particles. The acceleration of the 2.0 kg particle is

$$a_2 = \frac{F}{m_2} = \frac{5.3 \times 10^{-10} N}{2.0kg} = 2.65 \times 10^{-10} ms^{-2}$$

This acceleration is towards the 1.0 kg particle.

Illustration 2: Spheres of the same material and same radius r are touching each other. Show that gravitational force between them is directly proportional to r^4. **(JEE MAIN)**

Sol: The force of gravitation is directly proportional to the masses of the spheres. As the spheres are having the same masses, and mass $m \propto V \Rightarrow m \propto r^3$ thus the proportionality between the force and distance is easily established.

As the spheres are made of same material, and density so the mass of each sphere is $m_1 = m_2 =$ (volume) (destiny)

$$= \left(\frac{4}{3}\pi r^3\right)\rho$$

$$F = \frac{Gm_1m_2}{(2r)^2} = \frac{G\left(\frac{4}{3}\pi r^3\right)\left(\frac{4}{3}\pi r^3\right)\rho^2}{4r^2} \qquad \text{or} \qquad F \propto r^4$$

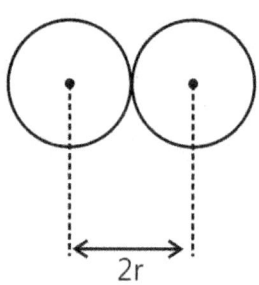

Figure 10.3

Illustration 3: Three particles each of mass m, are located at the vertices of an equilateral triangle of side a. At what speed will they move if they all revolve under the influence of their gravitational force of attraction in a circular orbit circumscribing the triangle while still preserving the equilateral triangle? **(JEE MAIN)**

Sol: The net force of gravitation on any one particle is due to other two particles. This gravitational force provides the necessary centripetal force to the particles to move in the circular orbit around the equilateral triangle.

$$\vec{F}_A = \vec{F}_{AB} + \vec{F}_{AC} = 2\left[\frac{GM^2}{a^2}\right]\cos 30° = \left[\frac{GM^2}{a^2}\sqrt{3}\right]$$

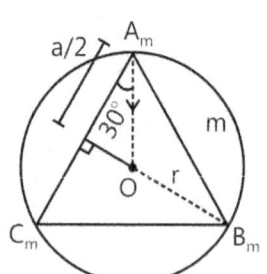

Figure 10.4

$$r = \frac{a}{\sqrt{3}}, \quad \text{Now} \quad \frac{mv^2}{r} = F; \text{ Or} \quad \frac{mv^2\sqrt{3}}{a} = \frac{GM^2}{a^2}\sqrt{3}; \quad \therefore v = \sqrt{\frac{GM}{a}}$$

3. GRAVITATIONAL FIELD

How would a particle interact with the surrounding or with other particles?

Every particle creates a field and when the other particle comes in to this particle's field, there would be an interaction between the particles.

The intensity of the field i.e. how intensely would it attract another particle in its field is called Gravitational field intensity or Gravitational field strength $\vec{E}$. It is defined as the force experienced by a unit mass placed at a distance

r due to mass M, i.e. $\vec{E} = \dfrac{\vec{F}}{M}$

NOMORECLASS CONCEPTS

Always remember, it is a vector quantity and should be added vectorially when calculating Gravitational field intensity at a point by one or more masses.

4. GRAVITATIONAL FIELD INTENSITY

(a) Due to a point mass M:

$$F = \dfrac{GMm}{r^2}; \qquad E = \dfrac{F}{m} = \dfrac{GM}{r^2}; \qquad E = \dfrac{GM}{r^2}$$

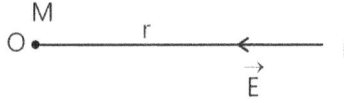

Figure 10.5

(b) Due to uniform ring of Mass M and radius a on its axis.

Consider any particle of mass dm on the ring, say at point A. The distance of this particle from P is

$AP = z = \sqrt{a^2 + r^2}$ .The gravitational field at P is dm is along $\overrightarrow{PA}$ and its magnitude is $dE = \dfrac{Gdm}{z^2}$

The component along PO is $\quad dE\cos\alpha = \dfrac{Gdm}{z^2}\cos\alpha$

The net gravitational field at P due to the ring is

$$E = \int \dfrac{Gdm}{z^2}\cos\alpha = \dfrac{G\cos\alpha}{z^2}\int dm = \dfrac{GM\cos\alpha}{z^2} = \dfrac{GMr}{(a^2 + r^2)^{3/2}}$$

The field is directed towards the center of the ring.

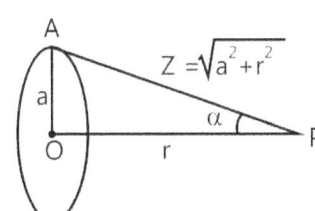

Figure 10.6

(c) Due to uniform disc of mass M and radius a on its axis.

Let us draw a circle of radius x with the center at O. We draw another concentric circle of radius x+dx. The part of the disc enclosed between these two circles can be treated as a uniform ring of radius x. The point P is on its axis at a distance r from the center. The area of this ring is $2\pi x dx$ .The area of the whole disc is πa^2 . As the disc is uniform, the mass of this ring is

$$dm = \dfrac{M}{\pi a^2}2\pi x dx = \dfrac{2Mx dx}{a^2}$$

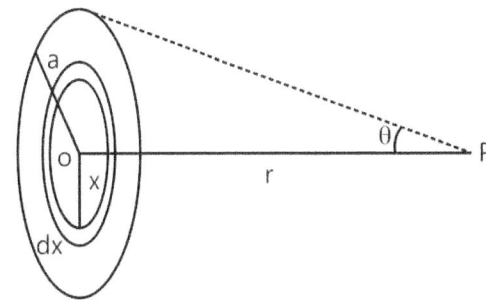

Figure 10.7

The gravitational field at P due to the ring is, by equation,

$$dE = \dfrac{G\left(\dfrac{2Mx dx}{a^2}\right)r}{\left(r^2 + x^2\right)^{3/2}} = \dfrac{2GMr}{a^2}\dfrac{x dx}{\left(r^2 + x^2\right)^{3/2}}$$

As x varies from 0 to a, the rings cover up the whole disc. The field due to each of these is in the same direction PO. Thus, the net field due to the whole disc is along PO and its magnitude is

$$E = \int_0^a \frac{2GMr}{a^2} \frac{x\,dx}{(r^2 + x^2)^{3/2}} = \frac{2GMr}{a^2} \int_0^a \frac{x\,dx}{(r^2 + x^2)^{3/2}} \qquad \dots(i)$$

Let $r^2 + x^2 = z^2$ then $2x\,dx = 2z\,dz$ and

$$\int \frac{x\,dx}{(r^2 + x^2)^{3/2}} = \int \frac{z\,dz}{z^3} = \int \frac{1}{z^2}\,dz = -\frac{1}{z} = -\frac{1}{\sqrt{r^2 - x^2}}$$

From (i) $E = \frac{2GMr}{a^2}\left[-\frac{1}{\sqrt{r^2 + x^2}}\right]_0^a = \frac{2GMr}{a^2}\left[\frac{1}{r} - \frac{1}{\sqrt{r^2 + a^2}}\right]$

Equation may be expressed in terms of the angle θ subtended by a radius of the disc at P as,

$$E = \frac{2GM}{a^2}(1 - \cos\theta).$$

(d) Due to uniform thin spherical shell of mass M and radius a from the triangle OAP,

$$z^2 = a^2 + r^2 - 2ar\cos\theta \qquad \text{or}$$

$$2z\,dz = 2ar\sin\theta\,d\theta$$

or $\quad \sin\theta\,d\theta = \frac{z\,dz}{ar}. \qquad \dots(ii)$

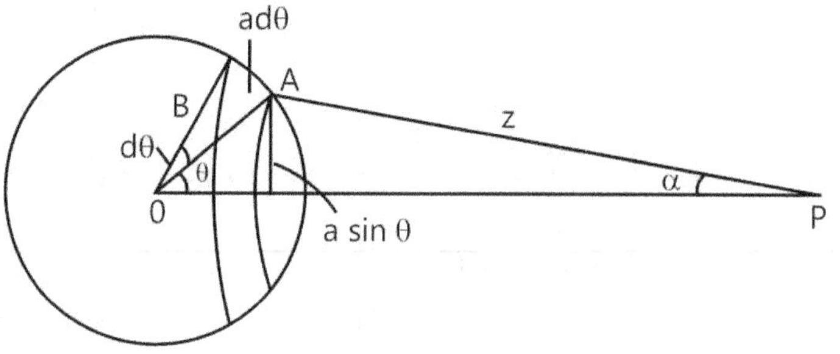

Also from the triangle OAP,

$$a^2 = z^2 + r^2 - 2zr\cos\alpha \qquad \text{or} \qquad \cos\alpha = \frac{z^2 + r^2 - a^2}{2zr}. \qquad \dots(iii)$$

Figure 10.8

Putting from (ii) and (iii) in (i), $\quad dE = \frac{GM}{4ar^2}\left(1 - \frac{a^2 - r^2}{z^2}\right)dz \qquad \text{or} \int dE = \frac{GM}{4ar^2}\left[z + \frac{a^2 - r^2}{z}\right]$

Case I: P is outside the shell (r > a)

In this case, z varies from $r - a$ to $r + a$. The field due to the whole shell is

$$E = \frac{GM}{4ar^2}\left[z + \frac{a^2 - r^2}{z}\right]_{r-a}^{r+a} = \frac{GM}{r^2}$$

We see that the shell may be treated as a point particle of the same mass placed at its center to calculate the gravitational field at an external point.

Case II: P is inside the shell

In this case, z varies from $a - r$ to $a + r$. The field at P due to the whole shell is $E = \frac{GM}{4ar^2}\left[z + \frac{a^2 - r^2}{z}\right]_{a-r}^{a+r} = 0$

Hence the field inside a uniform spherical shell is zero.

(e) Due to uniform solid sphere of mass M and radius a

(i) At an external point r (>a): Let us divide the sphere into thin spherical shells each centered at O. Let the mass of one such shell be dm. To calculate the gravitational field at P, we can replace the shell by a single particle of mass dm placed at the shell that is at O.

The field at P due to this shell is then $\quad dE = \frac{Gdm}{r^2}$

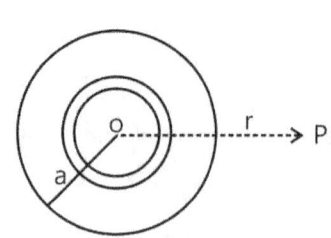

Figure 10.9

Towards PO. The field due to the whole sphere may be obtained by summing the fields of all the shells making the solid sphere.

Thus, $E = \int dE = \int \dfrac{Gdm}{r^2} = \dfrac{G}{r^2}\int dm = \dfrac{GM}{r^2}$

Thus, a uniform sphere may be treated as a single particle of equal mass placed at its center for calculating the gravitational field at an external point.

(ii) At an internal point r (<a):

Suppose the point P is inside the solid sphere (See Fig 10.10). In this case r<a. The sphere may be divided into thin spherical shells all centered at O.

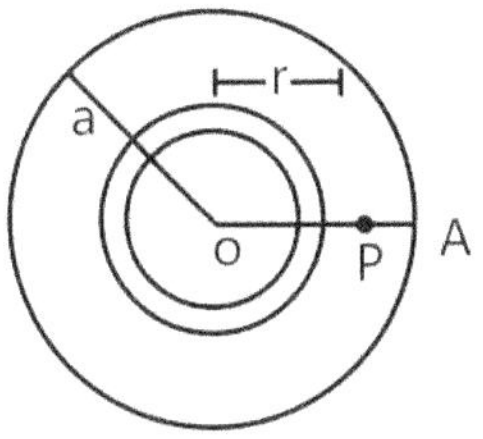

Suppose the mass of such a shell is dm. If the radius of the shell is less than r, the point is outside the shell. The field due to the shell is $dE = \dfrac{Gdm}{r^2}$ along PO.

Figure 10.10

If the radius of the shell considered is greater than r, the point P is internal and the field due to such a shell is zero. The total field due to the whole sphere is obtained by summing the fields due to all the shells. As all these fields are along the same direction, the net field is

$$E = \int dE = \int \dfrac{GdM}{r^2} = \dfrac{G}{r^2}\int dm \qquad \text{... (i)}$$

Only the masses of the shells with radii less than r should be added to get $z = \sqrt{a^2 + r^2}$. These shells form a solid sphere of radius r. The volume of this sphere is $\dfrac{4}{3}\pi r^3$. The volume of the whole sphere is $\dfrac{4}{3}\pi a^3$. As the given sphere is uniform, the mass of the sphere of radius r is $\dfrac{M}{\frac{4}{3}\pi a^3}\left(\dfrac{4}{3}\pi r^3\right) = \dfrac{Mr^3}{a^3}$

Thus, $\int dm = \dfrac{Mr^3}{a^3}$ and by (i) $E = \dfrac{G}{r^2}\dfrac{Mr^3}{a^3} = \dfrac{GM}{a^3}r$.

The gravitational field due to a uniform sphere at an internal point is proportional to the distance of the point from the center of the sphere.

NOMORECLASS CONCEPTS

One could assume the whole mass is concentrated at the center of mass (now assume it as point mass) for calculating the gravitation field at an external point for spherical shell, sphere nevertheless of mass distribution (uniformly/non-uniformly)

Mass distribution should be a function of radial distance only.

Remember the Gauss theorem in Electricity?

Equivalent Gauss theorem for gravitational field is $\oint \vec{E}.d\vec{S} = -4\pi G(m)$, m=enclosed mass I guess now you could deduce the note above. Can you?

Illustration 4: Three concentric shells of homogenous mass distribution of masses M_1, M_2 and M_3 having radii a, b and c respectively are situated as shown in Fig. 10.11. Find the force on a particle of mass m   **(JEE MAIN)**

(a) When the particle is located at Q.

(b) When the particle is located at P.

Sol: For a particle of mass m, lying at a distance r from the center of the spherical shell of mass M and radius r, the gravitational force of attraction is $\left(\dfrac{GMm}{r^2}\right)$. If the particle is lying inside the spherical shell then the force of gravitation on it is zero.

Attraction at an external point due to spherical shell of mass M is $\left(\dfrac{GMm}{r^2}\right)$ while at an internal point is zero.

(a) Point is external to shell M_1, M_2 and M_3,

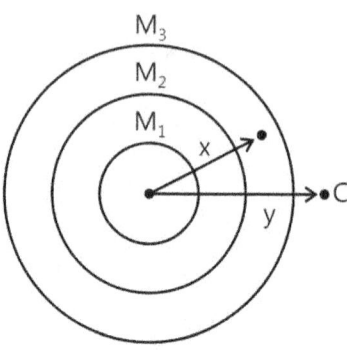

Figure 10.11

So, force at Q will be

$$Fq = \dfrac{GM_1m}{y^2} + \dfrac{GM_2m}{y^2} + \dfrac{GM_3m}{y^2} = \dfrac{Gm}{y^2}(M_1 + M_2 + M_3)$$

(b) Force at P will be

$$F_p = \dfrac{GM_1m}{x^2} + \dfrac{GM_2m}{x^2} + 0 = \dfrac{Gm}{x^2}(M_1 + M_2)$$

Illustration 5: A uniform ring of mass m and radius a is placed directly above a uniform sphere of mass M and of equal radius. The center of the ring is at a distance $\sqrt{3}a$ from the center of the sphere. Find the gravitational force exerted by the sphere on the ring. **(JEE ADVANCED)**

Sol: The field due to ring at the center of the sphere can be found easily, as the center of the sphere is lying at the axis of the ring. From Newton's third law of motion the force on the sphere due to the ring will be equal in magnitude to the force exerted by the sphere on the ring.

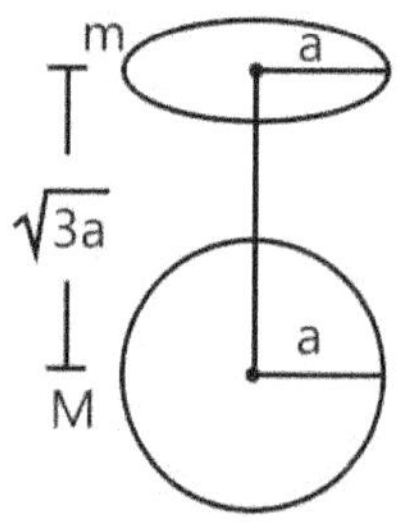

The gravitational field at any point on the ring due to the sphere is equal to the field due to a single particle of mass M placed at the center of the sphere. Thus, the force on the ring due to the sphere is also equal to the force on it by a particle of mass M placed at this point. By Newton's third law, it is equal to the force on the particle by the ring.

Now the gravitational field due to the ring at a distance $d = \sqrt{3}a$ on its axis is

Figure 10.12

$$E = \dfrac{G\,md}{(a^2 + d^2)^{3/2}} = \dfrac{\sqrt{3}\,Gm}{8a^2}$$

The force on a particle of mass M placed here is $F = ME = \dfrac{\sqrt{3}\,GMm}{8a^2}$. Thus we have used the formula for field due to a ring.

This is also the force due to the sphere on the ring.

5. EARTH'S GRAVITATIONAL FIELD

We have seen what gravitational field is and how an object would interact with other objects. Earth is no different as it creates a gravitational field and interacts with us.

$g = F/m$ (g should be written as g bar and F as F bar. Take care of that)

6. VARIATION IN THE VALUE OF ACCELERATION DUE TO GRAVITY (g)

Variation in the value of g: The value of g varies from place to place on the surface of earth. It also varies as we go above or below the surface of the earth. Thus, value of g depends on the following factors:-

(a) **Shape of the earth:** The earth is not a perfect sphere. It is somewhat flat at the two poles. The equatorial radius is approximately 21 km more than the polar radius. And since

$$g = \frac{GM}{R^2} \quad \text{Or} \quad g \propto \frac{1}{R^2}$$

The value of g is minimum at the equator and maximum at the poles.

(b) **Height above the surface of the earth:** The gravitational force on mass m due to Earth of mass M at height h above the surface of earth is

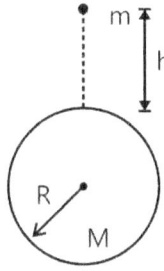

$$F = \frac{GMm}{(R+h)^2}$$

So the acceleration due to gravity is $g' = \frac{F}{m} = \frac{GM}{(R+h)^2}$

This can also be written as, $\quad g' = \dfrac{GM}{R^2\left(1+\dfrac{h}{R}\right)^2} \quad$ Or $\quad g' = \dfrac{g}{\left(1+\dfrac{h}{R}\right)^2} \quad$ as $\quad \dfrac{GM}{R^2} = g$

Figure 10.13

Thus, $g' < g$ i.e., the value of acceleration due to gravity g goes on decreasing as we go above the surface of earth. Further,

$$g' = g\left(1+\frac{h}{R}\right)^{-2} \quad \text{or} \quad g' \approx g\left(1-\frac{2h}{R}\right) \text{ if } h<<R$$

So on going above the surface of the earth, acceleration due to gravity decreases. Note that mass is always constant.

(c) **Depth below the surface of the earth:** Let an object of mass m is situated at a depth h below the earth's surface. Its distance from the center of earth is (R - h). This mass is situated at the surface of the inner solid sphere and lies inside the outer spherical shell. The gravitational force of attraction on a mass inside a spherical shell is always zero. Therefore, the object experiences gravitational attraction only due to inner solid sphere.

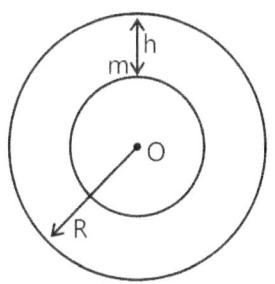

The mass of this sphere is $M' = \left(\dfrac{M}{4/3\,\pi R^3}\right)\dfrac{4}{3}\pi(R-h)^3$ or $M' = \dfrac{(R-h)^3}{R^3}M$

Figure 10.14

$$F = \frac{GM'm}{(R-h)^2} = \frac{GMm(R-h)}{R^3} \quad \text{and} \quad g' = \frac{F}{m}$$

Substituting the values, we get $g' = g\left(1-\dfrac{h}{R}\right)$ i.e., $\quad g' < g$

(d) **Axial rotation of the earth:** Let us consider a particle P at rest on the surface of the earth, in latitude ϕ. Then the pseudo force acting on the particles is $mr\omega^2$ in outward direction. The true acceleration g is acting towards the center O of the earth. Thus, the effective accelerating g' is the resultant of g and $r\omega^2$ or

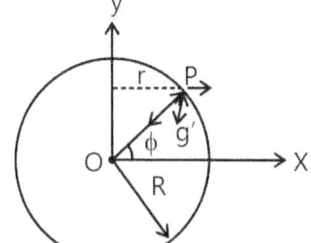

$$g' = \sqrt{g^2 + (r\omega^2)^2 + 2g(r\omega^2)\cos(180 - \phi)}$$

or $\quad g' = \sqrt{g^2 + r^2\omega^4 - 2gr\omega^2\cos\phi} \quad\quad\quad \dots \text{(i)}$

Figure 10.15

Here, the term $r^2\omega^4$ comes out to be too small as $\omega = \dfrac{2\pi}{T} = \dfrac{2\pi}{24 \times 3600}$

rad/s is small. Hence, this term can be ignored. Also, $r = R\cos\phi$. Therefore, Eq. (i) can be written as

$$g' = (g^2 - 2gR\omega^2 \cos^2 \phi)^{1/2}$$

$$= g\left(1 - \frac{2R\omega^2 \cos^2 \phi}{g}\right)^{1/2} = g\left(1 - \frac{R\omega^2 \cos^2 \phi}{g}\right)$$

Thus, $\qquad g' = g - R\omega^2 \cos^2 \phi \qquad\qquad R\omega^2$ is almost $0.03 \ m/s^2$

NOMORECLASS CONCEPTS

They is always a decrease in the value of acceleration due to gravity from that of g at the surface irrespective of the condition.

If earth were to rotate faster 'g' would decrease at all points except at the poles. Guessed it? ϕ is 90 at poles. Also remember ϕ is 0 at equator.

Illustration 6: Suppose the earth increases its speed of rotation. At what new time period will the weight of a body on the equator become zero? Take g = 10 m/s^2 and radius of earth R = 6400km. **(JEE MAIN)**

Sol: When rotational speed of earth is increased, the centrifugal force acting on the particle at rest at equator also increases. At the equator, the centrifugal force is opposite to the force of gravity. Thus the apparent value of g is

$g' = g - R\omega^2$. For mass of body to be zero at the equator, $g' = 0$ i.e. $\omega = \sqrt{\dfrac{g}{R}}$. The time period of rotation is $T = \dfrac{2\pi}{\omega}$.

The weight will become zero, when $g' = 0$ or $g - R\omega^2 = 0 \qquad$ (on the equator $g' = g - R\omega^2$)

or $\omega = \sqrt{\dfrac{g}{R}}; \quad \therefore \dfrac{2\pi}{T} = \sqrt{\dfrac{g}{R}}$ or $T = 2\pi\sqrt{\dfrac{R}{g}}$

Substituting the values, $\qquad T = \dfrac{2\pi\sqrt{\dfrac{6400 \times 10^3}{10}}}{3600}h \qquad$ or $\qquad T = 1.4 \ h$

Thus, the new time period should be 1.4 h instead of 24 h for the weight of a body to be zero on the equator.

Illustration 7: A simple pendulum has a time period exactly 2 s when used in a laboratory at North Pole. What will be the time period if the same pendulum is used in a laboratory at equator? Account for the earth's rotation only.

Take $g = \dfrac{GM}{R^2} = 9.8 m/s^2$ and radius of earth=6400 km. **(JEE ADVANCED)**

Sol: The time period of simple pendulum is given by $t = 2\pi\sqrt{\dfrac{\ell}{g}}$ where ℓ is the length of pendulum. At the equator

value of acceleration due to gravity 'g' is different than at the pole. The apparent value of g is $g' = g - R\omega^2$. Thus the time periods will be different.

Consider the pendulum in its mean position at the North Pole. As the pole is on the axis of rotation, the bob is

in equilibrium. Hence in the mean position, the tension T is balanced by earth's attraction. Thus, $T = \dfrac{GMm}{R^2} = mg$.

The time period t is $t = 2\pi\sqrt{\dfrac{\ell}{T/m}} = 2\pi\sqrt{\dfrac{\ell}{g}} \qquad\qquad$ ... (i)

At equator, the lab and the pendulum rotate with the earth at angular velocity $\omega = \dfrac{2\pi \text{ radian}}{24 \text{ hour}}$ in a circle of radius equal to 6400 km. Using Newton's second law,

$$\frac{GMm}{R^2} - T' = \omega^2 R \quad \text{or,} \quad T' = m(g - \omega^2 R)$$

Where T' is the tension in the string.

The time period will be

$$t' = 2\pi\sqrt{\frac{l}{(T'/m)}} = 2\pi\sqrt{\frac{l}{g - \omega^2 R}} \qquad \qquad \text{... (ii)}$$

By (i) and (ii)

$$\frac{t'}{t} = \sqrt{\frac{g}{g - \omega^2 R}} = \left(1 - \frac{\omega^2 R}{g}\right)^{-1/2} \quad \text{or,} \quad t' = t\left(1 + \frac{\omega^2 R}{2g}\right)$$

Putting the values, $t' = 2.004$ seconds.

7. GRAVITATIONAL POTENTIAL ENERGY

Suppose I would like to move a particle form another particle's field, work is either done against the gravitational field or extracted from it. This negative work is called as Gravitational Potential energy.

Gravitational force is a conservative in nature. Work done by gravitational field$= U_f - U_i = -\int\limits_i^f \vec{F}.\vec{dr}$.

Let a particle of mass m_1 be kept fixed at a point A (See Fig 10.16) and another particle of mass m_2 is taken from a point B to a point C. Initially, the distance between the particles is $AB = r_1$ and finally it becomes $AC = r_2$. We have to calculate the change in potential energy of the system of the two particles as the distance changes from r_1 to r_2.

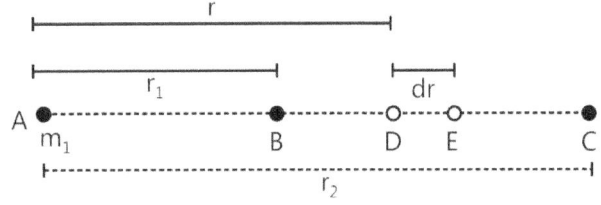

Figure 10.16

Consider a small displacement when the distance between the particles changes from r to r + dr. In the Fig 10.16, this corresponds to the second particle going from D to E.

The force on the second particle is $F = \dfrac{Gm_1 m_2}{r^2}$ along $\overrightarrow{DA}$

The work done by the gravitational force in the displacement is $dW = -\dfrac{Gm_1 m_2}{r^2}dr$.

The change in potential energy of the two-particle system during this displacement is $dU = -dW = \dfrac{Gm_1 m_2}{r^2}dr$.

The change in potential energy as the distance between the particles from r_1 to r_2 is

$$U(r_2) - U(r_1) = \int dU = \int\limits_{r_1}^{r_2}\frac{Gm_1 m_2}{r^2}dr = Gm_1 m_2 \int\limits_{r_1}^{r_2}\frac{1}{r^2}dr = Gm_1 m_2\left[-\frac{1}{r}\right]_{r_1}^{r_2} = Gm_1 m_2\left(\frac{1}{r_1} - \frac{1}{r_2}\right)$$

This is the change in potential energy of the particles when moved from B to C.

Suppose the same particles which are of mass m_1 and m_2 are very far from each other and we need to calculate the change in potential energy when the distance between them becomes r. Then using above formulae,

we get $\quad U(r) - U(\infty) = Gm_1 m_2\left[\dfrac{1}{\infty} - \dfrac{1}{r}\right] = -\dfrac{Gm_1 m_2}{r}$

1.9

We make a standard assumption that the potential energy of the two-particle system to be zero when the distance between them is infinity. This means that we choose $U(\infty) = 0$.

Note: Just as one assumed current to be in opposite direction with the flow of electrons, the potential at infinity is assumed to be zero.

8. GRAVITATIONAL POTENTIAL

The potential at a point may also be defined as the work done per unit mass by an external agent in bringing a particle slowly from the reference point to the given point. Generally the reference point is chosen at infinity so that the potential at infinity is zero.

We define the "change in potential" $V_B - V_A$ between the two points as $V_B - V_A = \dfrac{U_B - U_A}{m}$

Calculation of some Gravitational potentials:

(a) Potential due to point mass M at a point P which is at a distance r

(b) (ii) Potential due to Uniform ring of radius "a" and mass M at a point P on its axis.

(c) $V_{(r)} = \dfrac{U_{(r)} - U_{(\infty)}}{m}$

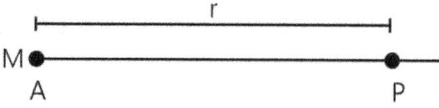

Figure 10.17

But $U(r) - U(\infty) = -\dfrac{GMm}{r}$ so that $V = -\dfrac{GM}{r}$

The gravitational potential due to a point mass M at a distance r is $-\dfrac{GM}{r}$

(d) Consider any small part of the ring of mass dm. The point P is at a distance $z = \sqrt{a^2 + r^2}$ from dm.

$dV = -\dfrac{GdM}{r} = -\dfrac{Gdm}{\sqrt{a^2 + r^2}}$;

$V = \int dV = \int -\dfrac{Gdm}{\sqrt{a^2 + r^2}} = -\dfrac{G}{\sqrt{a^2 + r^2}} \int dm = -\dfrac{GM}{\sqrt{a^2 + r^2}}$

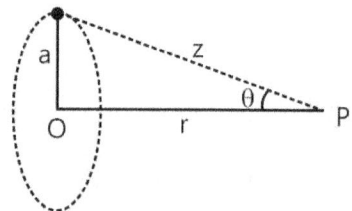

Figure 10.18

	Potential	Gravitational Field
Point Mass at a distance r	$\dfrac{-GM}{r}$	$\dfrac{-GM}{r^2}\vec{e}_r$
Uniform Ring at a point on its axis	$\dfrac{-GM}{\sqrt{a^2+r^2}}$	$\dfrac{GMr}{\left(a^2+r^2\right)^{3/2}}$ towards center of ring
Uniform Thin spherical shell	$\dfrac{-GM}{a}$ (inside) $\dfrac{-GM}{r}$ (outside)	0 (inside) $\dfrac{GM}{r^2}$ (outside)
Uniform Solid Sphere	$\dfrac{-GMr^2}{a^3}$ (Inside) $\dfrac{-GM}{2a^3}\left(3a^2-r^2\right)$ (outside)	$\dfrac{GMr}{a^3}$ (inside) $\dfrac{GM}{r^2}$ (outside)

Only the magnitudes of gravitational field are written. As the gravitational force is attractive in nature, the direction could be easily found out.

Gravitational force, potential and potential energy all are taken with negative sign because the gravitational force is always attractive in nature.

$$E_x = -\frac{\partial V}{\partial x}, \quad E_y = -\frac{\partial V}{\partial y} \text{ and } E_z = -\frac{\partial V}{\partial z}$$

Potential using the field for various cases $V\left(\vec{r_2}\right) - V\left(\vec{r_1}\right) = -\displaystyle\int_{r_1}^{r_2} \vec{E}.\overline{dr}$.

Illustration 8: A particle of mass 1 kg is kept on the surface of a uniform sphere of mass 20 kg and radius 1.0 m. Find the work to be done against the gravitational force between them to take the particle away from the sphere.

(JEE MAIN)

Sol: The work done in moving a particle away from the sphere will be equal to the change in gravitational potential energy of the particle in the gravitational field of the sphere.

Potential at the surface of sphere, $\quad V = -\dfrac{GM}{R} = -\dfrac{(6.67 \times 10^{-11})(20)}{1}$ J/kg $= -1.334 \times 10^{-9}$ J/kg i.e., 1.334×10^{-9} J work is obtained to bring a mass of 1 kg from infinity to the surface of sphere. Hence, the same amount of work will have to be done to take the particle away from the surface of sphere. Thus, $W = 1.334 \times 10^{-9}$ J

Illustration 9: A particle is fired vertically upward with a speed of 9.8 km/s. Find the maximum height attained by the particle. Radius of earth = 6400 km and g at the surface=9.8 m/s^2. Consider only earth's gravitation.

(JEE MAIN)

Sol: Particle initially moves with kinetic energy only in upwards direction opposite to the gravitation pull of earth. The loss in its kinetic energy is equal to the gain in the potential energy. At the highest point of its vertical motion, kinetic energy is converted completely into potential energy.

At the surface of the earth, the potential energy of the earth-particle system is $-\dfrac{GMm}{R}$ with usual symbols. The kinetic energy is $\dfrac{1}{2}mv_0^2$ where v_0 = 9.8 km/s. At the maximum height the kinetic energy is zero. If the maximum height reached is H, the potential energy of the earth-particle system at this instant is $-\dfrac{GMm}{R+H}$. Using conservation of energy, $-\dfrac{GMm}{R} + \dfrac{1}{2}mv_0^2 = -\dfrac{GMm}{R+H}$

Writing $GM = gR^2$ and dividing by m, $-gR + \dfrac{v_0^2}{2} = \dfrac{-gR^2}{R+H}$ or $\dfrac{R^2}{R+H} = R - \dfrac{v_0^2}{2g}$ or $R+H = \dfrac{R^2}{R - \dfrac{v_0^2}{2g}}$ Putting the values of R, v_0 and g on the right side,

$$R+H = \dfrac{(6400\text{km})^2}{6400\text{km} - \dfrac{(9.8\text{kms}^{-1})^2}{2 \times 9.8\text{ms}^{-2}}} = \dfrac{(6400\text{km})^2}{1500\text{km}} = 27300\text{km} \quad \text{or} \quad H = (27300 - 6400)\text{km} = 20900\text{km}.$$

Illustration 10: Two particles of equal masses go round a circle of radius R under the action of their mutual gravitational attraction. Find the speed of each particle.

(JEE MAIN)

Sol: As the particles go around the circle they always remain diametrically opposite to each other. To sustain their respective circular motion the necessary centripetal acceleration is provided by the gravitation force of attraction between them.

The particles will always remain diametrically opposite so that the force on each particle will be directed along the radius. Consider the motion of one of the particles. The force on the particle is $F = \dfrac{Gm^2}{4R^2}$. If Thus, by Newton's law,

$\dfrac{Gm^2}{4R^2} = \dfrac{mv^2}{R}$ or $v = \sqrt{\dfrac{Gm}{4R}}$

9. BINDING ENERGY

It is the energy due to which a system is bound. Suppose the mass m is placed on the surface of earth. The radius of the earth is R and its mass M. Then, the kinetic energy of the particle K=0 and potential energy of the particle is $U = -\dfrac{GMm}{R}$.

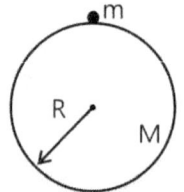

Therefore, the total mechanical energy of the particle is, $E = K + U = 0 - \dfrac{GMm}{R}$ or $E = -\dfrac{GMm}{R}$

Figure 10.19

It is due to this energy, the particle is attached to the earth. If this amount of energy is supplied to the particle in any form (normally kinetic), the particle no longer remains bound to the earth. It goes out of the gravitational field of earth.

Illustration 11: Assuming the earth to be a sphere of uniform mass destiny, calculate the energy needed to completely disassemble it against the gravitational pull amongst its constituent particles. Given the product of mass and radius of the earth $= 2.5 \times 10^{31}$ kgm, $g = 10 \text{m/s}^2$.

(JEE MAIN)

Sol: The work done to completely disassemble the earth will be equal to change in potential energy of the earth. Initial potential energy is negative and final will be zero.

If M and R are the mass and radius of the earth, then the density ρ of the earth is $\rho = \dfrac{3M}{4\pi R^3}$

The earth may be supposed to be made up of a large number of thin concentric spherical shells. It can be disassembled by removing such shells one by one. When a sphere of radius x is left, the energy needed to remove a shell of thickness lying between x and $x + dx$ is $dU = \dfrac{Gm_1 m_2}{x}$

Where m_1 = mass of the sphere of radius $x = \dfrac{4}{3}\pi x^3 \rho$,

and m_2 = mass of the spherical shell of radius x and thickness dx $= 4\pi x^2 dx\rho$

$$\therefore \quad dw = dU = \dfrac{G\left(\dfrac{4}{3}\pi x^3 \rho\right)\left(4\pi x^2 dx\rho\right)}{x} = \dfrac{16}{3}G\pi^2 \rho^2 x^4 dx$$

Total energy required $U = \int dU = \dfrac{16G\pi^2\rho^2}{3}\int_0^R x^4 dx = \dfrac{16G\pi^2\rho^2}{3}\dfrac{R^5}{5} = \dfrac{16}{15}G\pi^2\left(\dfrac{M}{(4\,3)\pi R^3}\right)^2 R^5 = \dfrac{3}{5}\dfrac{GM^2}{R}$

$= \dfrac{3}{5}gMR = \dfrac{3}{5}\times 10 \times 2.5\times 10^{31} = 1.5\times 10^{32}\,\text{J}.$

10. ESCAPE VELOCITY

The minimum velocity needed to take a particle infinitely away from the earth is called the escape velocity. On the surface of earth its value 11.2 km/s.

As we discussed the binding energy of a particle on the surface of earth kept at rest is $\dfrac{GMm}{R}$. If this much energy in the form of kinetic energy is supplied to the particle, it leaves the gravitational field of the earth. So, if v_e is the escape velocity of the particle, then

$$\dfrac{1}{2}mv_e^2 = \dfrac{GMm}{R} \quad \text{or} \quad v_e = \sqrt{\dfrac{2GM}{R}} \quad \text{or } v_e = \sqrt{2gR} \qquad \text{as} \qquad g = \dfrac{GM}{R^2}$$

NOMORECLASS CONCEPTS

Escape velocity is independent of angle of projection.

Illustration 12: Calculate the escape velocity from the surface of moon. The mass of the moon is 7.4×10^{22} kg and radius $= 1.74 \times 10^6$ m

(JEE MAIN)

Sol: Escape velocity of any object placed on moon is given by $v_e = \sqrt{\dfrac{2GM_m}{R_m}}$

Escape velocity from the surface of moon is $v_e = \sqrt{\dfrac{2GM_m}{R_m}}$

Substituting the values, we have $\quad v_e = \sqrt{\dfrac{2 \times 6.67 \times 10^{-11} \times 7.4 \times 10^{22}}{1.74 \times 10^6}} = 2.4 \times 10^3$ m/s or 2.4 km/s

11. SATELLITES

Satellites are generally of two types:

Natural Satellites: Moon is a natural satellite of the earth.

Artificial Satellite: These are launched in to space by humans and they help us in weather forecasting, telecommunications etc. The path of these satellites is elliptical with the center of earth at a focus.

Orbital Speed: The necessary centripetal force to the satellite is being provided by the gravitational force exerted by the earth on the satellite. Thus,

$\therefore \quad v_o = \sqrt{\dfrac{GM}{r}} \quad$ or $\quad v_o \propto \dfrac{1}{\sqrt{r}}$

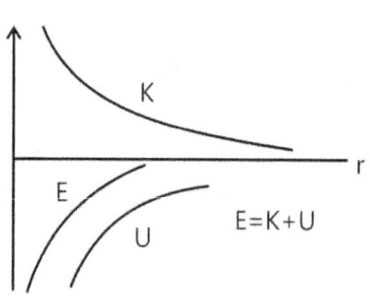

Hence, the orbital speed (v_o) of the satellite decreases as the orbital

radius (r) of the satellite increases. Further, the orbital speed of a satellite

close to the earth's surface $(r \approx R)$ is, $v_o = \sqrt{\dfrac{GM}{R}} = \sqrt{gR} = \dfrac{v_e}{\sqrt{2}}$;

Figure 10.20

Substituting $\quad v_e = 11.2$ km/s ; $v_o = 7.9$ km/s

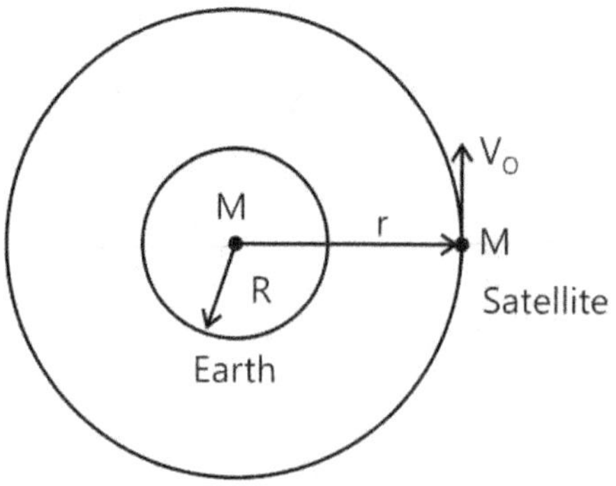

Period of Revolution: The period of revolution (T) is given by $T = \dfrac{2\pi r}{v_o}$ or $T = \dfrac{2\pi r}{\sqrt{\dfrac{GM}{r}}}$ or $\quad T = 2\pi \sqrt{\dfrac{r^3}{GM}}$

Or $T = 2\pi \sqrt{\dfrac{r^3}{gR^2}} \quad$ (as $GM = gR^2$)

Energy of Satellite: The potential energy of the system is $\quad U = -\dfrac{GMm}{r}$

The kinetic energy of the satellite is, $K = \dfrac{1}{2}mv_0^2 = \dfrac{1}{2}m\left(\dfrac{GM}{r}\right)$

or $\quad K = \dfrac{1}{2}\dfrac{GMm}{r}$

The total energy is, $E = K + U = -\dfrac{GMm}{2r}$ or $\quad E = -\dfrac{GMm}{2r}$

This energy is constant and negative, i.e., the system is closed. The farther the satellite from the earth the greater its total energy.

Figure 10.21

The velocity of a satellite is independent of its mass. It only depends upon the mass of the planet around which it revolves.

What if the time period of rotation of satellite is exactly 24 hours just as the time period of rotation of earth? Its position w.r.t earth is fixed right! Try calculating the distance from the earth's surface. By the way, these satellites are called Geo-stationary (stationary w.r.t earth) satellites.

Illustration 13: Consider an earth's satellite so positioned that it appears stationary to an observer on earth and serves the purpose of a fixed relay station for international transmission of TV and other communications. What would be the height at which the satellite should be positioned and what would be the direction of its motion? Given that the radius of the earth is 6400 km and acceleration due to gravity on the surface of the earth is 9.8 m/s². **(JEE ADVANCED)**

Sol: For any artificial satellite to appear stationary with respect to a point on earth, it must rotate with the same angular speed as that of the earth and in the direction of motion as of the earth. The angular velocity of the satellite at height h above earth surface is given by $\omega = \sqrt{GM/r^3}$ where r=R+h.

For a satellite to remain above a given point on the earth's surface, it must rotate with the same angular velocity as the point on earth's surface. Therefore the satellite must rotate in the equatorial plane from west to east with a time period of 24 hours.

Now as for a satellite orbital velocity is $v_0 = \sqrt{GM/r}$

$$T = \frac{2\pi r}{v_0} = 2\pi r\sqrt{\frac{r}{GM}} = 2\pi r\sqrt{\frac{r}{gR^2}} \ (as \ g = GM/R^2) \ or \ r = \left[gR^2\frac{3}{4\pi^2}\right]^{\frac{1}{3}} = 4.23\times107m = 42300km$$

So the height of the satellite above the surface of earth, h = r – R = 42300 – 6400 ≈ 36000km

[The speed of a geostationary satellite $v_0 = R\sqrt{g/r} = r\omega = 3.1\,km/s$]

Illustration 14: Two satellites S_1 and S_2 revolve round a planet in coplanar circular orbits in the same sense. Their periods of revolution are 1 h and 8 h respectively. The radius of the orbit of S_1 is 10^4 km. When S_2 is closest to S_1 find (a) the speed of S_2 relative to S_1 and (b) the angular speed of S_2 as observed by an astronaut in S_1.

(JEE ADVANCED)

Sol: According to Kepler's laws of planetary motion, $T^2 \propto R^3$. The orbital velocity of the satellite $v_0 = \frac{2\pi R}{T} = R\omega$ where ω is the angular velocity of revolution of satellite.

Let the mass of the planet be M, that of S_1 be m_1 and of S_2 be m_2.

Let the radius of the orbit of S_1 be $R_1(=10^4 km)$ and so S_2 be R_2.

Let v_1 and v_2 be the linear speeds of S_1 and S_2 with respect to the planet. The given Fig 10.22 shows the situation.

As the square of the time period is proportional to the cube of the radius,

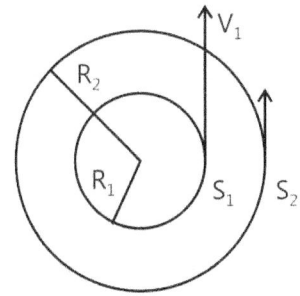

Figure 10.22

$$\left(\frac{R_2}{R_1}\right)^3 = \left(\frac{T_2}{T_1}\right)^2 = \left(\frac{8h}{1h}\right)^2 = 64 \quad or \quad \frac{R_2}{R_1} = 4 \ or \ R_2 = 4R_1 = 4\times10^4 km$$

Now the time period of S_1 is 1 h.

So, $\dfrac{2\pi R_1}{v_1} = 1h$ or $v_1 = \dfrac{2\pi R_1}{1h} = 2\pi \times 10^4 \, kmh^{-1}$

Similarly, $v_2 = \dfrac{2\pi R_2}{8h} = \pi \times 10^4 \, kmh^{-1}$

(a) At the closest separation, they are moving in the same direction. Hence the speed of S_2 with respect to S_1 is $|v_2 - v_1| = \pi \times 10^4 \, kmh^{-1}$

(b) As seen from S_1, the satellite S_2 is at a distance $R_2 - R_1 = 3 \times 10^4 \, km$ at the closest separation. Also, it is moving at $\pi \times 10^4 \, kmh^{-1}$ in a direction perpendicular to the line joining them.

Thus, the angular speed of S_2 as observed by S_1 is $\omega = \dfrac{\pi \times 10^4 \, kmh^{-1}}{3 \times 10^4 \, km} = \dfrac{\pi}{3} radh^{-1}$

Illustration 15: A spaceship is launched into a circular orbit close to the earth's surface. What additional velocity is now to be added to the spaceship in the orbit to overcome the gravitational pull? Radius of earth = 6400 km, g = 9.8 m/s². **(JEE MAIN)**

Sol: The potential energy of the spaceship close to the earth is negative (- mgR). The orbital speed close to the earth is $v = \sqrt{gR}$, so the kinetic energy is mgR/2. The total energy is - mgR/2. We need to provide the additional kinetic energy = mgR/2 such that the spaceship escapes the gravitational pull of the earth.

The extra kinetic energy to be given is $\dfrac{mv^2}{2} = \dfrac{mgR}{2}$, so that the extra velocity given is $v' = \sqrt{gR}$.

The velocity is $v' = \sqrt{9.8 \times 6400000} = 7.91 \times 10^3 \, m/s = 7.91 \, km/s$

Illustration 16: An artificial satellite is moving in a circular orbit around the earth with a speed equal to one fourth the magnitude of escape velocity from the earth.

(i) Determine the height of the satellite above the earth's surface.

(ii) If the satellite is stopped suddenly in its orbit and allowed to fall freely towards the earth, find the speed with which it hits the surface of the earth. **(JEE MAIN)**

Sol: For satellite the escape velocity is $v_e = \sqrt{2Rg}$. According to given data the satellite is moving in the orbit with one fourth the magnitude of this velocity. When satellite stops revolving, it falls freely under action of gravity from the height h above the surface of the earth. The loss in the gravitational potential energy in falling height h is equal to gain in the kinetic energy of the satellite.

(i) Let M and R be the mass and radius of the earth respectively. Let m be the mass of satellite. Here escape velocity from earth $v_e = \sqrt{(2Rg)}$

Velocity of satellite $v_g = \dfrac{v_e}{4} = \sqrt{(2Rg)}/4$ ...(i)

Further $v_c = \sqrt{\left(\dfrac{GM}{r}\right)} = \sqrt{\left(\dfrac{R^2 g}{R+h}\right)}$ $\therefore$ $v_g^2 = \dfrac{R^2 g}{R+h}$ ...(ii)

From equation (i) and (ii), we get $H = 7R = 44800 \, km$

(ii) Now, the total energy at height h = total energy on earth's surface (principle of conservation of energy). Let it reach earth's surface with velocity v.

$\therefore$ $0 - GM\dfrac{m}{R+h} = \dfrac{1}{2}mv^2 - GM\dfrac{m}{R}$ Or $\dfrac{1}{2}mv^2 = \dfrac{GMm}{R} - \dfrac{GMm}{7R}$ $(\because h = 7R)$

Solving we get $v = \sqrt{12Rg/7}$ $\therefore$ $v = \sqrt{(1.714 \times 6400 \times 10^3 \times 9.8)} = 10.368 \text{km/sec}$

12. PLANETS AND THEIR MOTION

12.1 Law of Orbits

All the planets move in elliptical orbits with the sun as one of its focii.

12.2 Law of Areas

The radius vector from the sun at the focus of elliptical orbit to the planet sweeps out equal areas in equal intervals of time.

If the radius vector R sweeps an angle $d\theta$ in time dt, area ASB

swept by radius vector in time $dt = dA = \frac{1}{2} \times R \times Rd\theta$

$\therefore$ $dA = \frac{1}{2}R^2 \frac{d\theta}{dt}dt = \frac{1}{2}\omega R^2 dt$

Areal velocity $= \dfrac{area}{time} = \dfrac{dA}{dt} = \dfrac{1}{2}\omega R^2$

So ωR^2 is constant for area SAB and area SCD. It shows that the angular momentum $mR^2\omega$ is conserved for planetary motion. When R decreases, ω increases so that ωR^2 is constant.

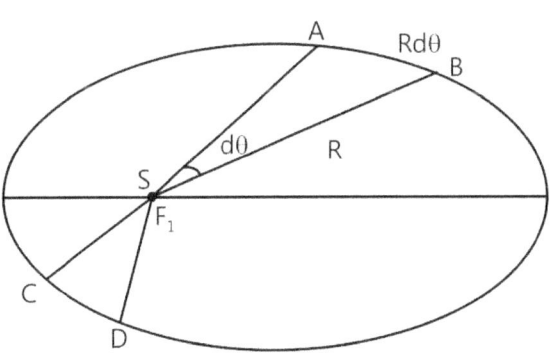

Figure 10.23

12.3 Laws of Periods

The square of the time period of revolution of a planet is proportional to the cube of the mean distance of the planet from the sun.

If a is the mean distance of sun from the planet, T^2 is proportional to a^3 or $T^2 = Ka^3$ where K is a constant.

If a_1 and a_2 are semi-major axis of the orbits of two planets around the sun with respective time periods T_1 and T_2,

then $\dfrac{T_1^2}{T_2^2} = \dfrac{a_1^3}{a_2^3}$

NOMORECLASS CONCEPTS

Observe the time period of rotation of satellite. Got it? (It follows Kepler's third law too)

When the planet is farthest from Sun, it is said to be at the Apogee of Aphelion.

When the planet is at nearest to the Sun, it is said to be at Perigee or Perhilion.

Illustration 17: The minimum and maximum distance of a satellite from the center of the earth are 2R and 4R respectively, where R is the radius of earth and M is the mass of the earth. Find:

(a) Its minimum and maximum speeds,

(b) Radius of curvature at the point of minimum distance.   **(JEE ADVANCED)**

Sol: The speed of the satellite is minimum when is at the maximum distance from the earth and vice versa. At the point of minimum or maximum distance from earth the velocity vector is perpendicular to the radius vector from the earth. Apply law of conservation of angular momentum and energy at the two points.

(a) Applying conservation of angular momentum

$$mv_1(2R) = mv_2(4R) \quad v_1 = 2v_2 \qquad \text{...(i)}$$

From conservation of energy

$$\frac{1}{2}mv_1^2 - \frac{GMm}{2R} = \frac{1}{2}mv_2^2 - \frac{GMm}{4R} \qquad \text{...(ii)}$$

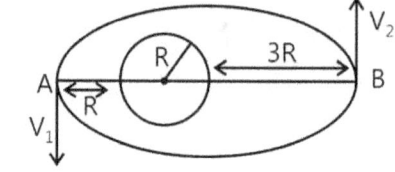

Figure 10.24

Solving Eqs. (i) and (ii), we get

$$v_2 = \sqrt{\frac{GM}{6R}}, \qquad v_1 = \sqrt{\frac{2GM}{3R}}$$

(b) If r is the radius of curvature at point A

$$\frac{mv_1^2}{r} = \frac{GMm}{(2R^2)}; \quad r = \frac{4v_1^2R^2}{GM} = \frac{8R}{3} \qquad \text{(Putting value of } v_1)$$

Illustration 18: The planet Neptune travels around the Sun with a period of 165 year. Show that the radius of its orbit is approximately thirty times that of Earth's orbit, both being considered as circular. **(JEE ADVANCED)**

Sol: According to the Kepler's laws of planetary motion $T^2 \propto R^3$ where T is the time period of revolution and R is the radius of the orbit of revolution of planet. Taking the ratio of time periods of revolution of Earth and Neptune, we get the ratio of radius of their orbits.

$$T_1 = T_{Earth} = 1 \text{ year}; T_2 = T_{Neptune} = 165 \text{ year} = 165 \, T_1$$

Let R_1 and R_2 be the radii of the circular orbits of Earth and Neptune respectively.

$$\frac{T_1^2}{T_2^2} = \frac{R_1^3}{R_2^3} \quad \therefore \quad R_2^3 = \frac{R_1^3 T_2^2}{T_1^2} \quad \text{or} \quad R_2^3 = \frac{R_1^3 \times 165^2}{1^2}$$

$$\therefore \quad R_2^3 = 165^2 R_1^3 \quad \text{or} \quad R_2 \approx 30 R_1$$

13. MOTION ABOUT THE CENTRE OF MASS

As shown in the Fig 10.25, for the case of circular orbits, two objects are moving about their common center of mass. If we consider the motion of the smaller body,

$$\frac{GMm}{(r+R)^2} = m\omega^2 r$$

The revised law of periods in

$$T^2 = \left(\frac{4\pi^2}{GM}\right) r^3 \left(1 + \frac{R}{r}\right)^2$$

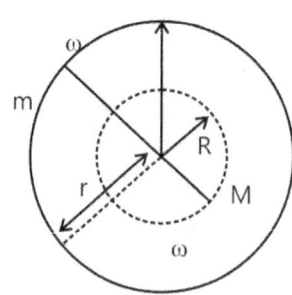

Two bodies moving in circular orbits under the influences of each other's gravitational attraction

Figure 10.25

Illustration 19: A pair of stars rotate about their common center of mass. One of them has mass m and the other 2m. Their centers are a distance d apart, d being large compared to the size of either star.

(a) Derive an expression for the period or rotation of the stars about their common center of mass in terms of d, m and G

(b) Compare the angular moments of the two stars about their common center of mass.

(c) Compare the kinetic energies of the two stars. **(JEE MAIN)**

Sol: The gravitational pull between two stars provides the necessary centripetal acceleration to make them revolve in a circular orbit. The time period of revolution of each star is $T = \dfrac{2\pi}{\omega}$. The angular momentum of the revolving body is given by $L = I\omega = m r^2 \omega$. And the kinetic energy is given by $E = \dfrac{I\omega^2}{2}$.

The center of mass O is at a distance 2d/3 from the star of mass m and d/3 from the star of mass 2m. Both the stars rotate with the same angular velocity ω.

(a) Since the gravitational force provides the centripetal force, then

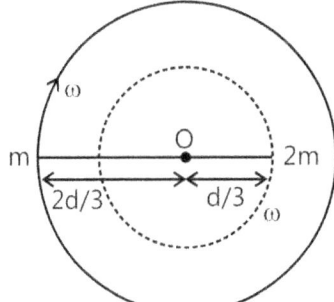

$$m\left(\frac{2d}{3}\right)\omega^2 = \frac{Gm.2m}{d^2} \Rightarrow \omega = \sqrt{3Gm/d^3} \quad \text{or} \quad T = \frac{2\pi}{\omega} = 2\pi\sqrt{d^3/3Gm}$$

(b) Ratio of angular momenta

$$\frac{L_{small}}{L_{large}} = \frac{m\left(2d/3\right)^2\omega}{2m\left(d/3\right)^2\omega} = 2$$

(c) Ratio of kinetic energies

$$\frac{E_{small}}{E_{large}} = \frac{\frac{1}{2}I_{small}\omega^2}{\frac{1}{2}I_{big}\omega^2} = 2$$

Figure 10.26

PROBLEM-SOLVING TACTICS

1. Most of the problems are easy, as gravitation and electrostatics are analogous to each other. Just be careful that gravitational force is always attractive, whereas electrostatic force can be attractive as well as repulsive and make changes as necessary.

2. Assumptions are appreciated in real cases of satellites and planetary motion.

3. Ideas and concepts of circular motion must be strong because they are generally applied here.

4. While dealing practical cases on Earth, be careful about Earth's rotation on its own axis.

5. Most questions are solved with ease by using work-energy theorem and laws of motion

FORMULAE SHEET

S. No.	Description	Formulae	
1	Magnitude of gravitational force between two particles of mass m_1 & m_2 placed at a distance r is	$F = \dfrac{Gm_1m_2}{r^2}$ $G = 6.67 \times 10^{-11} N-m^2/kg^2$ Note: It acts along the line joining two particles.	
2	Acceleration due to gravity (g)	$g = \dfrac{GM}{R^2}$ SI units:- m/s^2 M is the mass of the earth and its radius R.	
3		Gravitational force $= \dfrac{GMm}{(R+h)^2}$ Acceleration due to gravity $= g' = F/m = \dfrac{GM}{(R+h)^2}$ If $h \ll R$ $g' = g\left(1 - \dfrac{2h}{R}\right)$	
4		At a certain, Depth H, acceleration due to gravity g' is $g' = g\left(1 - \dfrac{h}{R}\right)$ g is acceleration due to gravity at surface of earth.	
5	Effect of g due to axial rotation of earth	$g' = g - R\omega^2 \cos^2\phi$ g' is the acceleration due to gravity on the particle on the earth surface in latitude ϕ.	
6	Gravitational field strength	$\vec{E} = \dfrac{\vec{F}}{m}$ SI unit is N/kg.	
		Gravitational Field	**Gravitational Potential**
7	Point Mass	$\dfrac{GM}{r^2}$	$-\dfrac{GM}{r}$
8	Uniform ring at point on its axis	$\dfrac{GMr}{\left(a^2 + r^2\right)^{3/2}}$ (towards center of ring)	$-\dfrac{GM}{\sqrt{a^2 + r^2}}$
9	Uniform thin spherical shell	Inside θ Outside $\dfrac{GM}{r^2}$	Inside $-\dfrac{GM}{a}$ Outside $-\dfrac{GM}{r}$

10	Uniform solid sphere	Inside $\dfrac{GMr}{a^3}$ Outside $\dfrac{GM}{r^2}$	Inside $-GMr^2/a^3$ Outside $-\dfrac{GM}{2a^2}\left(3a^2-r^2\right)$ Here, a is the radius and r is the location of point mass.
11	Gravitational potential	Note: It is a scalar; SI unit is J/kg.	
12		$\vec{E}=-\left[\dfrac{\partial y}{\partial x}\hat{i}+\dfrac{\partial V}{\partial y}\hat{j}+\dfrac{\partial V}{\partial z}\hat{k}\right]$ Note: It is partial derivative $dV=-\vec{E}.\vec{dr}$.	
13	Gravitational potential energy	$U=-\dfrac{Gm_1m_2}{r}$ System of particle $\left(m_1\,m_2\,m_3\,m_4\right)$ $U=-G\left[\dfrac{m_4m_3}{r_{43}}+\dfrac{m_4m_2}{r_{42}}+\dfrac{m_4m_1}{r_{41}}+\dfrac{m_3m_2}{r_{32}}+\dfrac{m_3m_1}{r_{31}}+\dfrac{m_2m_1}{r_{21}}\right]$ They are $\dfrac{4(4-1)}{2}=6$ Pairs	
14	For an n particle system, no. of pairs would be	$\dfrac{n(n-1)}{2}$ Pairs	
15	Binding Energy	$E=\dfrac{GMm}{R}$ It is due to this energy particle is bound to earth.	
16	Escape Velocity	$v_e=\sqrt{2gR}$	
17	Motion of Satellites	Orbital Speed $v_o=\sqrt{\dfrac{GM}{r}}$ Time period: $T=\dfrac{2\pi r}{v_o}=2\pi\sqrt{\dfrac{r^3}{GM}}$ Energy of satellite: $U=-\dfrac{GMm}{r}$; $K=\dfrac{GMm}{2r}$ U is The potential energy Total Energy "E" K is The kinetic energy $E=K+U=-\dfrac{GMm}{2r}=-K$	

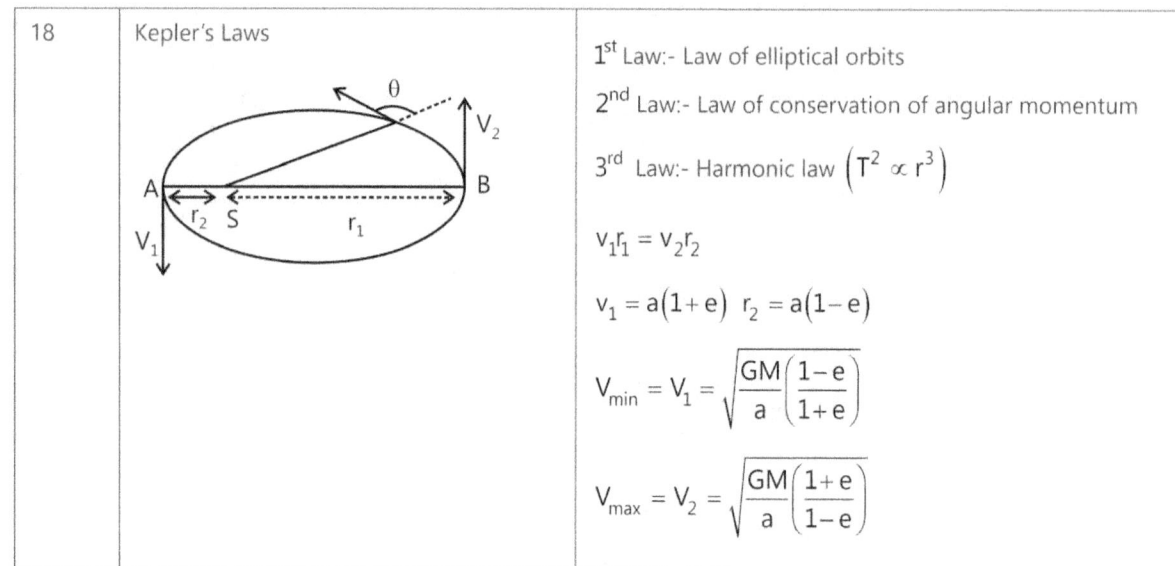

1st Law:- Law of elliptical orbits

2nd Law:- Law of conservation of angular momentum

3rd Law:- Harmonic law $\left(T^2 \propto r^3\right)$

$v_1 r_1 = v_2 r_2$

$v_1 = a(1+e) \quad r_2 = a(1-e)$

$V_{min} = V_1 = \sqrt{\dfrac{GM}{a}\left(\dfrac{1-e}{1+e}\right)}$

$V_{max} = V_2 = \sqrt{\dfrac{GM}{a}\left(\dfrac{1+e}{1-e}\right)}$

Solved Examples

JEE Main/Boards

Example 1: Two concentric shells of mass M_1 and M_2 are as shown. Calculate the gravitational force on m due to M_1 at points P, Q and R.

Sol: For a particle of mass m, lying at a distance r from the center of the spherical shell of mass M, the gravitational force of attraction is $\left(\dfrac{GMm}{r^2}\right)$. If the particle is lying inside the spherical shell then the force of gravitation on it is zero.

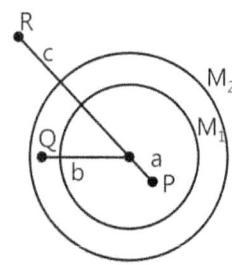

At P, F = 0 At Q, $F = \dfrac{GM_1 m}{b^2}$

At R, $F = \dfrac{G(M_1 + M_2)m}{c^2}$

Example 2: Find the potential energy of gravitational interaction of a point mass m and a thin uniform rod of mass M and length l, if they are located along a straight line at a distance a from each other.

Sol: The gravitational potential energy is given by $U = \dfrac{Gm_1 m_2}{r}$ where m_1 and m_2 are point masses.

Consider the gravitational potential energy of interaction between the point mass m and an infinitesimal element of the rod of mass dm. The total potential energy will be the summation of energy of interaction of all the small elements.

Consider small element dx of the rod whose mass $dm = \dfrac{M}{l}dx$

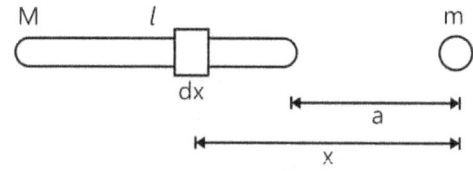

$\Rightarrow dU = -\dfrac{Gm\left(\dfrac{M}{l}dx\right)}{x}$

$\Rightarrow U = \int dU = -\dfrac{GmM}{l}\int_a^{a+l}\dfrac{dx}{x} = -\dfrac{GmM}{l}\left[\ln x\right]_a^{a+l}$

$\Rightarrow U = -\dfrac{GmM}{l}\log_e\left(\dfrac{a+l}{a}\right)$

Example 3: If the radius of the earth contracts to half of its present value without change in its mass, what will be the new duration of the day?

Sol: The angular momentum of the earth is given by $L = I\omega = \frac{2}{5}MR^2\omega$ since earth is considered to be sphere of uniform mass density. As there is no external force is acting on the earth, the angular momentum of the earth must remain constant after the radius of earth reduces to half of its original size. The time period of revolution is $T = \frac{2\pi}{\omega}$.

Present angular momentum of earth $L_1 = I\omega = \frac{2}{5}MR^2\omega$

New angular momentum because of change in radius

$$L_2 = \frac{2}{5}M\left(\frac{R}{2}\right)^2\omega'$$

If external torque is zero then angular momentum must be conserved

$L_1 = L_2$

$\frac{2}{5}MR^2\omega = \frac{1}{4}\times\frac{2}{5}MR^2\omega'$ i.e., $\omega' = 4\omega$

$T' = \frac{1}{4}T = \frac{1}{4}\times 24 = 6h$

Example 4: Two particles of equal mass go round a circle of radius R under the action of their mutual gravitational attraction. Find the speed of each particle.

Sol: As the particles go around the circle they always remain diametrically opposite to each other. To sustain their respective circular motion the necessary centripetal acceleration is provided by the gravitation force of attraction between them.

The particles will always remain diametrically opposite so that the force on each particle will be directed along the radius. Consider the motion of one of the particles.

The force on the particle is $F = \frac{Gm^2}{4R^2}$. If the speed is v, its acceleration is v^2/R.

Thus, by Newton's law,

$$\frac{Gm^2}{4R^2} = \frac{mv^2}{R} \quad \text{Or, } v = \sqrt{\frac{Gm}{4R}}$$

Example 5: Three particles A, B and C, each of mass m, are placed in a line with AB=BC=d. Find the gravitational force on a fourth particle P of same mass, placed at a distance d from the particle B on the perpendicular bisector of the line AC.

Sol: The gravitational force acting on the particle P due to each of other particles is given by $F = \frac{Gm^2}{(r)^2}$ where r is the separation between P and the other particle. As the force is vector quantity the resultant force on particle P has to be found by vector addition.

The force at P due to A is

$F_A = \frac{Gm^2}{(AP)^2} = \frac{Gm^2}{2d^2}$ along PA. The force at P due to C is

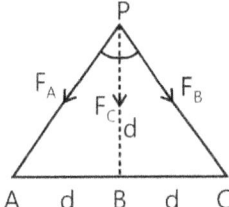

$F_C = \frac{Gm^2}{(CP)^2} = \frac{Gm^2}{2d^2}$ along PC. The force at P due to B is

$F_B = \frac{Gm^2}{d^2}$ along PB

The resultant of F_A, F_B and F_C will be along PB. Clearly $\angle APB = \angle BPC = 45°$

Components of F_A along $PB = F_A\cos 45° = \frac{Gm^2}{2\sqrt{2}d^2}$

Component of F_C along $PB = F_C\cos 45° = \frac{Gm^2}{2\sqrt{2}d^2}$

Component of F_B along $PB = \frac{Gm^2}{d^2}$

Hence, the resultant of the three forces is

$\frac{Gm^2}{d^2}\left(\frac{1}{2\sqrt{2}}+\frac{1}{2\sqrt{2}}+1\right) = \frac{Gm^2}{d^2}\left(1+\frac{1}{\sqrt{2}}\right)$ along PB.

Example 6: What is the fractional decrease in the value of free-fall acceleration g for a particle when it is lifted from the surface to an elevation h? (h<<R)

Sol: The gravitational acceleration g at height h is given by $g = \frac{GM}{(R+h)^2}$. As here R>>h then $g \approx \frac{GM}{R^2}$. The fractional decrease in g at height h above the surface of the earth is given by $\frac{\Delta g}{g}$.

1.23

The acceleration due to gravity is $g = \dfrac{GM}{R^2}$

$\therefore \dfrac{\Delta g}{\Delta R} = \dfrac{-2GM}{R^3}$ (Differentiating)

$\Rightarrow \dfrac{dg}{h} = \dfrac{-2GM}{R^2}\dfrac{1}{R} \Rightarrow \dfrac{dg}{g} = -2\left(\dfrac{h}{R}\right)$

Example 7: A double star is a system of two stars moving around the center of inertia of the system due to gravitation. Find the distance between the components of the double star, if its total mass equals M and period of revolution is T.

Sol: Each star is moving in circular orbit whose center is at the combined center of inertia. Find the radius of orbit of one of the stars in terms of the separation between them and find the orbital velocity of the star in terms of d.

Center of inertia

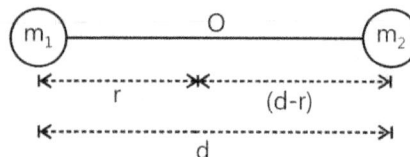

The situation is shown in the above figure

Here $m_1 r = m_2(d-r)$

$\therefore (m_1 + m_2)r = m_2 d$

$r = \dfrac{m_2 d}{(m_1 + m_2)}$

Also $M = (m_1 + m_2)$

As gravitational force provides the necessary centripetal force for rotation, we have

$G\dfrac{m_1 m_2}{d^2} = \dfrac{m_1 v_1^2}{r} = \dfrac{m_1 v_1^2 (m_1 + m_2)}{m_2 d}$

$\therefore v_1 = \left[\dfrac{Gm_2^2}{(m_1 + m_2)d}\right]^{1/2} = m_2\left[\dfrac{G}{Md}\right]^{1/2}$

Now $T = \dfrac{2\pi}{\omega} = \dfrac{2\pi r}{v_1} = \dfrac{2\pi r}{m_2\sqrt{G/Md}} = \dfrac{2\pi d^{3/2}}{\sqrt{GM}} \left(\text{as } r = \dfrac{m_2 d}{M}\right)$

$\therefore \dfrac{T}{2\pi} = \dfrac{d^{3/2}}{\sqrt{GM}}$ or $d = \sqrt[3]{\left(\dfrac{T}{2\pi}\right)^2 GM}$.

Example 8: Find the distance of a point from the earth's center where the resultant gravitational field due to the earth and the moon is zero. The mass of the earth is 6.0×10^{24} kg and that of the moon is 7.4×10^{22} kg. The distance between the earth and the moon is 4.0×10^5 km.

Sol: If a body is placed between moon and the earth then it is under action of gravitational force due to earth and moon simultaneously. When the gravitational field due to earth is equal in magnitude but opposite in direction to gravitational field due to moon then the net field is zero.

The point must be on the line joining the centers of the earth and the moon and in between them. If the distance of the point from the earth is x, the distance from the moon is $(4.0 \times 10^5 \text{ km} - x)$. The magnitude of the gravitational field due to the earth is

$E_1 = \dfrac{GM_e}{x^2} = \dfrac{G \times 6 \times 10^{24}\text{kg}}{x^2}$

and the magnitude of the gravitational field due to the moon is

$E_2 = \dfrac{GM_m}{\left(4.0 \times 10^5\text{km} - x\right)^2} = \dfrac{G \times 7.4 \times 10^{22}\text{kg}}{\left(4.0 \times 10^5\text{km} - x\right)^2}$

These fields are in opposite directions. For the resultant field to be zero $E_1 = E_2$,

Or, $\dfrac{6 \times 10^{24}\text{kg}}{x^2} = \dfrac{7.4 \times 10^{22}\text{kg}}{\left(4.0 \times 10^5\text{km} - x\right)^2}$

Or, $\dfrac{x}{4.0 \times 10^5\text{km} - x} = \sqrt{\dfrac{6 \times 10^{24}}{7.4 \times 10^{22}}} = 9$

Or, $x = 3.6 \times 10^5$ km

Example 9: A planet of mass m_1 revolves around the sun of mass m_2. The distance between the sun and the planet is r. Taking into consideration the motion of the sun, find the total energy of the system assuming the orbits to be circular.

Sol: The gravitational pull between sun and planet provides the necessary centripetal acceleration to make them revolve in circular orbits with same angular velocities. The center of each circular orbit will be at the combined center of mass but their radii will be different.

Both the planet and the sun revolve around their center of mass with same angular velocity (say ω)

COM

$$r = r_1 + r_2 \qquad \ldots \text{(i)}$$

$$m_1 r_1 \omega^2 = m_2 r_2 \omega^2 = \frac{Gm_1 m_2}{r^2} \qquad \ldots \text{(ii)}$$

Solving Eqs. (i) and (ii), we get

$$r_1 = r\left(\frac{m_2}{m_1 + m_2}\right)$$

$$r_2 = r\left(\frac{m_1}{m_1 + m_2}\right)$$

$$\omega^2 = \frac{G(m_1 + m_2)}{r^3}$$

And now, total energy of the system is E= P.E. + K.E.

$$\text{or } E = -\frac{Gm_1 m_2}{r} + \frac{1}{2}m_1 r_1^2 \omega^2 + \frac{1}{2}m_2 r_2^2 \omega^2$$

Substituting the values of r_1, r_2 and ω^2, we get

$$E = -\frac{Gm_1 m_2}{2r}.$$

Example 10: Two particles A and B of masses 1 kg and 2 kg respectively are kept 1 m apart and are released to move under mutual attraction. Find the speed of A when that of B is 3.6 cm/hour. What is the separation between the particles at this instant?

Sol: As the particles A and B are initially at rest, the system has potential energy only, but as they move towards each other the loss in potential energy is equal to gain in kinetic energy. As particle is moving under their mutual interaction, the linear momentum system must be conserved.

The linear momentum of the pair A+B is initially zero. As only mutual attraction is taken into account – which is internal when A+B is taken as the system – the linear momentum will remain zero. The particles move in opposite directions. If the speed of A is v when the speed of B is $3.6 \text{ cm/hour} = 10^{-5} \text{ m/s}$,

$$(1\text{kg})v = (2\text{kg})(10^{-5}\text{ms}^{-1})$$

$$\text{or, } v = 2 \times 10^{-5} \text{ms}^{-1}$$

The potential energy of the pair is $-\dfrac{Gm_A m_B}{R}$ with usual symbols. Initial potential energy

$$= -\frac{6.67 \times 10^{-11}\text{N} - \text{m}^2/\text{kg}^2 \times 2\text{kg} \times 1\text{kg}}{1\text{m}}$$

$$= -13.34 \times 10^{-11}\text{J}.$$

If the separation at the given instant is d, using conservation of energy,

$$-13.34 \times 10^{-11}\text{J} + 0$$

$$= -\frac{13.34 \times 10^{-11}\text{J} - \text{m}}{d} + \frac{1}{2}(2\text{ kg})\left(10^{-5}\text{m/s}\right)^2$$

$$+ \frac{1}{2}(1\text{kg})\left(2 \times 10^{-5}\text{m/s}\right)^2$$

Solving this, d = 0.31m.

Example 11: The gravitational field in a region is given by $\vec{E} = \left(10\text{Nkg}^{-1}\right)\left(\vec{i} + \vec{j}\right)$. Find the work done by an external agent to slowly shift a particle of mass of 2 kg from the point (0,0) to a point (5m, 4m).

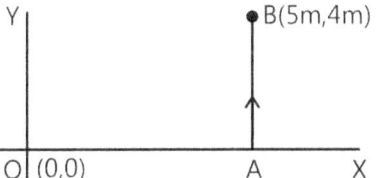

Sol: As the particle is moving slowly, the kinetic energy of the particle remains zero during its motion. The work done by the external agent to move the particle is given by $W = -\Delta U = \int_i^f \vec{F}.\vec{dr}$

As the particle is slowly shifted, its kinetic energy remains zero. The total work done on the particle is thus zero. The work done by the external agent should be negative of the work done by the gravitational field.

The work done by the field is $dW = -dU = \int_i^f \vec{F}.\vec{dr}$

Consider the figure. Suppose the particle is taken from O to A and then from A to B. The force on the particle is

$$\vec{F} = m\vec{E} = (2\text{kg})\left(10\text{Nkg}^{-1}\right)\left(\vec{i} + \vec{j}\right) = (20\text{N})\left(\vec{i} + \vec{j}\right)$$

The work done by the field during the displacement OA is

$$W_1 = \int_0^{5m} F_x dx = \int_0^{5m} (20\text{N})dx = 20\text{N} \times 5\text{m} = 100\text{J}.$$

Similarly, the work done in displacement AB is

$$W_2 = \int_0^{4m} F_y dy = \int_0^{4m} (20\text{N})dy = (20\text{N})(4\text{m}) = 80\text{J}$$

Thus, the total work done by the field, as the particle is shifted from O to B, is 180 J.

The work done by the external agent is -180 J.

Note that the work is independent of the path so that we can choose any path convenient to us from O to B.

1.25

Example 12: A uniform solid sphere of mass M and radius 'a' is surrounded symmetrically by a uniform thin and spherical shell of equal mass and radius 2a. Find the gravitational field at a distance

(a) $\dfrac{3}{2}$ a from the center, (b) $\dfrac{5}{2}$ a from the center.

Sol: If the particle is inside the spherical shell then the gravitation field due to the shell is zero. The gravitational field at distance r from the center of the sphere is given by $E = \dfrac{GM}{r^2}$.

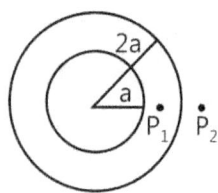

Given figure shows the situation. The point p_1 is at a distance $\dfrac{3}{2}$ a from the center and p_2 is at a distance $\dfrac{5}{2}$ a from the center. As p_1 is inside the cavity of the thin spherical shell, the field here due to the shell is zero. The field due to the solid sphere is

$$E = \frac{GM}{\left(\dfrac{3}{2}a\right)^2} = \frac{4GM}{9a^2}$$

This is also the resultant field. The direction is towards the center. The point p_2 is outside the sphere as well as the shell. Both may be replaced by single particles of the same mass at the center. The field due to each of them is

$$E' = \frac{GM}{\left(\dfrac{5}{2}a\right)^2} = \frac{4GM}{25a^2}$$

The resultant field is $E = 2E' = \dfrac{8GM}{25a^2}$ towards the center.

Example 13: A planet of mass m revolves in an elliptical orbit around the sun so that its maximum and minimum distance from the sun are equal to r_a and r_p respectively. Find the angular momentum of this planet relative to the sun.

Sol: At the apogee and perigee the radius vector is perpendicular to the velocity vector of the plane. Use the law of conservation of angular momentum and energy at these two points.

Using conservation of angular momentum
$$mv_p r_p = mv_a r_a$$

As velocities are perpendicular to the radius, vectors at apogee and perigee, $v_p r_p = v_a r_a$

Using conservation of energy,

$$-\frac{GMm}{r_p} + \frac{1}{2}mv_p^2 = \frac{-GMm}{r_a} + \frac{1}{2}mv_a^2$$

By solving, the above equations,

$$v_p = \sqrt{\frac{2GMr_a}{r_p\left(r_p + r_a\right)}} \; ; \;\; L = mv_p r_p = m\sqrt{\frac{2GMr_p r_a}{\left(r_p + r_a\right)}}$$

JEE Advanced/Boards

Example 1: The distance between the centers of two stars is 10 a. The masses of these stars are M and 16 M and their radii, 'a' and '2a' respectively. A body of mass m is fired straight from the surface of the larger star towards the smaller star. What should be its minimum initial speed to reach the surface of the smaller star? Obtain the expression in terms of G, M and a.

Sol: At a certain distance from the centers of the stars, the gravitational fields due to the stars are equal in magnitude but opposite in direction. As the body of mass m is projected from the surface of larger star towards the surface of smaller star, the kinetic energy lost by the body is equal to gain of its potential energy when it reaches at the point of zero field.

Let O be the point along O_1O_2 where gravitational intensities due to both the stars balance each other.

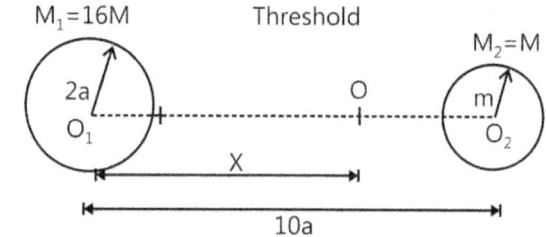

Let $O_1O = x$

$$\therefore \frac{GM_1}{x^2} = \frac{GM_2}{\left(10a - x\right)^2}$$

Or $16\left(10 - x\right)^2 = x^2$ or $x = 8a$

Potential energy of the body on surface of larger star,

$$U_1 = -\frac{Gm(16M)}{2a} - \frac{GmM}{8a} = -\frac{65GMm}{8a}$$

Potential energy at

$$O = -\frac{GMm}{2a} - \frac{G(16M)m}{8a} = -\frac{5GMm}{2a} = U_O$$

As $U_1 + \frac{1}{2}mv_{min}^2 = U_O$

$$\therefore -\frac{65GMm}{8a} + \frac{1}{2}mv_{min}^2 = -\frac{5GMm}{2a}$$

or $\frac{1}{2}mv_{min}^2 = \frac{65GMm}{8a} - \frac{5GMm}{2a} = \frac{45}{8}\frac{GMm}{a}$

$$v_{min}^2 \frac{45}{4}\frac{GM}{a} = \frac{9 \times 5}{4} \times \frac{GM}{a}$$

$$v_{min} = \frac{3}{2} \times \sqrt{\frac{5GM}{a}}$$

Example 2: Two masses m_1 and m_2, at an infinite distance from each other are initially at rest, start interacting gravitationally. Find their velocity of approach when they are at a distance r apart.

Sol: As the masses move towards each other gain in kinetic energy is equal to loss in gravitational potential energy. This problem is best solved in center of mass frame where the total kinetic energy of masses depends on the square of velocity of approach.

Let v_r be their velocity of approach. From conservation of energy:

Increase in kinetic energy=decrease in gravitational potential energy

Or $\frac{1}{2}\mu v_r^2 = \frac{Gm_1m_2}{r}$ ... (i)

Here, μ = reduced mass $= \frac{m_1m_2}{m_1 + m_2}$

Substituting in Eq. (i), we get

$$v_r = \sqrt{\frac{2G(m_1 + m_2)}{r}}$$

Example 3: A satellite is revolving round the earth in a circular orbit of radius e and velocity v_0. A particle is projected from the satellite in forward direction with relative velocity $v = \left(\sqrt{5/4} - 1\right)v_0$. Calculate its

minimum and maximum distance from earth's center during subsequent motion of the particle.

Sol: As the particle is projected from the satellite while the satellite is still in circular motion, the net velocity of the particle is sum of velocity relative to satellite and the velocity of the satellite. As the particle is still bound to the gravitational attraction of the earth, the orbit of the particle will be ellipse. The point of projection is perigee. Conserve the angular momentum at the apogee and perigee.

The orbital speed of satellite is

$$v_o = \sqrt{\frac{GM}{r}}$$ ... (i)

Where M=mass of earth

Absolute velocity of particle would be:

$$v_p = v + v_o = \sqrt{\frac{5}{4}}v_o = \sqrt{1.25}\,v_o$$ ... (ii)

Since, v_p lies between orbital velocity and escape velocity, path of the particle would be an ellipse with r being the minimum distance.

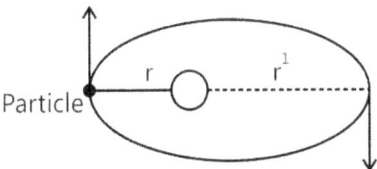

Let r' be the maximum distance and v_p' its velocity at the moment. $v_p = \sqrt{\frac{5}{4}}v_o$

Then, from the conservation of angular momentum and conservation of mechanical energy, we get

$$mv_p r = mv_p' r'$$ ... (iii)

$$\frac{1}{2}mv_p^2 - \frac{GMm}{r} = \frac{1}{2}mv_p'^2 - \frac{GMm}{r'}$$ ... (iv)

Solving the above Eqs. (i), (ii), (iii) and (iv), we get $r' = \frac{5r}{3}$ and r.

Hence, the maximum and minimum distance are $\frac{5r}{3}$ and r respectively.

Example 4: An earth satellite is revolving in a circular orbit of radius 'a' with velocity v_o. A gun is in the satellite and is aimed towards the earth. A bullet is fired from the gun with muzzle velocity $\dfrac{v_o}{2}$. Neglecting resistance offered by cosmic dust and recoil of gun, calculate maximum and minimum distance of bullet from the center of earth during its subsequent motion.

Sol: Conserve the angular momentum and energy of the particle between the points, the point of projection and at perigee. At perigee velocity is perpendicular to radius.

The orbital speed of the satellite is

$$v_o = \sqrt{\dfrac{GM}{a}} \qquad \dots (i)$$

From conservation of angular momentum at P and Q we have

$$mav_o = mvr$$

Or $v = \dfrac{av_o}{r}$ $\qquad \dots(ii)$

From conservation of mechanical energy at P and Q. we have

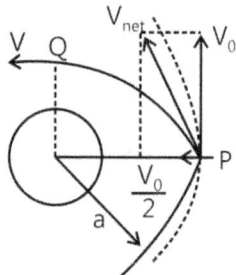

$$\dfrac{1}{2}m\left(v_o^2 + \dfrac{v_o^2}{4}\right) - \dfrac{GMm}{a} = \dfrac{1}{2}mv^2 - \dfrac{GMm}{r}$$

or $\dfrac{5}{8}v_o^2 - \dfrac{GM}{a} = \dfrac{v^2}{2} - \dfrac{GMm}{r}$

Substituting values of v and v_o from Eqs. (i) and (ii), we get

$$\dfrac{5}{8}\dfrac{GM}{a} - \dfrac{GM}{a} = \dfrac{a^2}{r^2}\left(\dfrac{GM}{2a}\right) - \dfrac{GM}{r}$$

or $-\dfrac{3}{8a} = \dfrac{a}{2r^2} - \dfrac{1}{r}$ or $-3r^2 = 4a^2 - 8ar$

or $3r^2 - 8ar + 4a^2 = 0$

or $r = \dfrac{8a \pm \sqrt{64a^2 - 48a^2}}{6}$

or $r = \dfrac{8a \pm 4a}{6}$ or $r = 2a$ and $\dfrac{2a}{3}$

Hence, the maximum and minimum distance are $2a$ and $\dfrac{2a}{3}$ respectively.

Example 5: Binary stars of comparable masses m_1 and m_2 rotate under the influence of each other's gravity with a time period T. If they are stopped suddenly in their motion, find their relative velocity when they collide with each other. The radii of the stars are R_1 and R_2 respectively. G is the universal constant of gravitation.

Sol: They rotate about center of mass, such that the necessary centripetal acceleration for the rotational motion is provided by the gravitational force of attraction. As the stars start approaching each other and collide, the loss in the gravitational energy of system is equal to the gain in the kinetic energy of the system. Find the initial separation in terms of the time period.

Both the stars rotate about their center of mass (COM).

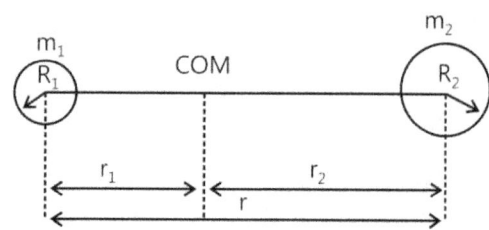

For the position of COM

$$\dfrac{r_1}{m_2} = \dfrac{r_2}{m_1} = \dfrac{r_1 + r_2}{m_1 + m_2} = \dfrac{r}{m_1 + m_2} \quad (r = r_1 + r_2)$$

Also, $m_1 r_1 \omega^2 = \dfrac{Gm_1m_2}{r^2}$ or $\omega^2 = \dfrac{Gm_2}{r_1 r^2}$ $\left(\omega = \dfrac{2\pi}{T}\right)$

But, $r_1 = \dfrac{m_2 r}{m_1 + m_2}$

$\therefore \omega^2 = \dfrac{G(m_1 + m_2)}{r^3}$

Or $r = \left\{\dfrac{G(m_1 + m_2)}{\omega^2}\right\}^{1/3}$ $\qquad \dots(i)$

Applying conservation of mechanical energy we have

$$-\dfrac{Gm_1m_2}{r} = -\dfrac{Gm_1m_2}{(R_1 + R_2)} + \dfrac{1}{2}\mu v_r^2 \qquad \dots(ii)$$

1.28

Here, μ = reduced mass = $\dfrac{m_1 m_2}{m_1 + m_2}$ and

v_r = relative velocity between the two stars.

From Eq. (ii), we find that

$$v_r^2 = \dfrac{2Gm_1 m_2}{\mu}\left(\dfrac{1}{R_1 + R_2} - \dfrac{1}{r}\right)$$

$$= \dfrac{2Gm_1 m_2}{\dfrac{m_1 m_2}{m_1 + m_2}}\left(\dfrac{1}{R_1 + R_2} - \dfrac{1}{r}\right)$$

$$= 2G(m_1 + m_2)\left(\dfrac{1}{R_1 + R_2} - \dfrac{1}{r}\right)$$

Substituting the value of r from Eq. (i), we get

$$v_r = \sqrt{2G(m_1 + m_2)\left[\dfrac{1}{R_1 + R_2} - \left\{\dfrac{4\pi^2}{G(m_1 + m_2)T^2}\right\}^{1/3}\right]}$$

Example 6: Find the maximum and minimum distance of the planet A from the sun S, if at a certain moment of time it was a distance r_0 and travelling with the velocity v_0 with the angle between the radius vector and velocity vector being equal to ϕ.

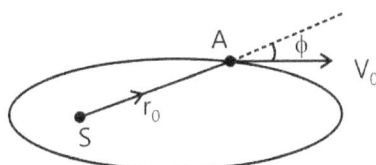

Sol: As the planet revolves around the sun, the mechanical energy of the system is conserved. Conserve the angular momentum between the given point and apogee.

At minimum and maximum distance velocity vector $\left(\vec{v}\right)$ makes an angle of $90°$ with radius vector. Hence, from conservation of angular momentum,

$$mv_0 r_0 \sin\phi = mrv \qquad \text{...(i)}$$

Here, m is the mass of the planet.

From energy conservation law it follows that.

$$\dfrac{mv_0^2}{2} - \dfrac{GMm}{r_0} = \dfrac{mv^2}{2} - \dfrac{GMm}{r} \qquad \text{...(ii)}$$

Here, M is the mass of the sun.

Solving Eqs. (i) and (ii) for r, we get we values of r, one is r_{max} and another is r_{min}. So,

$$r_{max} = \dfrac{r_0}{2 - K}\left(1 + \sqrt{1 - K(2 - K)\sin^2\phi}\right)$$

and $r_{min} = \dfrac{r_0}{2 - K}\left(1 - \sqrt{1 - K(2 - K)\sin^2\phi}\right)$

Here, $K = \dfrac{r_0^2 v_0^2}{GM}$

Example 7: The density inside a solid sphere of radius 'a' is given by $\rho = \rho_0\, a/r$, where ρ_0 is the destiny at the surface and r denotes the distance from the center. Find the gravitational field due to this sphere at a distance of '2a' from its center.

Sol: The given mass distribution is having spherical symmetry. Any spherically symmetrical body can be replaced by a point particle of the same mass situated at the center of the spherical body. The gravitational field due to the sphere at the point 2a from the center of the sphere is given by $E = \dfrac{GM}{(2a)^2}$

The field is required at a point outside the sphere. Dividing the sphere in concentric shells, each shell can be replaced by a point particle at its center having mass equal to the mass of the shell. Thus, the whole sphere can be replaced by a point particle at its center having mass equal to the mass of the given sphere. If the mass of the sphere is M, the gravitational field at the given point is

$$E = \dfrac{GM}{(2a)^2} = \dfrac{GM}{4a^2} \qquad \text{... (i)}$$

The mass M may be calculated as follows: Consider a concentric shell of radius r and thickness dr. Its volume is

$$dV = \left(4\pi^2\right)dr$$ and its mass is

$$dM = \rho dV = \left(\rho_0\dfrac{a}{r}\right)\left(4\pi^2 dr\right) = 4\rho_0 ar\, dr.$$

The mass of the whole sphere is

$$M = \int_0^a 4\rho_0 ar\, dr = 2\pi\rho_0 a^3$$

Thus, by (i) the gravitational field is

$$E = \dfrac{2\pi G\rho_0 a^3}{4a^2} = \dfrac{1}{2} \times G\rho_0 a.$$

Example 8: Two satellites of same mass are launched in the same orbit round the earth so as to rotate opposite to each other. They collide solidly and stick together as wreckage. Obtain the total energy of the system before and just after the collision. Describe the subsequent motion of the wreckage.

Sol: Both the satellites are moving in the same orbit so their orbital velocity will be same. As the masses of the satellites are equal, and they are moving in the opposite direction their total momentum before and after the collision is zero.

The two satellites round the earth are shown in figure

Potential energy of the satellite in its orbit $= -GMm/r$

Kinetic energy of satellite in its orbit is

$K = GMm/2r$

Where m is mass of satellite, M is the mass of the earth and r is the orbital radius.

Total energy $= \dfrac{GMm}{2r} - \dfrac{GMm}{r} = -\dfrac{GMm}{2r}$

When there are two satellites, the total energy would be

$$\left(-\dfrac{GMm}{2r}\right) + \left(-\dfrac{GMm}{2r}\right) = \left(-\dfrac{GMm}{r}\right)$$

Let after collision, v' be the velocity of wreckage by the law of conservation of momentum $mv - mv = (m+m)v'$

$\therefore v' = 0$

The wreckage of mass (2m) has no kinetic energy, but it has only potential energy,

So, energy after collision $= -\dfrac{GM(2m)}{r}$

Now the combined mass has zero velocity just after collision and therefore, the wreckage stops rotating and falls down under gravity.

JEE Main/Boards

Exercise 1

Q.1 Why Newton's law of gravitation is called a universal law?

Q.2 On earth value of $G = 6.67 \times 10^{-11} Nm^2kg^2$. What is its value on moon, where g is nearly one-sixth than that of earth?

Q.3 An artificial satellite is revolving around the earth at a height 200 km from the earth's surface. If a packet is released from the satellite, what will happen to it? Will it reach the earth?

Q.4 A spring balance is suspended inside an artificial satellite revolving around the earth. If a boy of mass 2 kg is suspended from it, what would be its reading?

Q.5 The escape velocity from earth for a piece of 10 gram is $11.2\,kms^{-1}$. What would it be for a piece of mass 100 gram?

Q.6 Where will the true weight of the body be zero?

Q.7 If the force of gravity acts on all bodies in proportion to their masses, why does not a heavy body fall correspondingly faster than a light body.

Q.8 The gravitational potential energy of a body at a distance from the center of earth is U. What is the weight of the body at that point?

Q.9 The distance of two planets from the sun are 10^{11} and 10^{10} meters respectively. What is the ratio of time periods of these two planets?

Q.10 For a satellite, escape speed is $11\,kms^{-1}$. If the satellite is launched at an angle of 60° with the vertical, what will be the escape speed?

Q.11 Prove that the value of acceleration due to gravity at a point above the surface of the earth is inversely proportional to the square of the distance of that point from the center of the earth.

Q.12 Gravitational force between two bodies is 1 newton. If the distance between them is made twice, what will be the force?

Q.13 If a person goes to a height equal to radius of earth from its surface, what would be his weight relative to that on the earth?

Q.14 If the change in the value of g at a height h above the surface of the earth is the same as at a depth x below it, both x and h being much smaller than the radius of the earth, find the relation between x and h.

Q.15 The gravitational force acting on a rocket at a height h from the surface of earth is 1/3 of the force acting on a body at sea level. What is the ratio of h and R (radius of earth)?

Q.16 Does the gravitational force of attraction of the earth becomes zero at some height above the surface of earth? Explain.

Q.17 What do you understand by gravity and acceleration due to gravity. Establish a relation between g and G.

Q.18 Explain how the knowledge of g helps us to find (i) mass of earth and (ii) mean density of earth?

Q.19 What do you understand by 'Escape velocity'? Derive an expression for it in terms of parameters of given planet.

Q.20 What do you understand by Gravitational field, Intensity of gravitational field. Prove that gravitational intensity at a point is equal to the acceleration due to gravity at that point.

Q.21 Explain Kepler's laws of planetary motion and deduce Newton's law of gravitation from them.

Q.22 Explain Newton's law of gravitation. Define gravitational constant, and give its dimensional formula. Give the evidences in support of the Newton's law of gravitation.

Q.23 Let the speed of the planet at the perihelion P in the given figure be v_p and the sun-planet distance SP be r_p. Relate r_p, v_p to the corresponding quantities at the aphelion (r_A, v_A). Will the planet take equal take equal times to traverse BAC and CPB?

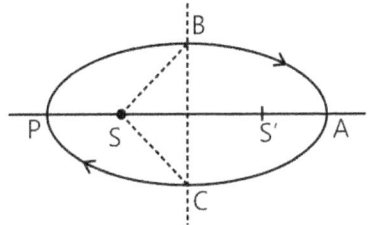

Q.24 Two satellites of a planet have period 32 days and 256 days. If the radius of the orbit of former is R, find the orbital radius of the latter.

Q.25 If the distance of the earth from the sun were half the present value, how many days will make one year? Given, 1 year= 365 days.

Q.26 Estimate the mass of the sun, assuming the orbit of earth round the sun to be a circle. The distance between the sun and the earth is $1.49 \times 10^{11} m$, and $G = 6.66 \times 10^{-11} Nm^2 kg^{-2}$.

Q.27 If the mass of the sun is $2 \times 10^{30} kg$, the distance of the earth from sun is $1.5 \times 10^{11} m$ and period of revolution of the earth around sun is one year (= 365.3 days), calculate the value of gravitational constant.

Q.28 Calculate the mass and mean density of earth from the following data:

Radius of earth $= 6.37 \times 10^6 m$, acceleration due to gravity $= 9.8 ms^{-2}$ and

Gravitational constant $= 6.6 \times 10^{-11} Nm^2 kg^{-2}$

Q.29 If the radius of the earth shrinks by 2.5%, mass remaining constant, then how would the value of acceleration due to gravity change?

Q.30 At what altitude the acceleration due to gravity above the earth's surface would be half of its value on the surface of the earth? Radius of earth is 6400 km.

Q.31 The radius of earth is approximately 6000 km. What will be your weight at 600 km above the surface of earth? At 12000 km above? At 18000 km above? Your weight on earth is 80 kg wt.

Q.32 At what height from the surface of earth, the acceleration due to gravity is the same at a depth 160 km below the surface of earth. Radius of earth is 6400 km.

1.31

Q.33 What is the minimum energy required to launch a satellite of mass m from the surface of earth of mass M, radius R in a circular orbit at an attitude 2 R.

Q.34 A rocket is launched vertically from the surface of the earth with an initial velocity $10 \, kms^{-1}$. How far above the surface of the earth would it go? Radius of the earth $= 6400 \, km; g = 9.8 \, ms^{-2}$.

Q.35 A remote sensing satellite of the earth revolves in a circular orbit at a height of 250km above the earth's surface. What is the (a) orbital speed, and (b) period of revolution of satellite? Radius of the earth $= 6.38 \times 10^6$ m, and acceleration due to gravity at the surface of earth $= 9.8 \, ms^{-2}$

Q.36 A satellite revolves round a planet in an orbit just above the surface of planet. Taking $G = 6.67 \times 10^{-11} \, Nm^2kg^{-2}$ and the means density of the planet $= 5.51 \times 10^3 \, kgm^{-3}$, find the period of satellite.

Q.37 Find the speed of escape at the moon given that its radius 1.7×10^6 m and the value of g at its surface is $1.63 \, ms^{-2}$.

Q.38 If the earth has a mass nine times and radius twice that of the planet Mars, calculate the maximum speed required by a rocket to pull out of the gravitational force of Mars. Given escape speed on the surface of earth is $11.2 \, kms^{-1}$

Exercise 2

Single Correct Choice Type

Q. 1 At what altitude will the acceleration due to gravity be 25% of that at the earth's surface (given radius of earth is R)?

(A) R/4   (B) R   (C) 3R/8   (D) R/2

Q.2 Let ω be the angular velocity of the earth's rotation about its axis. Assume that the acceleration due to gravity on the earth's surface has the same value at the equator and the poles. An object weighed at the equator gives the same reading as a reading taken at a depth d below earth's surface at a pole (d<<R). The value of d is

(A) $\dfrac{\omega^2 R^2}{g}$ (B) $\dfrac{\omega^2 R^2}{2g}$ (C) $\dfrac{2\omega^2 R^2}{g}$ (D) $\dfrac{\sqrt{Rg}}{g}$

Q.3 If the radius of the earth be increased by a factor of 5, by what factor its density be changed to keep the value of g the same?

(A) 1/25   (B) 1/5   (C) $1/\sqrt{5}$ (D) 5

Q.4 The mass and diameter of a planet are twice those of earth. What will be the period of oscillation of a pendulum on this planet if it is a second's pendulum on earth?

(A) $\sqrt{2}$ second   (B) $2\sqrt{2}$ second

(C) $\dfrac{1}{\sqrt{2}}$ second   (D) $\dfrac{1}{2\sqrt{2}}$ second

Q.5 A particle of mass M is at a distance a form surface of a thin spherical shell of equal mass and having radius a.

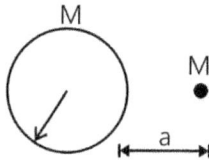

(A) Gravitational field and potential both are zero at center of the shell.

(B) Gravitational field is zero not only inside the shell but at a point outside the shell also.

(C) Inside the shell, gravitational field alone is zero.

(D) Neither gravitational field nor gravitational potential is zero inside the shell.

Q.6 A spherical uniform planet is rotating about its axis. The velocity of a point on its equator is v. Due to the rotation of planet about its axis the acceleration due to gravity g at equator is $\dfrac{1}{2}$ of g at poles. The escape velocity of a particle on the pole of planet in terms of V.

(A) $v_e = 2v$ (B) $v_e = v$

(C) $v_e = v\sqrt{2}$ (D) $v_e = \sqrt{3}v$

Q.7 Two planets A and B have the same material density. If the radius of A is twice that of B, then the ratio of the escape velocity $\dfrac{v_A}{v_B}$ is.

(A) 2   (B) $\sqrt{2}$ (C) $1/\sqrt{2}$ (D) $1/2$

Q.8. The escape velocity for a planet is v_e. A tunnel is dug along a diameter of the planet and a small body is dropped into it at the surface. When the body reaches the center of the planet, its speed will be

(A) v_e (B) $\dfrac{v_e}{\sqrt{2}}$ (C) $\dfrac{v_e}{2}$ (D) Zero

Q.9 A hollow spherical shell is compressed to half its radius. The gravitational potential at the center

(A) Increases

(B) Decreases

(C) Remains same

(D) During the compression increases then returns at the previous value.

Q.10 A (nonrotating) star collapses onto itself from an initial radius R_i with its mass remaining unchanged. Which curve in the following figure best gives the gravitational acceleration a_g on the surface of the star as function of the radius of the star during the collapse?

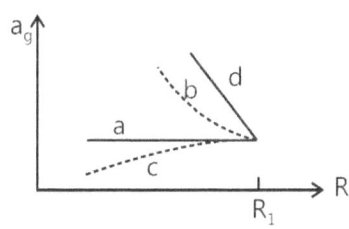

(A) a   (B) b   (C) c   (D) d

Q.11 A mass is at the center of a square, with four masses at the corners as shown

Rank the choices according to the magnitude of the gravitational force on the center mass.

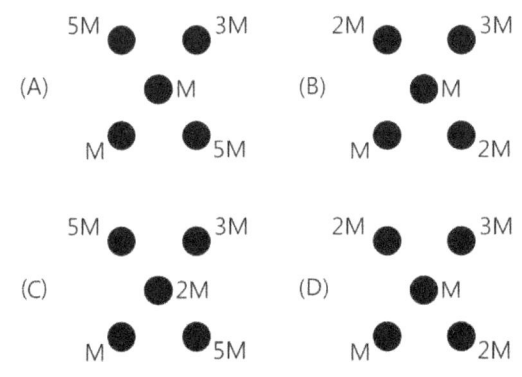

(A) $F_A = F_B < F_C = F_D$ (B) $F_A > F_B < F_D < F_C$

(C) $F_A = F_B > F_C = F_D$ (D) None

Q.12 A satellite of the earth is revolving in circular orbit with a uniform velocity V. If the gravitational force suddenly disappears, the satellite will

(A) Continue to move with the same velocity in the same orbit.

(B) Move tangentially to the original orbit with velocity V.

(C) Fall down with increasing velocity.

(D) Come to a stop somewhere in its original orbit.

Q.13 A satellite revolves in the geostationary orbit but in a direction east to west. The time interval between its successive passing about a point on the equator is:

(A) 48 hrs   (B) 24 hrs

(C) 12 hrs   (D) Never

Q.14 Two point masses of mass 4m and m respectively separated by d distance are revolving under mutual force of attraction. Ratio of their kinetic energies will be:

(A) 1:4   (B) 1:5   (C) 1:1   (D) 1:2

Q.15 Select the correct choice(s)

(A) The gravitational field inside a spherical cavity, within a spherical planet must be non-zero and uniform.

(B) When a body is projected horizontally at an appreciable large height above the earth, with a velocity less than for a circular orbit, it will fall to the earth along a parabolic path

(C) A body of zero total mechanical energy placed in a gravitational field if it is travelling away from source of field will escape the field.

(D) Earth's satellite must be in equatorial plane.

Q.16 A satellite of mass m, initially at rest on the earth, is launched into a circular orbit at a height equal to the radius of the earth. The minimum energy required is.

(A) $\dfrac{\sqrt{3}}{4}mgR$ (B) $\dfrac{1}{2}mgR$

(C) $\dfrac{1}{4}mgR$ (D) $\dfrac{3}{4}mgR$

Q.17 The following figure shows the variation of energy with the orbit radius of a circular planetary motion. Find the correct statement about the curves A, B and C

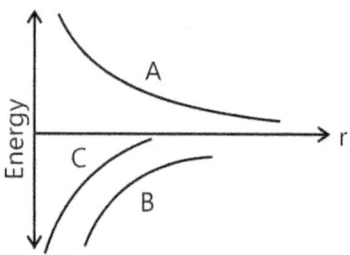

(A) A shows the kinetic energy, B the total energy and C the potential energy of the system.

(B) C shows the total energy, V the kinetic energy and A the potential energy of the system.

(C) C and A are kinetic and potential energies respectively and B is the total energy of the system.

(D) A and B are kinetic and potential energies and C is the total energy of the system.

Q.18 When a satellite moves around the earth in a certain orbit, the quantity which remains constant is:

(A) Angular velocity   (B) Kinetic energy

(C) Aerial velocity   (D) Potential energy

Q.19 A satellite of mass 5 M orbits the earth in a circular orbit. At one point in its orbit, the satellite explodes into two pieces, one of mass M and the other of mass 4M. After the explosion the mass M ends up travelling in the same circular orbit, but in opposite direction. After explosion the mass 4M is.

(A) In a circular orbit

(B) Unbound

(C) Elliptical orbit

(D) Data is insufficient to determine the nature of the orbit.

Q.20 A satellite can be in a geostationary orbit around earth at a distance r from the center. If the angular velocity of earth about its axis doubles, a satellite can now be in a geostationary orbit around earth if its distance from the center is

(A) $\dfrac{r}{2}$ (B) $\dfrac{r}{2\sqrt{2}}$ (C) $\dfrac{r}{(4)^{1/3}}$ (D) $\dfrac{r}{(2)^{1/3}}$

Q.21 A planet of mass m is in an elliptical orbit around the sun $(m \ll M_{sun})$ with an orbital period T. If A be the area of orbit, then its angular momentum would be:

(A) $\dfrac{2mA}{T}$ (B) mAT   (C) $\dfrac{mA}{2T}$ (D) 2mAT

Q.22 Satellite A and B are orbiting around the in orbits of ratio R and 4R respectively. The ratio of their aerial velocities is:

(A) 1:2   (B) 1:4   (C) 1:8   (D) 1:16

Q.23 In older times, people used to think that the Earth was flat. Imagine that the earth is indeed not a sphere of radius R, but an infinite plate of thickness H. What value of H is needed to allow the same gravitational acceleration to be experienced as on the surface of the actual Earth? (Assume that the Earth's density is uniform and equal in the two models.)

(A) $\dfrac{2R}{3}$ (B) $\dfrac{4R}{3}$ (C) $\dfrac{8R}{3}$ (D) $\dfrac{R}{3}$

Q.24 A planet revolves about the sun in elliptical orbit. The aerial velocity $\left(\dfrac{dA}{dt}\right)$ of the planet is $4.0 \times 10^{16}\,\text{m}^2/\text{s}$. The least distance between planet and the sun is $2 \times 10^{12}\,\text{m}$. Then the maximum speed of the planet in km/s is:

(A) 10   (B) 20   (C) 40   (D) None of these

Previous Years' Questions

Q.1 If the radius of x the earth were to shrink by one per cent, its mass remaining the same, the acceleration due to gravity on the earth's surface would   *(1981)*

(A) Decrease   (B) Remain unchanged

(C) Increase   (D) Be zero

Q.2 If g is the acceleration due to gravity on the earth's surface, the gain in the potential energy of an object of mass m raised from the surface of the earth to a height equal to the radius R of the earth, is   *(1983)*

(A) $\dfrac{1}{2}mgR$ (B) 2mgR

(C) mgR   (D) $\dfrac{1}{4}mgR$

Q.3 Imagine a light planet revolving around a very massive star in a circular orbit of radius R with a period of revolution T. If the gravitational force of attraction between the planet and the star is proportional to $R^{-5/2}$, then   *(1989)*

(A) T^2 is proportional to R^2

(B) T^2 is proportional to $R^{7/2}$

(C) T^2 is proportional to $R^{3/2}$

(D) T^2 is proportional to $R^{3.75}$

Q.4 If the distance between the earth and the sun were half its present value, the number of days in a year would have been *(1996)*

(A) 64.5   (B) 129   (C) 182.5   (D) 730)

Q.5 A simple pendulum has a time period T_1 when on the earth's surface and T_2 when taken to a height R above the earth's surface, where R is the radius of the earth. The value of T_2/T_1 is *(2001)*

(A) 1   (B) $\sqrt{2}$ (C) 4   (D) 2

Q.6 A geostationary satellite orbits around the earth in a circular orbit of radius 36,000km. Then, the time period of a spy satellite orbiting a few hundred km above the earth's surface $(R_e = 6400km)$ will approximately be *(2002)*

(A) 1/2h   (B) 1 h   (C) 2 h   (D) 4 h

Q.7 A double star system consists of two stars A and B which have time periods T_A and T_B. Radius R_A and R_B and mass M_A and M_B. Choose the correct option. *(2006)*

(A) $T_A > T_B$ then $R_A > R_B$

(B) if $T_A > T_B$ then $M_A > M_B$

(C) $\left(\dfrac{T_A}{T_B}\right)^2 = \left(\dfrac{R_A}{R_B}\right)^3$

(D) $T_A = T_B$

Q.8 A satellite is moving with a constant speed v in a circular orbit about the earth. An object of mass m is ejected from the satellite such that it just escapes from the gravitational pull of the earth. At the time of its ejection, the kinetic energy of the object is *(2011)*

(A) $\dfrac{1}{2}mv^2$ (B) mv^2 (C) $\dfrac{3}{2}mv^2$ (D) $2mv^2$

Q.9 A planet in a distant solar system is 10 times more massive than the earth and its radius is 10 times smaller. Given that the escape velocity from the earth is 11 kms^{-1}, the escape velocity from the surface of the planet would be *(2008)*

(A) 1.1 kms^{-1} (B) 11 kms^{-1}

(C) 110 kms^{-1} (D) 0.11 kms^{-1}

Q. 10 Statement-I: For a mass M kept at the centre of a cube of side 'a', the flux of gravitational field passing through its sides is 4π GM.

Statement-II: If the direction of a field due to a point source is radial and its dependence on the distance 'r' for the source is given as $1/r^2$, its flux through a closed surface depends only on the strength of the source enclosed by the surface and not on the size or shape of the surface *(2008)*

(A) Statement-I is false, statement-II is true.

(B) Statement-I is true, statement-II is true; statement-II is correct explanation for statement-I.

(C) Statement-I is true, statement-II is true; statement-II is not a correct explanation for statement-I.

(D) Statement-I is true, statement-II is false.

Q.11 The height at which the acceleration due to gravity becomes $\dfrac{g}{9}$ (where g = the acceleration due to gravity on the surface of the earth) in terms of R, the radius of the earth is *(2009)*

(A) 2R   (B) $\dfrac{R}{\sqrt{2}}$ (C) $\dfrac{R}{2}$ (D) $\sqrt{2}R$

Q.12 Two bodies of masses m and 4 m are placed at a distance r. The gravitational potential at a point on the line joining them where the gravitational field is zero is: *(2011)*

(A) $-\dfrac{4Gm}{r}$ (B) $-\dfrac{6Gm}{r}$

(C) $-\dfrac{9Gm}{r}$ (D) Zero

Q.13 The mass of a spaceship is 1000 kg. It is to be launched from the earth's surface out into free space. The value of 'g' and 'R' (radius of earth) are 10 m/s^2 and 6400km respectively. The required energy for this work will be: *(2012)*

(A) 6.4×10^{11} Joules   (B) 6.4×10^8 Joules

(C) 6.4×10^9 Joules   (D) 6.4×10^{10} Joules

Q.14 What is the minimum energy required to launch a satellite of mass m from the surface of a planet of mass M and radius R in a circular orbit at an altitude of 2R? *(2013)*

(A) $\dfrac{5GmM}{6R}$

(B) $\dfrac{2GmM}{3R}$

(C) $\dfrac{GmM}{2R}$

(D) $\dfrac{GmM}{3R}$

Q.15 From a solid sphere of mass M and radius R, a spherical portion of radius R/2 is removed, as shown in the figure. Taking gravitational potential V = 0 at r = ∞, the potential at the centre of the cavity thus formed is: (G = gravitational constant) *(2015)*

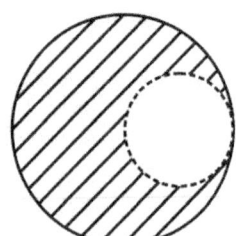

(A) $\dfrac{-GM}{2R}$

(B) $\dfrac{-GM}{R}$

(C) $\dfrac{-2GM}{3R}$

(D) $\dfrac{-2GM}{R}$

Q.16 A satellite is revolving in a circular orbit at a height 'h' from the earth's surface (radius or earth R; h<<R). The minimum increase in its orbital velocity required, so that the satellite could escape from the earth's gravitational field, is close to: (Neglect the effect of atmosphere.) *(2016)*

(A) $\sqrt{gR}$

(B) $\sqrt{gR/2}$

(C) $\sqrt{gR}(\sqrt{2}-1)$

(D) $\sqrt{2gR}$

JEE Advanced/Boards

Exercise 1

Q.1 A small mass and a thin uniform rod each of mass 'm' are positioned along the same straight line as shown. Find the force of gravitational attraction exerted by the rod on the small mass.

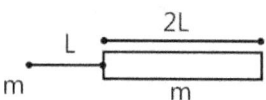

Q.2 A particle is forced vertically form the surface of the earth with a velocity kv_e where v_e is the escape velocity and k<1. Neglecting air resistance and assuming earth's radius as R_e, calculate the height to which it will rise from the surface of the earth.

Q.3 A point P lies on the axis of a fixed ring of mass M and radius a, at a distance a from its center C. A small particle starts from P and reaches C under gravitational attraction only. Its speed at C will be_____.

Q.4 Calculate the distance from the surface of the earth at which above and below the surface acceleration due to gravity is the same.

Q.5 An object is projected vertically upward from the surface of the earth of mass M with a velocity such that the maximum height reached is eight times the radius R of the earth. Calculate:

(i) The initial speed of projection

(ii) The speed at half the maximum height.

Q.6 A sphere of radius R has it center at the origin. It has a uniform mass density ρ_0 except that there is a spherical hole of radius r = R/2 whose center is at x = R/2 as in the given figure. (a) Find gravitational field at points on the axis for x>R

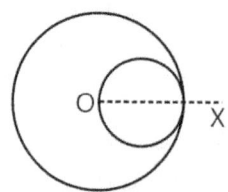

(b) Show that the gravitational field inside the hole is uniform. Find its magnitude and direction.

Q.7 A small body of mass is projected with a velocity just sufficient to make it reach from the surface of a planet (of radius 2R and mass 3M) to the surface of another planet (of radius R and mass M). The distance between the centers of the two spherical planet is 6R. The distance of the body from the center of bigger planet is 'x' at any moment. During the journey, find the distance x where the speed of the body is (a) maximum (b) minimum. Assume motion of body along the line joining centers of planets.

Q.8 A man can jump over b=4m wide trench on earth. If mean density of an imaginary planet is twice that of the earth, calculate its maximum possible radius so that he may escape from it by jumping. Given radius of earth=6400km.

Q.9 A satellite P is revolving around the earth at a height h = radius of earth (R) above equator. Another satellite Q is at a height 2h revolving in opposite direction. At an instant the two are at same vertical line passing through center of sphere. Find the least time after which again they are in this situation.

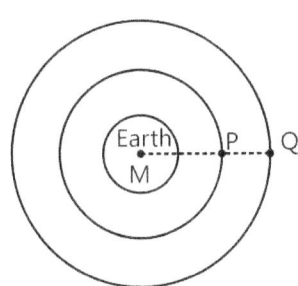

Q.10 Two small dense stars rotate about their common center of mass as a binary system with the period 1 year for each. One star is of double the mass of the other and the mass of the lighter one is 1/3 of the mass of the sun. Find the distance between the stars if distance between the earth & the sun is R.

Q.11 Four masses (each of m) are placed at the vertices of a regular pyramid (triangular base) of side 'a'. Find the work done by the system while taking them apart so that they form the pyramid of side '2a'

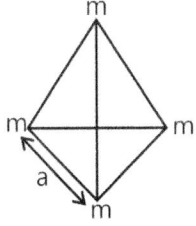

Q.12 A thin spherical shell of total mass M and radius R is held fixed. There is a small hole in the shell. A mass m is released from rest at a distance R from the hole along a line that passes through the hole and also through the center of the shell. This mass subsequently moves under the gravitational force of the shell. How long does the mass take to travel from the hole to the point diametrically opposite?

Q.13 A satellite close to the earth is in orbit above the equator with a period of rotation of 1.5 hours. If it is above a point P on the equator at some time, it will be above P again after time_____.

Q.14 A satellite is moving in a circular orbit around the earth. The total energy of the satellite is $E = -2 \times 10^5 \, J$. The amount of energy to be imparted to the satellite to transfer it to a circular orbit where its potential energy is equal to_____.

Q.15 A satellite of mass m is orbiting the earth in a circular orbit radius r. It starts losing energy due to small air resistance at the rate of C J/s. Then the time taken for the satellite to reach the earth is____.

Q.16 A satellite is orbiting the Earth of mass M in equatorial plane in a circular orbit having radius 2R and same sense of rotation as that of the Earth. Find duration of time for which a man standing on the equator will be able to see the satellite continuously. Assume that the man can see the satellite when it is above horizon. Take Earth's angular velocity = ω

Q.17 A launching pad with a spaceship is moving along a circular orbit of the moon, whose radius R is triple that of moon Rm. The ship leaves the launching pad with a relative velocity equal to the launching pad's initial orbital velocity v_0 and the launching pad then falls to the moon. Determine the angle θ with the horizontal at which the launching pad crashes into the surface if its mass is twice that of the spaceship m.

Q.18 A body moving radially away from a planet of mass M, when at distance r from planet, explodes in such a way that two of its many fragments move in mutually perpendicular circular orbits around the planet. What will be

(a) Their velocity in circular orbits

(b) Maximum distance between the two fragments before collision and

(c) Magnitude of their relative velocity just before they collide.

1.37

Q.19 A cord of length 64 m is used to connect a 100 kg astronaut to spaceship whose mass is much larger than that of the astronaut. Estimate the value of the tension in the cord. Assume that the spaceship is orbiting near earth's surface. Assume that the spaceship and the astronaut fall on a straight line from the earth's center. The radius of the earth is 6400km.

Q.20 Imagine a planet of mass M with a small moon of mass m and radius a orbiting it and keeping the same face toward it. If the moon now approaches the planet, there will be a critical distance from the planet's center at which loose material lying on the moon's surface will be lifted off. Show that this distance is given by $r_e = a(3M/m)^{1/3}$. This critical distance is called Roche's limit.

Q.21 A hypothetical planet of mass M has three moons each of equal mass 'm' revolving in the same circular orbit of radius R. The masses are equally spaced and thus form an equilateral triangle. Find:

(i) The total P.E. of the system

(ii) The orbital speed of each moon such that they maintain this configuration.

Q.22 A remote sensing satellite is revolving in an orbit of radius x over the equator of earth. Find the area on earth's surface in which satellite cannot send message.

Q.23 A pair of stars rotate about a common center of mass. One of the stars has a mass M which is twice as large as the mass m of the other. Their centers are a distance d apart, d being large compared to the size of either star.

(a) Derive an expression for the period of rotation of the star about their common center of mass in terms of d, m, G.

(b) Compare the angular momentum of the two stars about their common center of mass by calculating the ratio L_m/L_M.

(c) Compare the kinetic energies of the two stars by calculating the ratio K_m/K_M.

Q.24 Assume that a geosynchronous communications satellite is in orbit at the longitude of Mumbai. You are in Mumbai and want to pick up its signals. In what direction should you point the axis of your parabolic antenna? The latitude of Mumbai 30° N.

Q.25 The fastest possible rate of rotation of a planet such that for which the gravitational force on material at the equator barely provides the centripetal force needed for the rotation. Show that the corresponding shortest period of rotation is given by $T = \sqrt{\dfrac{3\pi}{G\rho}}$, where ρ is the density of the planet, assumed to be homogeneous.

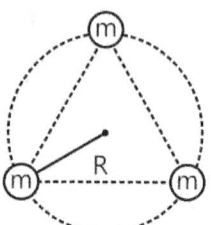

Exercise 2

Multiple Correct Choice Type

Q.1 Assuming the earth to be a sphere of uniform density the acceleration due to gravity

(A) At a point outside the earth is inversely proportional to the square of its distance from the center.

(B) At a point outside the earth in inversely proportional to its distance from the center.

(C) At a point inside is zero.

(D) At a point inside is proportional to its distance from the center.

Q.2 Two masses m_1 and $m_2 (m_1 < m_2)$ are released from rest from a finite distance. They start under their mutual gravitational attraction.

(A) Acceleration of m_1 is more than that of m_2.

(B) Acceleration of m_1 is more than that of m_1.

(C) Center of mass of system will remain at rest in all reference frames.

(D) Total energy of system remains constant.

Q.3 Inside a hollow isolated spherical shell

(A) Everywhere gravitational potential is zero.

(B) Everywhere gravitational field is zero.

(C) Everywhere gravitational potential is same.

(D) Everywhere gravitational field is same.

Q.4 When a satellite in a circular orbit around the earth centers the atmospheric region, it encounters small air resistance to its motion. Then

(A) Its kinetic energy increases.

(B) Its kinetic energy decreases.

(C) Its angular momentum about the earth decreases.

(D) Its period of revolution around the earth increases.

Q.5 A communications Earth satellite

(A) Goes round the earth from east to west.

(B) Can be in the equatorial plane only.

(C) Can be vertically above any place on the earth.

(D) Goes round the earth from west to east.

Q.6 An earth satellite is moved from one stable circular orbit to another larger and stable circular orbit. The following quantities increase for the satellite as a result of this change:-

(A) Gravitational potential energy

(B) Angular velocity

(C) Linear orbital velocity

(D) Centripetal acceleration

Q.7 A satellite S is moving in an elliptical orbit around the earth. The mass of the satellite is very small compared to the mass of the earth.

(A) The acceleration of S is always directed towards the center of the earth.

(B) The angular momentum of S about the center of the earth changes in direction, but its magnitude remains constant.

(C) The total mechanical energy of S varies periodically with time.

(D) The linear momentum of S remains constant in magnitude.

Q.8 If a satellite orbits as close to the earth's surface as possible,

(A) Its speed is maximum

(B) Time period of its rotation is minimum

(C) The total energy of the 'earth plus satellite' system is minimum

(D) The total energy of the 'earth plus satellite' system is maximum

Q.9 For a satellite to orbit around the earth, which of the following must be true?

(A) It must be above the equator at some time.

(B) Its cannot pass over the poles at any time

(C) Its height above the surface cannot exceed 36,000 km

(D) Its period of rotation must be $> 2\pi\sqrt{R/g}$ where R is radius of earth

Q.10 Two satellites s_1 & s_2 of equal masses revolve in the same sense around a heavy planet in coplanar circular orbit of radii R & 4R.

(A) The ratio of period of revolution s_1 & s_2 is 1:8

(B) Their velocities are in the ratio 2:1

(C) Their angular momentum about the planet are in the ratio 2:1

(D) The ratio of angular velocities of s_1 w.r.t. s_2 when all three are in same line is 9:5

Assertion Reasoning Type

(A) Statement-I is true, statement-II is true and statement-II is correct explanation for statement-I.

(B) Statement-I is true, statement-II is true and statement-II is NOT the correct the explanation for statement-I.

(C) Statement-I is true, statement-II is false.

(D) Statement-I false, statement-II is true.

Q.11 Statement-I: Moon revolving around earth does not come despite earth's gravitational attraction.

Statement-II: A radially outward force balances earth's force of attraction during revolution of moon.

Q.12 Statement-I: Time period of simple pendulum in an orbiting geostationary satellite in infinite.

Statement-II: Earth's gravitational field becomes negligible at large distance from it.

Q.13 Statement-I: Geostationary satellite may be setup in equatorial plane in orbits of any radius more than earth's radius.

Statement-II: Geostationary satellite have period of revolution of 24 hrs.

Q.14 Statement-I: For the calculation of gravitational force between any two uniform spherical shells, they can always be replaced by particles of same mass placed at respective centers.

Statement-II: Gravitational field of a uniform spherical shell out side it is the same as that of particle of same mass placed at its center of mass.

Q.15 Statement-I: It takes more fuel for a spacecraft to travel from the earth to moon than for the return trip.

Statement-II: Potential energy of spacecraft at moon's surface is greater than that at earth surface.

Comprehension Type

Paragraph 1:

Two uniform spherical stars made of same material have radii R and 2R. Mass of the smaller planet is m. They start moving from rest towards each other from a large distance under mutual force of gravity. The collision between the stars is inelastic with coefficient of restitution ½.

Q.16 Kinetic energy of the system just after the collision is:

(A) $\dfrac{8Gm^2}{3R}$

(B) $\dfrac{2Gm^2}{3R}$

(C) $\dfrac{4Gm^2}{3R}$

(D) Cannot be determined

Q.17 The maximum separation between their centers after their first collision

(A) 4 R   (B) 6 R   (C) 8 R   (D) 12 R

Paragraph 2:

The given figure shows the orbit of a planet P round the sun S, AB and CD are the minor and major axes of the ellipse.

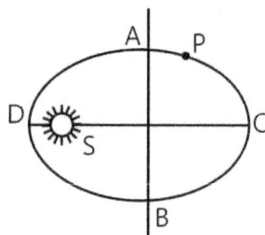

Q.18 If t_1 is the time taken by the planet to travel along ACB and t_2 the time along BDA, then

(A) $t_1 = t_2$

(B) $t_1 > t_2$

(C) $t_1 < t_2$

(D) Nothing can be concluded

Q.19 If U is the potential energy and K kinetic energy then $|U| > |K|$ at

(A) Only D   (B) Only C

(C) Both D & C   (D) Neither D nor C

Paragraph 3:

During the formation of stars from clouds of hydrogen gas in space, due to gravitational force of attraction, volume of gas decreases, which in turn heats the gas. Specific heat capacity of gas is S, universal gravitational constant is G and mass in a hydrogen cloud is M.

Q.20 If radius of gas cloud decreases from R to R/2, the increment in temperature of gas is (assume No loss of energy outside due to radiations, and clouds are spherical in shape)

(A) $\dfrac{GM}{RS}$ (B) $\dfrac{3GM}{5RS}$ (C) $\dfrac{3GS}{5MR}$ (D) $-\dfrac{3GM}{RS}$

Q.21 Assume the initial temperature of gas is 0 K and thermonuclear reactions will start at T_0 K temperature, the minimum mass of gas required so that thermonuclear reactions start when radius of cloud becomes half of initial radius (R). Assume uniform temperature in entire volume of gas.

(A) $\dfrac{5}{3}\dfrac{SRT_0}{G}$ (B) $\dfrac{3}{5}\dfrac{SRT_0}{G}$

(C) $\dfrac{SRT_0}{G}$ (D) None

Paragraph 4:

In some parts of universe, it is found that acceleration produced in a body is inversely proportional to the square of its mass and directly proportional to the net force (F) according to equation $a = c\dfrac{F}{m^2}$ where c is constant, whose magnitude is 1, if m is measured in kg, a is measured in m/s² and F is in Bose. Also action and reaction force are equal and opposite and on different interacting bodies.

Q.22 In the given figure shown, two blocks of mass m_1 = 2 kg and m_2 = 4 kg are attached via an ideal massless string over frictionless mass less pulley. If acceleration due to gravity g = 5 m/s².The tension in the string is

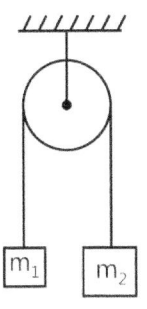

(A) 6 Bose   (B) 1.67 Bose

(C) 3 Bose   (D) 32 Bose

Q.23 In the given figure a block of mass m=2 kg is placed on smooth inclined plane. The minimum value of force F needed to support the block is $\left(g = 5m/s^2\right)$

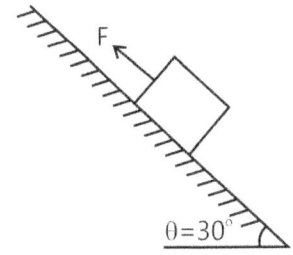

(A) Zero, Newton   (B) 10 Bose

(C) 20 Bose   (D) 10 Newton

Previous Years' Questions

Q.1. Statement-I: An astronaut in an orbiting space station above the earth experiences weightlessness.

Statement-II: An object moving around the earth under the influence of earth's gravitational force is in a state of 'free-fall'   **(2008)**

(A) If statement-I is true, statement-II is true; statement-II is the correct explanation for statement-I

(B) If statement-I is true; statement-II is true; statement-II is not a correct is true; statement-I

(C) If statement-I is true; statement-II is false

(D) If statement-I is false; statement-II is true

Q.2 A solid sphere of uniform density and radius 4 units is located with its center at the origin O of coordinates. Two spheres of equal radii 1 unit, with their centers at A (-2, 0, 0) and B (2, 0, 0) respectively, are taken out of the solid leaving behind spherical cavities as shown in the given figure.   **(1993)**

Then

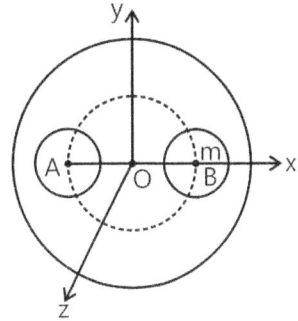

(A) the gravitational field due to this object at the origin is zero.

(B) the gravitational field at the point B (2, 0, 0) is zero

(C) the gravitational potential is the same at all points of circle $y^2 + z^2 = 36$

(D) the gravitational potential is the same at all points on the circle $y^2 + z^2 = 4$

Q.3 The magnitude of the gravitational field at distance r_1 and r_2 from the center of a uniform sphere of radius R and mass M are F_1 and F_2 respectively. Then   **(1994)**

(A) $\dfrac{F_1}{F_2} = \dfrac{r_1}{r_2}$ if $r_1 < R$ and $r_2 < R$

(B) $\dfrac{F_1}{F_2} = \dfrac{r_2^2}{r_1^2}$ if $r_1 > R$ and $r_2 > R$

(C) $\dfrac{F_1}{F_2} = \dfrac{r_1^3}{r_2^3}$ if $r_1 < R$ and $r_2 < R$

(D) $\dfrac{F_1}{F_2} = \dfrac{r_1^2}{r_2^2}$ if $r_1 < R$ and $r_2 < R$

Q.4 Two satellites S_1 and S_2 revolve round a planet in coplanar circular orbits in the same sense. Their periods of revolution are 1 h and 8 h respectively. The radius of the orbit of S_1 is 10^4 km. when S_2 is closed to S_1. Find

(a) the speed of S_2 relative to S_1,

(b) the angular speed of S_2 as actually observed by an astronaut in S_1 **(1986)**

Q.5 Three particles, each of mass m, are situated at the vertices of an equilateral triangle of side length a. The only forces acting on the particles are their mutual gravitational forces. It is desired that each particle moves in a circle while maintaining the original mutual separation a. Find the initial velocity that should be given to each particle and also the time period of circular motion **(1988)**

Q.6 An artificial satellite is moving in circular orbit around the earth with a speed equal to half the magnitude of escape velocity from the earth. **(1990)**

(a) Determine the height of the satellite above the earth's surface.

(b) If the satellite is stopped suddenly in its orbit and allowed to fall freely onto the earth, find the speed with which it hits surface of the earth.

Q.7 Distance between the centers of two stars is 10a. The masses of these stars are M and 16 M and their radii a and 2a respectively. A body of mass m is fired straight from the surface of the larger star towards the surface of the smaller star. What should be its minimum initial speed to reach the surface of the smaller star? Obtain the expression in terms of G, M and a. **(1996)**

Q.8 There is a crater of depth $\dfrac{R}{100}$ on the surface of the moon (radius R). A projectile is fired vertically upward from the crater with velocity, which is equal to the escape velocity v from the surface of the moon. Find the maximum height attained by the projectile. **(2003)**

Q.9 Gravitational acceleration on the surface of a planet is $\dfrac{\sqrt{6}}{11}$ g, where g is the gravitational acceleration on the surface of the earth. The average mass density of the planet is $\dfrac{2}{3}$ times that of the earth. If the escape speed on the surface of the earth is taken to be 11 km/s the escape speed on the surface of the planet in km/s will be? **(2010)**

Q.10 A spherically symmetric gravitational system of particles has a mass density $\rho = \begin{cases} \rho_0 \text{ for } r \le R \\ 0 \text{ for } r > R \end{cases}$ where ρ_0 is a constant. A test mass can undergo circular motion under the influence of the gravitational field of particles. Its speed v as a function of distance r from the center of the system is represented by **(2008)**

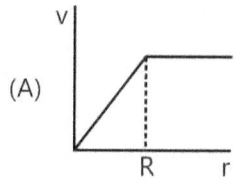 (A)

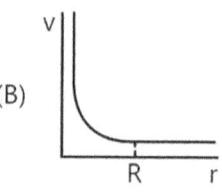 (B)

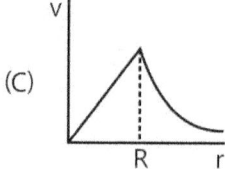 (C)

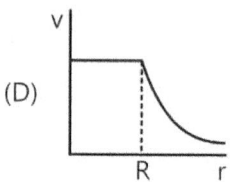 (D)

Q.11 Two spherical planets P and Q have the same uniform density ρ, masses M_P and M_Q, an surface areas A and 4A, respectively. A spherical planet R also has uniform density ρ and its mass is $(M_P + M_Q)$. The escape velocities from the planets P, Q and R, are V_P, V_Q and V respectively. Then **(2012)**

(A) $V_Q > V_R > V_P$

(B) $V_R > V_Q > V_P$

(C) $\dfrac{V_R}{V_P} = 3$

(D) $\dfrac{V_P}{V_Q} = \dfrac{1}{2}$

Q.12 A planet of radius R = $\dfrac{1}{10}$ × (radius of Earth) has the same mass density as Earth. Scientists dig a well of depth $\dfrac{R}{5}$ on it and lower a wire of the same length and of linear mass density 10^{-3} kgm^{-1} into it. If the wire is not touching anywhere, the force applied at the top of the wire by a person holding it in place is (take the radius of Earth = 6×10^6 m and the acceleration due to gravity of Earth is 10 ms^{-2}) **(2014)**

(A) 96 N   (B) 108 N   (C) 120 N   (D) 150 N

Q.13 In an experiment to determine the acceleration due to gravity g, the formula used for the time period of a periodic motion is $T = 2\pi = \sqrt{\dfrac{7(R-r)}{5g}}$. The values of R and r are measured to be (60 ± 1) mm and (10 ± 1) mm, respectively. In five successive measurements, the time period is found to be 0.52 s, 0.56 s, 0.57 s, 0.54 s and 0.59 s. The least count of the watch used for the measurement of time period is 0.01 s. Which of the following statement(s) is (are) true? **(2016)**

(A) The error in the measurement of r is 10%

(B) The error in the measurement of T is 3.57%

(C) The error in the measurement of T is 2%

(D) The error in the determined value of g is 11%

Important Questions

JEE Main/Boards

Exercise 1

Q.23   Q.31   Q.33

Q.35

Exercise 2

Q. 2   Q.13   Q.17

Q.20

JEE Advanced/Boards

Exercise 1

Q.1   Q.5   Q.6

Q.11   Q.16   Q.21

Exercise 2

Q.1   Q.2   Q.5

Q.6   Q.7   Q.8

Q.10   Q.11   Q.12

Q.16

Answer Key

JEE Main /Boards

Exercise 1

Q.3 No

Q.4 Zero

Q.9 $10\sqrt{10}$

Q.10 11 km/s

Q.13 One-Fourth

Q.14 $x=2h$

Q.15 0.732

Q.23 $v_A/v_P = r_P/r_A$; No, time for path BAC is greater than time for path CPB

Q.24 4 R

Q.25 129 days

Q.26 1.972×10^{30} kg

Q.27 $6.69 \times 10^{-11} m^2 kg^{-2}$

Q.28 6.025×10^{24} kg; 5.56×10^3 kg/m^3

Q.29 Increase by 5%

Q.30 2649.6 km

Q.31 66.12 kg wt; 8.89 kg wt; 5 kg wt.

Q.32 80 km

Q.33 $\dfrac{5GmM}{6R}$

Q.34 2.56×10^4 km

Q.35 (a) 7756.6 m/s;   (b) 5373 s

Q.36 5064 s

Q.37 $2.354 \times 10^3 ms^{-1}$

Q.38 $5.28 kms^{-1}$

Exercise 2

Single Correct Choice Type

Q.1 B	Q.2 A	Q.3 B	Q.4 B	Q.5 C	Q.6 A
Q.7 A	Q.8 B	Q.9 B	Q.10 B	Q.11 A	Q.12 B
Q.13 C	Q.14 A	Q.15 C	Q.16 D	Q.17 D	Q.18 C
Q.19 B	Q.20 C	Q.21 A	Q.22 A	Q.23 A	Q.24 C

Previous Years' Questions

Q.1 C	Q.2 A	Q.3 B	Q.4 B	Q.5 D	Q.6 C
Q.7 D	Q.8 B	Q.9 C	Q.10 B	Q.11 A	Q.12 C
Q.13 D	Q.14 A	Q.15 B	Q.16 C		

JEE Advanced/Boards

Exercise 1

Q.1 $\dfrac{GM^2}{3L^2}$

Q.2 $\dfrac{R_e k^2}{1-k^2}$

Q.3 $\sqrt{\dfrac{2GM}{a}\left(1-\dfrac{1}{\sqrt{2}}\right)}$

Q.4 $h=\dfrac{\sqrt{3}-1}{2}$

Q.5 (i) $\dfrac{4}{3}\sqrt{\dfrac{GM}{R}}$ (ii) $\dfrac{2}{3}\sqrt{\dfrac{2GM}{5R}}$

Q.6 (a) $E=\dfrac{\pi G\rho_0 R^3}{6}\left[\dfrac{1}{\left(x-(R/2)\right)^2}-\dfrac{8}{x^2}\right]$, (b) $E=\dfrac{GM}{2R^2}$

Q.7 $2R,\ 3R\left[3-\sqrt{3}\right]$

Q.8 $\sqrt{6.4}\,$km

Q.9 $\dfrac{2\pi R^{3/2}\left(6\sqrt{6}\right)}{\sqrt{GM}\left(2\sqrt{2}+3\sqrt{3}\right)}$

Q.10 R

Q.11 $\dfrac{6GM^2}{2a}$

Q.12 $2\times\sqrt{R^3/GM}$

Q.13 1.6 hours if it is rotating from west to east, 24/17 hours if it is rotating from east to west

Q.14 $1\times10^5\,$J

Q.15 $t=\dfrac{GMm}{2C}\left(\dfrac{1}{R_e}-\dfrac{1}{r}\right)$

Q.16 $\dfrac{2\pi}{3\left(\sqrt{\dfrac{Gm}{8R^3}}-\omega_e\right)}$

Q.17 $\cos\theta=\dfrac{3}{\sqrt{10}}$

Q.18 (a) $\sqrt{\dfrac{GM}{r}}$; (b) $r\sqrt{2}$; (c) $\sqrt{\dfrac{2GM}{R}}$

Q.19 $T=3\times10^{-2}\,$N

Q.20 $r_e=a\left(3M/m\right)^{1/3}$

Q.21 (i) $-\dfrac{3GM}{R}\left(\dfrac{m}{\sqrt{3}}+m\right)$, (ii) $\left(\sqrt{\dfrac{GM}{R}\dfrac{(2\sqrt{3}+R)}{2\sqrt{3}}}\right)$

1.44

Q.22 $2\pi R^2\left(1 + \dfrac{R}{x}\right)$

Q.24 $\cot^{-1}\left(\sqrt{3} - \dfrac{32}{105}\right)$ to vertical

Q.23 (a) $T = \dfrac{2\pi d^{3/2}}{\sqrt{3GM}}$ (b) 2 (c) 2

Q.25 $T = \sqrt{\dfrac{3\pi}{G\rho}}$

Exercise 2

Multiple Correct Choice Type

Q.1 A, D   **Q.2** A, D   **Q.3** B, C, D   **Q.4** A,C   **Q.5** B, D   **Q.6** A, C, D

Q.7 A, D   **Q.8** A, B, C   **Q.9** A, D   **Q.10** A, B, D

Assertion Reasoning Type

Q.11 C   **Q.12** B   **Q.13** D   **Q.14** D   **Q.15** A

Comprehension Type

Paragraph 1:   **Q.16** B   **Q.17** A

Paragraph 2:   **Q.18** B   **Q.19** C

Paragraph 3:   **Q.20** B   **Q.21** A

Paragraph 4:   **Q.22** D   **Q.23** B

Previous Years' Questions

Q.1 A   **Q.2** A   **Q.3** B   **Q.4** (a) $-\pi \times 10^4$ km/h (b) 3×10^{-4} rad/s

Q.5 $v = \sqrt{\dfrac{Gm}{a}}, T = 2\pi\sqrt{\dfrac{a^3}{3Gm}}$ **Q.6** (a) 6400 km (b) 7.9 km/s

Q.7 $\dfrac{3\sqrt{5}}{2}\sqrt{\dfrac{GM}{a}}$ **Q.8** 99.5 R   **Q.9** 3   **Q.10** C

Q.11 B, D   **Q.12** B   **Q.13** A

JEE Main/Boards

Exercise 1

Sol 1: Newton's law of gravitation is called a universal law because it is applicable anywhere in the universe.

Sol 2: The value is same on moon. G is called universal gravitational constant, which is constant anywhere in the universe

Sol 3: The packet doesn't reach the earth (theoretically). Because in a satellite, centrifugal force balances the gravitational force on it. The same will happen with the packet, which has same initial velocity as that of satellite.

Sol 4: Its reading will be zero. A spring balance shows the net force the hanging body exerts on it net force by body = Mass × Acceleration

∴ The net acceleration is zero as the centrifugal and gravitational forces balance each other, it reads zero.

Sol 5: Escape velocity is always constant for a given celestial body. Escape velocity of earth is 11.2 km s⁻¹. It is same irrespective of mass.

Sol 6: Weight = Mass × Net acceleration It will be zero, if net acceleration is zero like a satellite.

Sol 7: Gravitational force (f) $= \dfrac{GMm}{r^2}$

∴ $f \propto m$

∴ Acceleration due to gravity $g = \dfrac{f}{m}$

$g = \dfrac{GM}{r^2}$

It is independent of m.

Here both fall at same time

Sol 8: $U = \dfrac{GMm}{r}$

weight $W = mg$

$= m.\left(\dfrac{GM}{r^2}\right) = \dfrac{1}{r}\left(\dfrac{GMm}{r}\right)$

$W = \dfrac{U}{r}$

Hence weight $= \dfrac{U}{r}$

Sol 9: $T \propto r^{3/2}$

$\dfrac{T_1}{T_2} = \left(\dfrac{r_1}{r_2}\right)^{\frac{3}{2}} = \left(\dfrac{10^{11}}{10^{10}}\right)^{\frac{3}{2}} = 10\sqrt{10}$

Sol 10: Escape speed is still 11 km s⁻¹ because escape speed is irrespective of angle of launch (of course not towards ground). We calculate escape velocity by

$\dfrac{1}{2}mV^2 = \dfrac{GMm}{r}$

(Kinetic energy + Potential energy)

= it Irrespective of angle of each.

Sol 11: Acceleration due to gravity

$g = \dfrac{GM}{r^2}$

$g \propto \dfrac{1}{r^2}$

Hence it is inversely proportional to r²

Sol 12: $f \propto \dfrac{1}{r^2}$

$f_2 = f_1 . \left(\dfrac{r_1}{r_2}\right)^2$

$\dfrac{r_1}{r_2} = \dfrac{1}{2}$

∴ $f_2 = \dfrac{f_1}{4} = \dfrac{1}{4}N$

∴ Gravitational force 0.25 N

Sol 13: If he goes to a height r, his distance from center is 2r

i.e., $\dfrac{r_1}{r_2} = \dfrac{1}{2}$

$w \propto \dfrac{1}{r^2}$ (w = Weight)

$$\frac{w_2}{w_1} = \left(\frac{r_1}{r_2}\right)^2$$

$$\Rightarrow \frac{w_2}{w_1} = \left(\frac{1}{2}\right)^2 = \frac{1}{4},$$

i.e., his weight quadrates

Sol 14: $g = \dfrac{GM}{r^2}$ for $r \geq r_0$ (r_0 is radius of earth)

$$\frac{dg}{dr} = -\frac{GM}{2r^3}$$

$$\Rightarrow \Delta g = -\frac{GM}{2r^3} \cdot \Delta r$$

$$\Rightarrow \Delta g = -\frac{GM}{2r_0^3} \cdot h$$

$$g = \frac{GM}{r_0^3} r_1, \qquad r_1 \leq r_0$$

$$\Delta g = \frac{GM}{r_0^3}(r_1 - r_0)$$

$$r_1 - r_0 = -x$$

$$\Rightarrow \Delta g = \frac{GM}{r_0^3} \cdot (-x)$$

$$\Rightarrow \frac{GM}{r_0^3}(-x) = \frac{GM}{r_0^3}(-2h)$$

$$\Rightarrow x = 2h$$

Sol 15: $\dfrac{f_1}{f_2} = \left(\dfrac{r_2}{r_1}\right)^2$

$r_1 = r_0,\ r_2 = $ Distance of rocket from center of earth

$$\frac{f_2}{f_1} = \frac{1}{3}$$

$$\Rightarrow 3 = \left(\frac{r_2}{r_1}\right)^2 \Rightarrow \frac{r_2}{r_1} = \sqrt{3}$$

Height of rocket $n = r_2 - r_1$

$$\frac{r_2}{r_1} = \sqrt{3}$$

$$\frac{r_2}{r_1} - 1 = \sqrt{3} - 1$$

$$\Rightarrow \frac{r_2 - r_1}{r_1} = \sqrt{3} - 1$$

$$\Rightarrow \frac{h}{r_1} = 0.732$$

Sol 16: No. gravitational force $f = \dfrac{GMm}{r^2}$

$$\Rightarrow f \propto \frac{1}{r^2}$$

$\dfrac{1}{r^2}$ will never become zero.

So force doesn't become zero.

Sol 17: Gravity is the force with which a body pulls another body towards its center.

Acceleration due to gravity is the acceleration which it produces in the body due to force of gravity

$$g = \frac{GM}{r^2}$$

Sol 18: $g = \dfrac{GM}{r^2}$

m = Mass of earth

r = Radius of earth

$$m = \frac{r^2 g}{G}$$

= We can calculate g by physical means, r is known, hence M can be calculated

$$\text{Mean density of earth} = \frac{M}{\frac{4}{3}\pi r^3}$$

∵ M can be calculated, mean density can also be calculated.

Sol 19: Escape velocity is the velocity with which when a body is projected from the surface of a celestial body, it crosses its potential barrier and escapes into out space for bodies to escape total energy ≥ 0

i.e. $K.E + P.E \geq 0$

At escape velocity

$K.E + P.E = 0$

$$\Rightarrow \frac{1}{2}mv^2 - \frac{GMm}{r} = 0$$

$$\Rightarrow v = \sqrt{\frac{2GM}{r}}$$

Sol 20: Gravity field is a field in which a body produces a force on another body.

Intensity of gravitational force is the force which a body attracts a body of unit mass

Intensity $= \dfrac{GM}{r^2}$ ... (i)

$= \dfrac{GM}{r^2} = g$

Sol 21: Kepler's law $T^2 \propto r^3$

$T = \dfrac{2\pi r}{v}$; $\Rightarrow \dfrac{r^2}{v^2} \propto r^3$

$\Rightarrow v^2 \propto \dfrac{1}{r}$ ...... (i)

For a planetary motion

$PE + KE = 0$

$\Rightarrow \dfrac{1}{2} mv^2 + PE = 0$

$\Rightarrow PE = -\dfrac{1}{2} mv^2$; $\Rightarrow PE \propto -\dfrac{m}{r}$

Gravitational force $= \dfrac{d}{dr} P.E$

$\Rightarrow F \propto \dfrac{d}{dr}\left(-\dfrac{m}{r}\right)$

$\Rightarrow f \propto \dfrac{m}{r^2} \Rightarrow f = \dfrac{km}{r^2}$ (k-some constant)

Let two bodies m_1, m_2 exert gravity on each other.

$f_{12} = \dfrac{k_2 m_2}{r^2} \Rightarrow f_{12} \propto m_1$ ...... (i)

$f_{21} = \dfrac{k_1 m_1}{r^2} \Rightarrow f_{21} \propto m_2$ ...... (ii)

Where f_{12} is force on body 1 by body 2 similarly f_{21} defined

but $|f_{12}| = |f_{21}|$ newton's third law

$\Rightarrow t_{12} \propto m_2$

$\Rightarrow f_{12} = k_3 \dfrac{m_1 m_2}{r^2}$ (k_3-some constant)

$\Rightarrow f \propto \dfrac{m_1 m_2}{r^2}$

Hence newton's law of gravity is deduced.

Sol 22: Newton's law of universal gravitation states that every point mass in the universe attracts every other point mass with a force that is directly proportional to the product of their masses and inversely proportional to the square of distance between them.

$f \propto \dfrac{m_1 m_2}{r^2}$

Gravitational constant (G) is a empirical physical constant involved in the calculation of gravitational

force between two bodies. In simple terms it can be sides the proportionality constant for Newton's law of gravitation

$G = 6.67 \times 10^{-4}\, N \left(\dfrac{m}{kg}\right)^2$

Dimensional formula $[G] = \dfrac{fL^2}{m^2}$

$f = \dfrac{GMm}{r^2}$

$[f] = \dfrac{[G][M][m]}{[r]^2} = \dfrac{FL^2}{m^2} \cdot m.m.L^{-2} = f$

It supports Newton's law empirically

Sol 23: The distance travelled $\vec{dx}$ in time dt is $\vec{dv}$

$= \vec{v} . dt$

Area swept by radius vector

$dA = \dfrac{1}{2} \vec{r} \times \vec{v}\, dt$

$dA_A = \dfrac{1}{2} r_A v_A\, dt$ and

$dA_P = \dfrac{1}{2} r_p v_p\, dt$

$\because \dfrac{dA}{dt}$ = Constant, (Kepler's 2nd law)

$\Rightarrow r_A v_A = r_p v_p$

The time taken is different as

area of SBAC $\pm$ SCPB

Sol 24: $T^2 \propto r^3$

$\Rightarrow \dfrac{r_1}{r_2} = \left(\dfrac{T_1}{T_2}\right)^{\frac{2}{3}}$; $\Rightarrow r_1 = r_2 \left(\dfrac{T_1}{T_2}\right)^{\frac{2}{3}}$

gen rL = R, T_1 = 32, T_2 = 256

$\Rightarrow r_1 = 4R$

Sol 25: $\because T_1 = T_2 = \left(\dfrac{r_1}{r_2}\right)^{\frac{3}{2}}$ (kepler's 3rd law)

$\dfrac{r_1}{r_2} = \dfrac{1}{2}$

$\therefore T_1 = 365 \left(\dfrac{1}{2}\right)^{\frac{3}{2}} = 129$ days

1 year would have 129 days.

Sol 26: For planetary motion,

$$\frac{1}{2}mv^2 = \frac{GMm}{r} \; ; \Rightarrow V = \sqrt{\frac{2GM}{r}}$$

$$V = \frac{2\pi r}{T} \; ; \Rightarrow T = 2\pi\sqrt{\frac{\pi^3}{2GM}}$$

$$\Rightarrow M = \frac{(2\pi)^2 . r^3}{2GT^2}$$

$$= \frac{4.\pi^2.(1.49\times10^{11})^3}{2\times6.66\times10^{-11}\times(365\times24\times3600)^2}$$

$$= 1.972 \times 10^{30} \text{ kg}$$

Sol 27: $M = \dfrac{(2\pi)^2 r^3}{2GT^2} \; ; \Rightarrow G = \dfrac{(2\pi)^2 r^3}{2MT^2}$

$r = 1.5 \times 10^{11}$ m

T = 365.3 days = 365.3 × 24 × 3600 seconds

$M = 2 \times 10^{30}$ Kg

$$\Rightarrow G = \frac{4(\pi)^2.(1.5\times10^{11})^3}{2\times2\times10^{30}\times(365.3\times24\times3600)^2}$$

$$= 6.69\times 10^{-11} \text{ m}^2/\text{Kg}^2$$

Note: For calculation purpose, you may take $\pi^2 = 10$

Sol 28: $g = \dfrac{GM}{r^2}$

$$M = \frac{gr^2}{G}$$

$$= \frac{9.8\times(6.37\times10^6)^2}{(6.66\times10^{-11})} = 6.025 \times 10^{24} \text{ kg}$$

Mean density $= \dfrac{M}{\dfrac{4}{3}\pi r^3} = \dfrac{gr^2}{G} \cdot \dfrac{1}{\dfrac{4}{3}\pi r^3}$

$$= \frac{3}{4}\frac{g}{Gr} = \frac{3}{4}\times\frac{9.8}{6.66\times10^{-11}\times6.37\times10^6}$$

Mean density (e) = 5.56 × 10³ kg/m³

Sol 29: $g = \dfrac{GM}{r^2} \; ; \dfrac{dg}{dr} = -\dfrac{2GM}{r^3}$

$$\Rightarrow \Delta g = -\frac{2GM}{r^3}\Delta r \; ; \frac{\Delta g}{g} = \frac{-\dfrac{2GM}{r^3}}{\dfrac{GM}{r^2}}\Delta r$$

$$\frac{\Delta g}{g} = -2\left(\frac{\Delta r}{r}\right) \; ; \frac{\Delta g}{g}\times100 = -2\frac{\Delta r}{r}\times100$$

$$\frac{\Delta r}{r}\times100 = 2.5\% \; ; \Rightarrow \frac{\Delta g}{g}\times100 = -2(-2.5)$$

$$= 5\%$$

∴ Acceleration due to gravity increases by 5%

Note :- Try focusing on the sign convention. It you get confused, use common sense which implies when body gets denser, its g increases like a black hole, etc.

Sol 30: $g \propto \dfrac{1}{r^2}$

$$\frac{g_1}{g_2} = \left(\frac{r_2}{r_1}\right)^2 \; ; \Rightarrow r_2 = r_1\sqrt{\frac{g_1}{g_2}}$$

$$\frac{g_1}{g_2} = 2 \; ; \Rightarrow r_2 = r_1\sqrt{2}$$

Height $= r_2 - r_1 = r_1 (\sqrt{2}-1)$

$$= 2649.6 \text{ km}$$

Sol 31: wt $\propto \dfrac{1}{r^2}$

$$w_2 = w_1.\left(\frac{r_1}{r_2}\right)^2$$

r_1 = 6000 km

r_2 = 6600, 18000, 2400 ($r_2 = r_1 + h$)

w_1 = 80 kg wt.

for r_2 = 6600

$$w_2 = 80\left(\frac{6000}{6600}\right)^2 = 66.12 \text{ kg wt}$$

for r_2 = 18000

$$w_2 = 80\left(\frac{6000}{18000}\right)^2 = 8.89 \text{ kg wt}$$

for r_2 = 24000

$$w_2 = 80\left(\frac{6000}{24000}\right)^2 = 5 \text{ kg.wt}$$

Sol 32: g at a depth x, $g_x = \dfrac{GM}{r_0^3}(r_0 - x)$

g at a height h, $g_h = \dfrac{GM}{(r_0 + h)^2}$

$g_x = gh$

By substituting we get solution

But for some intelligent manipulation

$$gx = \frac{gm}{r_0^3}(r_0 - x) \Rightarrow \Delta g = \frac{-GMx}{r_0^3}$$

$$g = \frac{GM}{r_0^3}$$

$$\Delta g = -\frac{2GM}{x^2}\Delta r \text{ (Differentiation)}$$

$\therefore$ g is equal $\Rightarrow \Delta g$ is equal

$$\Rightarrow -\frac{2GM}{r_0^3}\Delta r = -\frac{GM}{r_0^3}x \; ; \Rightarrow \Delta r = \frac{x}{2}$$

$\Delta r = n$

$x = 160km \; ; \Rightarrow h = \frac{160}{2} = 80 \text{ km}$

$\Rightarrow$ It is same at a height 80 km

Note: $h << r_0$ is assumed hence we could apply this method of differentiation

Sol 33: Energy required = Total change in energy Initial energy $= -\frac{GMm}{R}$

Find energy = P. E + K. E

$$= -\frac{GMm}{3R} + \frac{1}{2}mv^2$$

(r = 3R because altitude = 2R)

For orbital motion, centrifugal force = Gravitational for

$$\frac{mv^2}{r} = \frac{GMm}{r^2}$$

$$\frac{1}{2}mv^2 = \frac{GMm}{2r}$$

$$= \frac{GMm}{6R} \text{ (substitute } r = 3R)$$

$$\Rightarrow \text{Final energy} = \frac{GMm}{6R} - \frac{GMm}{3R} = -\frac{GMm}{6R}$$

$$\Rightarrow \text{Energy required} = -\frac{GMm}{6R} - \left(\frac{GMm}{R}\right)$$

$$= \frac{5GMm}{6R}$$

Sol 34: Kinetic energy = Change in potential energy

$$\Rightarrow \frac{1}{2}mv^2 = -\frac{GMm}{r} - \left(\frac{GMm}{R}\right)$$

$$\Rightarrow \frac{1}{r} = \frac{1}{R} - \frac{1}{2}\frac{v^2}{GM}$$

$$\frac{1}{r} = \frac{1}{R} - \frac{1}{2}\frac{v^2}{gR^2}$$

$$\frac{1}{r} = \frac{1}{6400} - \frac{1}{2} \times \frac{10^2}{9.8 \times 10^{-3} \times (6400)^2}$$

$r = 2.56 \times 10^4$ km

Sol 35: Orbital velocity $v = \sqrt{\frac{GM}{r}}$

$$= \sqrt{\frac{GM}{R+H}} = \sqrt{\frac{gR^2}{R+H}}$$

$g = 9.8 \text{ ms}^{-1}$

$R = 6.38 \times 10^{-6}$

$H = 250 \text{ km} = 2.5 \times 10^5$

$v = 7756.6 \text{ ms}^{-1}$

$$T = \frac{2\pi r}{v} = \frac{2\pi(R+H)}{v}$$

$T = 5373$ s

Sol 36: Let orbital velocity = v

$$\Rightarrow v = \sqrt{\frac{GM}{R}} \text{ where m = Mass, R = Radius of point}$$

$$T = \frac{2\pi R}{v} = 2\pi\sqrt{\frac{R^3}{GM}} \qquad\qquad .... \text{(i)}$$

$$M = \frac{4}{3}\pi R^3 \rho$$

$$\Rightarrow T = 2\pi\sqrt{\frac{R^3}{G.\frac{4}{3}\pi R^3 \rho}}$$

$$= 2\pi\sqrt{\frac{3}{4\pi G\rho}} = 5064 \text{ s}$$

Sol 37: For escape velocity

$$\frac{1}{2}mv_e^2 = \frac{GMm}{R}$$

$$v_e = \sqrt{\frac{2GM}{R}}$$

$$v_e = \sqrt{2gR}$$

$$= \sqrt{2 \times 1.63 \times 17 \times 10^6}$$

$$= 2.354 \times 10^3 \text{ ms}^{-1}$$

Sol 38: $v_e \propto \sqrt{\dfrac{M}{R}}$

$$\frac{v_m}{v_e} = \sqrt{\frac{M_m}{R_m} \times \frac{R_e}{M_e}}$$

$$v_m = v_e \sqrt{\left(\frac{M_m}{M_e}\right) \times \left(\frac{D_e}{R_m}\right)}$$

given $= \dfrac{M_m}{M_e} = \dfrac{1}{9}$

$\dfrac{R_e}{R_m} = 2$

$\Rightarrow v_m = v_e \sqrt{\dfrac{2}{9}} = 2\sqrt{\dfrac{2}{9}} = 5.28 \text{ kms}^{-1}$

M_m = Mass of planet mass

R_m = Radius of planet mass

Exercise 2

Single Correct Choice Type

Sol 1: (B) $g \propto \dfrac{1}{r^2}$

$$\frac{g_1}{g_2} = \left(\frac{r_2}{r_1}\right)^2 ; \Rightarrow \frac{r_2}{r_1} = \sqrt{\frac{g_1}{g_2}} = \sqrt{4}$$

$\Rightarrow r_2 = 2R ; \Rightarrow$ altitude $h = r_2 - R$

$h = R$

Sol 2: (A) Net acceleration at equator

$g' = g - R\omega^2$

($R\omega^2$ is radial acceleration)

$\therefore$ Weight at equator $mg' = mg - mR\omega^2$

acceleration at a depth d

$\Rightarrow g_d = \dfrac{g(R-d)}{R} = g - \dfrac{g}{R}d$

given $mg_d = mg'$

$\Rightarrow mg - mR\omega^2 = mg - \dfrac{mgd}{R} \Rightarrow d = \dfrac{R^2\omega^2}{g}$

Sol 3: (B) $g = \dfrac{GM}{r^2}$

$m = \dfrac{4}{3}\pi r^3 \rho ; \Rightarrow g = \dfrac{4}{3}\pi r r$

$r\rho = $ constant

$\therefore \dfrac{r_2}{r_1} = \dfrac{\rho_1}{\rho_2} = \dfrac{1}{5}\left(\dfrac{\rho_2}{\rho_1} = 5 \text{ given}\right)$

$\therefore$ Radius to be changed by a factor of $\dfrac{1}{5}$

Sol 4: (B) $T = 2\pi\sqrt{\dfrac{\ell}{g}}$

$T \propto \dfrac{1}{\sqrt{g}} ; g = \dfrac{GM}{r^2}$

$\dfrac{1}{\sqrt{g}} = \dfrac{r}{\sqrt{M}} ; \dfrac{T_1}{T_2} = \dfrac{r_1}{r_2}\sqrt{\dfrac{M_2}{M_1}}$

$T_2 = T_1\left(\dfrac{r_2}{r_1}\sqrt{\dfrac{M_1}{M_2}}\right)$

$\dfrac{r_2}{r_1} = 2; \dfrac{M_1}{M_2} = \dfrac{1}{2}; T_1 = 2\text{seconds}$

$\therefore T_2 = (2 \times 2)\left(\dfrac{1}{\sqrt{2}}\right) = 2\sqrt{2}$ second

Note: Time period of a seconds pendulum is 2 seconds.

Sol 5: (C) Inside the shell, the gravitational field due to sphere is zero, but there is gravity due to particle.

Sol 6: (A) $g - \dfrac{v^2}{R} = \dfrac{g}{2} ; \Rightarrow \dfrac{v^2}{R} = \dfrac{g}{2}$

$$v_e = \sqrt{\frac{2GM}{R}} = \sqrt{\frac{1}{2}\frac{GM}{R^2}(4R)}$$

$$= \sqrt{\left(\frac{g}{2}\right)(4R)} = \sqrt{\frac{v^2}{R}(4R)}$$

$v_e = 2v$

Sol 7: (A) $v \propto \sqrt{\dfrac{M}{R}}$

$$\frac{M}{R} = \frac{\frac{4}{3}\pi R^3 \rho}{R} \propto R^2 ; \Rightarrow v \propto R$$

$\Rightarrow \dfrac{v_A}{v_B} = \dfrac{R_A}{R_B} = 2$

1.51

Sol 8: (B) $v_e = \sqrt{\dfrac{2GM}{R}}$

P.E at surface $= \dfrac{GMm}{R}$

P.E at centre of earth $= \dfrac{3GMm}{2R}$

$KE = \Delta PE = \dfrac{1}{2}\dfrac{GMm}{R}$

$\dfrac{1}{2}mv^2 = \dfrac{1}{2}\dfrac{GMm}{R}$; $\Rightarrow v = \sqrt{\dfrac{GM}{R}}$

$= \dfrac{v_e}{\sqrt{2}}$

Sol 9: (B) Potential at surface = Potential at center for hollow sphere

Potential $P = -\dfrac{GM}{r}$

r = radius

Let $P_0 = -\dfrac{GM}{r_0}$

new potential $P = -\dfrac{GM}{\dfrac{r_0}{2}} = -\dfrac{2GM}{r_0}$

$P < P_0$

$\therefore$ Decreases

Sol 10: (B) $g \propto \dfrac{1}{r^2}$

Sol 11: (A) In A, both 5M forces, cancel each other hence net force is proportional to (3M – M) and M (at center)

The same is for B.

$\therefore F_A = F_B$

Similarly $f_c \propto (3m-m)2m$ $\therefore f_c > f_B$

same is for F_D

$\therefore F_C = f_D$

$\because F_A = F_B < F_C = F_D$

Sol 12: (B) It moves tangentially as there is no centripetal force.

Sol 13: (C) Time period of a geo stationary satellite is 24 hrs but due to the given situation, it moves twice above same point in one day

$\therefore$ Time for successful interval interval $= \dfrac{24}{2} = 12$ Hrs

Sol 14: (A) Let centre of mass be at a distance ℓ from 4m

$\Rightarrow 4\,m\,\ell = m(d - \ell)$

$\Rightarrow \ell = \dfrac{d}{5}$

$\Rightarrow$ Orbital radius of 4 m $= \dfrac{d}{5}$, m $= \dfrac{4d}{5}$

Both bodies have same angular velocities

$\Rightarrow \dfrac{v}{r}$ = constant

$\Rightarrow v = Kr$ (K = constant)

$v_{am} = = \dfrac{kd}{5}$, $v_m = \dfrac{4kd}{5}$

$\dfrac{KE_{am}}{KE_m} = \dfrac{\dfrac{1}{2}(4m)(v_{4m})^2}{\dfrac{1}{2}m(v_m)^2} = 4\left(\dfrac{1}{4}\right)^2 = \dfrac{1}{4}$

Sol 15: (C) C-options defines the information about the escape velocities

B-option it is elliptical path

Sol 16: (D) Change in potential energy

$\Delta E = \dfrac{GMm}{R} - \dfrac{GMm}{2R} = \dfrac{GMm}{2R}$

Final velocity or orbital velocity

$v_0 = \sqrt{\dfrac{GM}{2R}}$

Change in P.E = Change in K.E.

$\dfrac{GMm}{2R} = \dfrac{1}{2}mv^2 - \dfrac{1}{2}mv_0^2$

$\dfrac{1}{2}mv^2 = \dfrac{GMm}{2R} + \dfrac{1}{4}m \cdot \dfrac{GM}{R}$

$\dfrac{1}{2}mv^2 = \dfrac{3}{4}\dfrac{GMm}{R}$

$\dfrac{1}{2}mv^2 = \dfrac{3}{4}mgR$

$\therefore$ Energy required is $\dfrac{3}{4}mgR$

Sol 17: (D) K.E $\propto \dfrac{1}{r}$

$\therefore$ A is K.E.

Total energy > Potential energy,

total energy, potential energy $\propto -\dfrac{1}{r}$

$\therefore$ C is total energy

B is potential energy

Sol 18: (C) Kepler's 2nd law, areal velocity is constant.

Sol 19: (B) Let final velocity of 4 M be V_1

$5MV = 4MV_1 - MV$

(Conservation of linear momentum)

$\Rightarrow V_1 = \dfrac{3}{2}V$

now $V = \sqrt{\dfrac{GM}{r}}$

$V_e = \sqrt{\dfrac{2GM}{r}} = \sqrt{2}\,V = V_1 > \sqrt{2}\,V$

i.e., $V_1 > V_e$

$\Rightarrow$ Body gets unbound

Sol 20: (C) $T \propto r^{3/2}$

$\dfrac{T_2}{T_1} = \dfrac{1}{2}$; $\left(\dfrac{r_2}{r_1}\right)^{\frac{3}{2}} = \dfrac{1}{2}$

$\Rightarrow r_2 = r_1 \left(\dfrac{1}{2}\right)^{\frac{2}{3}} = \dfrac{r}{(4)^{\frac{1}{3}}}$

Sol 21: (A) Consider the planet to be at one of the vertex.

Let its distance from sun be r, velocity be v.

Area covered in time dT

$dA = \dfrac{1}{2}rv.\,dT$; $\dfrac{2dA}{dt} = vr$

$mvr = \dfrac{2mdA}{dt}$; $\dfrac{dA}{dt} = \dfrac{A}{T}$

$\therefore$ Angular momentum $= \dfrac{2mA}{T}$

Sol 22: (A) $\dfrac{dA}{dt} \propto rv$

$V \propto \dfrac{1}{\sqrt{r}}$; $\therefore \dfrac{dA}{dt} \propto \sqrt{r}$

$\therefore$ ratio of their area velocity

$= \sqrt{\dfrac{r_1}{r_2}} = \sqrt{\dfrac{1}{4}} = 1:2$

Sol 23: (A) Field due to gravity $E_1 = \dfrac{GM}{R^2}$

$m = \dfrac{4}{3}\pi R^3$; $\therefore E_1 = \dfrac{4}{3}\pi G\rho R$

Field due to infinite plate $E_2 = 2\pi\rho tG$

$(t = H) = 2\pi\rho HG$

$E_1 = E_2$

$\therefore \dfrac{4}{3}\pi G\rho R = 2\pi\rho HG$; $\Rightarrow H = \dfrac{2R}{3}$

Sol 24: (C) Maximum speed occurs at least distance

$\dfrac{dA}{dt} = \dfrac{1}{2}r_{min}\,v_{max.}$

$4 \times 10^{16} = \dfrac{1}{2} \times 2 \times 10^{12} \times v$

$v = 4 \times 10^4\ ms^{-1}$

$\therefore v = 40\ kms^{-1}$

Previous Years' Questions

Sol 1: (C) $g = \dfrac{GM}{R^2}$

or $g \propto \dfrac{1}{R^2}$

g will increase if R decreases

Sol 2: (A) $\Delta U = \dfrac{mgh}{1+\dfrac{h}{R}}$

Given, h = R

$\Delta U = \dfrac{mgR}{1+\dfrac{R}{R}} = \dfrac{1}{2}mgR$

Sol 3: (B) $\dfrac{mv^2}{R} \propto R^{-5/2}$ $\therefore v \propto R^{-3/4}$

Now, $T = \dfrac{2\pi R}{v}$

or $T^2 \propto \left(\dfrac{R}{v}\right)^2$

or $T^2 \propto \left(\dfrac{R}{R^{-3/4}}\right)^2$

or $T^2 \propto R^{7/2}$

Sol 4: (B) From Kepler's third law

$T^2 \propto r^3$ or $T \propto (r)^{3/2}$

$\therefore \dfrac{T_2}{T_1} = \left(\dfrac{r_2}{r_1}\right)^{3/2}$

or $T_2 = T_1\left(\dfrac{r_2}{r_1}\right)^{3/2} = (365)\left(\dfrac{1}{2}\right)^{3/2}$

$T_2 \approx 129$ days

Sol 5: (D) $T \propto \dfrac{1}{\sqrt{g}}$ i.e., $\dfrac{T_2}{T_1} = \sqrt{\dfrac{g_1}{g_2}}$

where g_1 = Acceleration due to gravity on

Earth's surface

$= g$

g_2 = Acceleration due to gravity at a height

h = R from earth's surface = g/4

$\left[\text{Using } g(h)\dfrac{g}{\left(1+\dfrac{h}{R}\right)^2}\right]\dfrac{T_2}{T_1} = \sqrt{\dfrac{g}{g/4}} = 2$

Sol 6: (C) Time period of a satellite very close to earth's surface is 84.6 min. Time period increases as the distance of the satellite from the surface of earth increase. So, time period of spy satellite orbiting a few 100 km above the earth's surface should be slightly greater than 84.6 min. Therefore. The most appropriate option is (C) or 2 h.

Sol 7: (D) In case of binary star system angular velocity and hence the time period of both the stars are equal.

Sol 8: (B) In circular orbit of a satellite, potential energy

$= - 2 \times$ (kinetic energy)

$= - 2 \times \dfrac{1}{2}mv^2 = - mv^2$

Just to escape from the gravitational pull, its total mechanical energy should be zero therefore, its kinetic energy should be $+ mv^2$

Sol 9: (C)

$V_{esc} = \sqrt{\dfrac{2GM}{R}} = \sqrt{\dfrac{2G \times 10M}{R/10}} = 10 \times 11 = 110 \text{km}/s$

Sol 10: (B) $g = GM/r^2$

Sol 11: (A) $g' = \dfrac{GM}{(R+h)^2}$, acceleration due to gravity at height h

$\Rightarrow \dfrac{g}{9} = \dfrac{GM}{R^2}\dfrac{R^2}{(R+h)^2} = g\left(\dfrac{R}{R+h}\right)^2$

$\Rightarrow \dfrac{1}{9} = \left(\dfrac{R}{R+h}\right)^2 \Rightarrow \dfrac{R}{R+h} = \dfrac{1}{3}$

$\Rightarrow 3R = R+h \Rightarrow 2R = h$

Sol 12: (C) Position of the null point from mass m,

$x = \dfrac{r}{1+\sqrt{\dfrac{4m}{m}}} = \dfrac{r}{3}$

$V = -Gm\left(\dfrac{3}{r} + \dfrac{12}{2r}\right) = -9\dfrac{Gm}{r}$

Sol 13: (D) To launch the spaceship out into free space, from energy conservation,

$\dfrac{-GMm}{R} + E = 0$

$E = \dfrac{GMm}{R} = \left(\dfrac{GM}{R^2}\right)mR = mgR$

$= 6.4 \times 10^{10}$ J

Sol 14: (A) $E_f = \dfrac{1}{2}mv_0^2 - \dfrac{GmM}{3R} = \dfrac{1}{2}m\dfrac{GM}{3R} - \dfrac{GmM}{3R}$

$= \dfrac{GmM}{3R}\left(\dfrac{1}{2} - 1\right) = \dfrac{-GMm}{6R}$

$E_i = \dfrac{-GMm}{R} + K$

$E_i = E_f$

$K = \dfrac{5GmM}{6R}$

Sol 15: (B) Potential at point P due to complete solid sphere

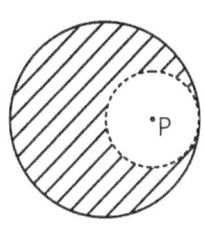

$$= -\frac{GM}{2R^3}\left(3R^2 - \left(\frac{R}{2}\right)^2\right)$$

$$= -\frac{GM}{2R^3}\left(3R^2 - \frac{R^2}{4}\right)$$

$$= -\frac{GM}{2R^3}\left(\frac{11R^2}{4}\right) = -\frac{11GM}{8R}$$

Potential at point P due to cavity part

$$= -\frac{3}{2}\frac{G\frac{M}{8}}{\frac{R}{2}} = \frac{-3GM}{8R}$$

So potential due to remaining part at point P

$$= \frac{-11GM}{8R} - \left(\frac{-3GM}{8R}\right)$$

$$= \frac{-11GM + 3GM}{8R} = \frac{-GM}{R}$$

Sol 16: (C) $\dfrac{GmM}{(R+h)^2} = \dfrac{GMm}{R}$

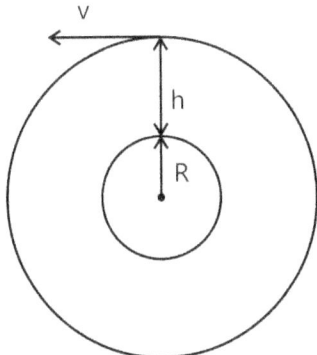

$$v = \sqrt{\frac{GM}{R}}$$

$$\frac{1}{2}mv_1^2 - \frac{GMm}{R} = 0$$

$$v_1 = \sqrt{\frac{2GM}{R}}$$

$$\Delta V = \sqrt{\frac{GM}{R}}(\sqrt{2} - 1) = \sqrt{gR}(\sqrt{2} - 1)$$

JEE Advanced/Boards

Exercise 1

Sol 1:

consider a small strip of rod of length $d\ell$ at a distance ℓ from the small mass. Let the mass of strip be dm

$df = \dfrac{GM}{\ell^2}dm$ (df= force of attraction between the strip and small mass)

$dm = \sigma\, dl$

where σ is linear density of rod

$$\sigma = \frac{M}{2L}$$

$$\Rightarrow df = \frac{GM}{\ell^2}.\sigma dl$$

integrating from L to 3L

$$\int_0^f df = \int_L^{3L} \frac{GM}{\ell^2}\sigma\, d\ell$$

$$f = GM\sigma. -\frac{1}{L}\Big|_2^{32}$$

$$f = GM\sigma.\frac{2}{3L}$$

$$= GM.\frac{M}{2L}.\frac{2}{3L}$$

$$F = \frac{GM^2}{3L^2}$$

Note: Try understanding the boundary conditions. It is most important aspect of physics. Here it is integrated from L to 3L because the rod starts from distance L till distance 3L from the small mass.

Sol 2: $v_c = \sqrt{\dfrac{2GM}{R}}$

Kinetic energy = Change in potential energy

$$\frac{1}{2}mv^2 = -\frac{GMm}{r} - \left(-\frac{GMm}{R}\right)$$

$$\frac{1}{2}m.(k.v_e)^2 = \frac{GMm}{R} - \frac{GMm}{r}$$

$\Rightarrow \dfrac{1}{R} - \dfrac{1}{r} = \dfrac{1}{2} \cdot m \cdot k^2 \cdot 2 \dfrac{GM}{R} \cdot \dfrac{1}{GMm}$

$\Rightarrow \dfrac{1}{R} - \dfrac{1}{r} = \dfrac{k^2}{R}$

$\Rightarrow r = \dfrac{R}{1-k^2}$

height = $r - R$

$= \dfrac{R}{1-R^2} - R = \dfrac{k^2 R}{1-k^2}$

Hence it will rise to a height of $\dfrac{k^2 R}{1-k^2}$

Sol 3: Consider a small path on the ring of length $d\ell$, which subtends an angle $d\theta$ at the center. Let its mass be dM

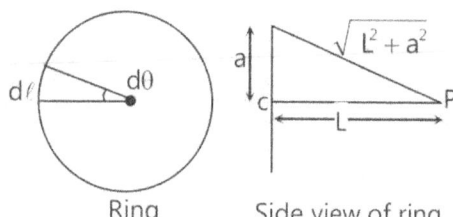

Ring   Side view of ring

$d\ell = a\, d\theta$

$dM = \sigma\, d\ell$

r = Linear density of ring $= \dfrac{M}{2\pi a}$

$\Rightarrow dM = \dfrac{M}{2\pi a} \cdot a\, d\theta = \dfrac{M}{2\pi} d\theta$

Let the particle be at a distance < along the axis from center.

Potential energy due to → Mass patch

$d_E = \dfrac{GM}{r} \cdot dM$ (m = mass of particles)

$= \dfrac{GM}{r} \cdot \dfrac{M}{2\pi} d\theta$

$r = \sqrt{L^2 + a^2}$

$\displaystyle\int_0^E dE = \int_0^{2\pi} \dfrac{GMm}{\sqrt{L^2 + a^2}} \cdot \dfrac{1}{2\pi} d\theta$

$E = \dfrac{GMm}{\sqrt{K^2 + a^2}} \times \dfrac{2\pi}{2\pi} = \dfrac{GMm}{L^2 + a^2}$

Kinetic energy = Change in potential energy

$\dfrac{1}{2}mv^2 = \dfrac{GMm}{a} - \dfrac{GMm}{\sqrt{a^2 + L^2}}$

$\dfrac{1}{2}mv^2 = \dfrac{GMm}{a\sqrt{(a^2 + \ell^2)}} \left(\sqrt{a^2 + L^2} - a\right)$

$v = \sqrt{\dfrac{2GM}{a\sqrt{a^2 + \ell^2}}\left(\sqrt{a^2 + \ell^2} - a\right)}$

Here given $\ell = a$

$\Rightarrow v = \sqrt{\dfrac{2GM\,(\sqrt{2}-1)}{a}\dfrac{}{\sqrt{2}}}$

$v = \sqrt{\dfrac{2GM}{a}\left(1 - \dfrac{1}{\sqrt{2}}\right)}$

Sol 4: Let the height be h

$g = \dfrac{GM}{(R+h)^2}$ (above the surface)

g below the surface

$g = \dfrac{GM(R-h)}{R^3}$

$\Rightarrow \dfrac{GM}{(R+h)^2} = \dfrac{GM(R-h)}{R^3}$

$\Rightarrow h^3 + h^2 R - hR^2 = 0$

$h = 0$ (which is an obvious solution)

$h^2 + hR - R^2 = 0$

$h = -\dfrac{R \pm \sqrt{3R^2}}{2}$

$h = \dfrac{\sqrt{3}-1}{2}$ ($\because h > 0$)

Sol 5: (i) Maximum height = 8R

$\Rightarrow$ Distance from center of earth (r)

$= 8R + R$

$= 9 R$

Kinetic energy = Change in potential energy

$\dfrac{1}{2}mv^2 = -\dfrac{GMm}{9R} - \left(-\dfrac{GMm}{R}\right)$

$\dfrac{1}{2}mv^2 = GMm\left(\dfrac{8}{9R}\right)$

$v = \sqrt{\dfrac{16GM}{9R}}$; $v = \dfrac{4}{3}\sqrt{\dfrac{GM}{R}}$

(ii) Half minimum height = 4R

$\Rightarrow$ r = 4R + R = 5R

$\Rightarrow \dfrac{1}{2}mv^2 = -\dfrac{GMm}{9R} - \left(-\dfrac{GMm}{5R}\right)$

$\Rightarrow v = \sqrt{\dfrac{8GM}{45R}}$; $v = \dfrac{2}{3}\sqrt{\dfrac{2GM}{5R}}$

Sol 6: We use principal of superposition gravitation field due to sphere I

$E_1 = -\dfrac{GM}{x^2}$; x > R,

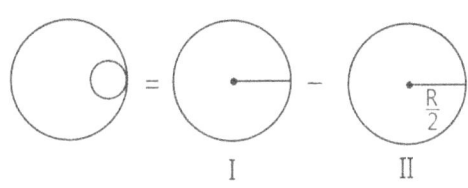

I   II

$= -\dfrac{GM}{R^3}.x$; x < R

Let M_2 mass of sphere II

Here the center is at $\dfrac{R}{2}$ hence distance function is $R - \dfrac{R}{2}$.

Assume sphere is uniform

$\Rightarrow M = \dfrac{4}{3}\pi R^3$

$m_2 = \dfrac{4}{3}\pi\left(\dfrac{R}{2}\right)^3 = \dfrac{1}{8}\left(\dfrac{4}{3}\pi R^3\right) = \dfrac{M}{8}$

$E_2 = -\dfrac{GM_2}{\left(x - \dfrac{R}{2}\right)^2}$ x < 0, x > R

$-\dfrac{GM_2}{\left(\dfrac{R}{2}\right)^3}\left(x - \dfrac{R}{2}\right)$ 0 < x < R

$= -\dfrac{GM}{8\left(R - \dfrac{R}{2}\right)^2}$ x < 0, x > R

$-\dfrac{GM_2}{R^3}\left(x - \dfrac{R}{2}\right)$

for x > R

E = $E_1 - E_2$

$= -\dfrac{GM}{x^2} + \dfrac{GM}{8\left(x - \dfrac{R}{2}\right)^2}$

$= \dfrac{GM}{8}\left(\dfrac{1}{\left(k - \dfrac{R}{2}\right)^2} - \dfrac{8}{x^2}\right)$

$M = \dfrac{4}{3}\pi R^3 \rho_0$

$\Rightarrow E = \dfrac{\pi G\rho_0 R^3}{6}\left(\dfrac{1}{\left(x - \dfrac{R}{2}\right)^2} - \dfrac{8}{x^2}\right)$

for x < R

E = $E_1 - E_2$

$= -\dfrac{GM}{R^3}x + \dfrac{GM}{R^3}\left(x - \dfrac{R}{2}\right)$

$E = \dfrac{GM}{2R^2}$

If is independent of x, hence uniform

Sol 7: Potential energy due to planet, at a distance r from its cents

$P.E_1 = -\dfrac{G(3M)m}{r}$; 2R < r < 5R

Potential due to plant 2

$P.E_2 = -\dfrac{GMm}{(6R - r)}$; R < r < 4R

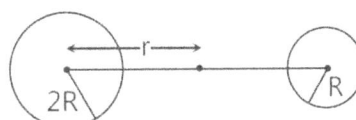

P.E = $P.E_1 + P.E_2$

$= -\dfrac{3GMm}{r} - \dfrac{GMm}{6R - r}$

$E(r) = -GMm\left(\dfrac{3}{r} + \dfrac{1}{6R - r}\right)$

Differentiating

$$\frac{dE}{dr} = -GMm\left[-\frac{3}{r^2} - \frac{1}{(6R-r)}(-1)\right]$$

For maximum E, $\frac{dE}{dr} = 0$

$$\Rightarrow -\frac{3}{r^2} + \frac{1}{(6R-r)^2} = 0$$

$$\Rightarrow r = \frac{6\sqrt{3}R}{\sqrt{3}+1}$$

For particle to reach other side it is sufficient if its velocity is zero at

$$r_0 = \frac{6\sqrt{3}R}{\sqrt{3}+1}$$

i.e., K.E = 0 at this point

∴ Speed is minimum at $r_0 = \dfrac{6\sqrt{3}R}{\sqrt{3}+1}$

$$= 3R(3-\sqrt{3})$$

Potential energy at x = 2R

∴ $PE_1 + PE_2$

$$= \frac{G.3M}{2R} - \frac{GMm}{6R-R} = -\frac{17GMm}{10R}$$

Potential energy at x = 5 R

$$\Rightarrow PE_1 + PE_2$$

$$= -\frac{G(3M)m}{5R} - \frac{GMm}{6R-R}$$

$$= -\frac{8}{5}\frac{GMm}{R} \quad PE(2r) < PE(5R)$$

$PE(r_0) - PE(2r) > PE(r_0) - PE(5R)$

Hence it has maximum speed at x = 2R

Sol 8: Maximum range $= \dfrac{v^2}{g}$

$$b = \frac{v^2}{g}$$

$$v = \sqrt{bg} = \sqrt{\frac{bGM}{r_e^2}}$$

$$M = \frac{4}{3}\pi r^3 \rho$$

$$\Rightarrow v = \sqrt{\frac{bG}{r_e^2}.\frac{4}{3}\pi r_e^3 \rho}$$

$$v = \sqrt{\frac{4\pi b G\, r_e \rho}{3}}$$

Escape velocity $V_e = \sqrt{\dfrac{2GM}{r}}$

$$= \sqrt{\frac{2G}{r_P}.\frac{4}{3}\pi r_P^3.2\rho} = \sqrt{\frac{16\pi\rho r_P^2}{3}}$$

v= ve

$$\Rightarrow \sqrt{\frac{4\pi\, bG\, r_e \rho}{3}} = \sqrt{\frac{16\pi\rho r_P^2}{3}}$$

$$\Rightarrow r_P = \sqrt{\frac{b r_e}{2}}$$

$$= \frac{\sqrt{4\times10^{-3}\times6400}}{2}$$

$$r_P = \sqrt{6.4} \text{ km}$$

Maximum radius of planet is $\sqrt{6.4}$ km

Sol 9: $V = \sqrt{\dfrac{GM}{r}}$ where V = Orbital velocity

$$V_1 = \sqrt{\frac{GM}{2R}}, \quad V_2 = \sqrt{\frac{GM}{3R}} \quad (r = R + \text{height})$$

Angular velocity $\omega = \dfrac{V}{R}$

$$\omega_1 = \frac{V}{2R}\sqrt{\frac{GM}{2R}}; \omega_2 = \frac{1}{3R}\sqrt{\frac{GM}{3R}}$$

Relative angular velocity $\omega_R = \omega_1 + \omega_2$

$$= \sqrt{\frac{GM}{R^3}}\left(\frac{1}{2\sqrt{2}} + \frac{1}{3\sqrt{3}}\right)$$

$$t = \frac{2\pi}{wR} = \frac{2\pi}{\sqrt{\dfrac{GM}{R^3}}\dfrac{2\sqrt{2}+3\sqrt{3}}{6\sqrt{6}}}$$

$$\therefore t = \frac{2\pi R^{\frac{3}{2}}(6\sqrt{6})}{\sqrt{GM}(2\sqrt{2}+3\sqrt{3})}$$

Sol 10: Let d be distance between them. Distance of centre of mass from m

$$r_1 = \frac{Md}{m+M} ; r_1 = \frac{2d}{3}$$

$$F = \frac{GMm}{d^2} = \frac{2GM^2}{d^2}$$

Gravitational force = Centrifugal force

$$\frac{2GM^2}{d^2} = \frac{mv_1^2}{\frac{2d}{3}} \; ; \Rightarrow v_1 = \sqrt{\frac{4GM}{3d}}$$

$$T = \frac{2\pi}{v_1} r_1 = \frac{2\pi \frac{2d}{3}}{\sqrt{\frac{4GM}{3d}}} = \frac{2\pi d^{\frac{3}{2}}}{\sqrt{3GM}}$$

$$m = \frac{M_s}{3} \quad (M_s = \text{Mass of surfs})$$

$$\therefore T = \frac{2\pi d^{\frac{3}{2}}}{\sqrt{GM_s}}$$

Time period of earth $T_e = \frac{2\pi d^{\frac{3}{2}}}{\sqrt{GM_s}}$

given $T = T_e$; $\Rightarrow d = R$

Sol 11: Total energy $= \sum_{i<j} \frac{-G.M_i M_j}{r_{ij}}$

$$= -G\left(\frac{M^2}{a}\right) \times (\Sigma 3) = -\frac{6GM^2}{a}$$

Final energy $= -\frac{6GM^2}{2a}$

Change in energy $= -\frac{6GM^2}{2a} + \frac{6GM^2}{a}$

$$= \frac{6GM^2}{2a}$$

$\therefore$ Work done is $\frac{6GM^2}{2a}$

Sol 12: Potential $= \frac{GM}{r}$

$\therefore$ Change in potential $= GMm\left(\frac{1}{R} - \frac{1}{2R}\right)$

$$= \frac{GMm}{2R}$$

K.E. = Charge in P.E

$$\frac{1}{2}mv^2 = \frac{GMm}{2R} \; ; v = \sqrt{\frac{GM}{R}}$$

Inside sphere v is constant

$\therefore$ Time $= \frac{2R}{v} = 2\sqrt{\frac{R^3}{GM}}$

Sol 13: $\omega_1 = \frac{2\pi}{T}$

Angular velocity of earth

$$\omega_e = \frac{2\pi}{T_0} \quad (T_0 = 24 \text{ Hz})$$

$$\omega_r = \omega_1 + \omega_e \text{ or } \omega_1 - \omega_e$$

$$T = \frac{2\pi}{\omega_r} = \frac{2\pi}{\frac{2\pi}{T} \pm \frac{2\pi}{T_0}} = \left(\frac{T_0 \pm T}{T_0 T}\right)^{-1}$$

$$= \left(\frac{24 \pm 1.5}{24 \times 1.5}\right)^{-1} = \left(\frac{17}{24}\right)^{-1} \text{ hrs}, \left(\frac{5}{8}\right)^{-1} \text{ hrs}$$

Sol 14: For a satellite

$$|K.E| = \frac{1}{2}|P.E.| = |\text{total energy}|$$

$$E_1 = -2 \times 10^5 \text{ J}$$

$$\Rightarrow U_1 = -4 \times 10^{-5}, K_1 = 2 \times 10^5$$

$$U_2 = -2 \times 10^{-5}$$

$$\Rightarrow E_1 = -1 \times 10^{-5} \text{ J}$$

$$\Delta E = 1 \times 10^5 \text{ J}$$

$\therefore$ Energy required is 10^5 J

Sol 15: Total energy $= -\frac{GMm}{2r}$, r = Radius

$\therefore$ Change is energy $= \frac{GMm}{2}\left(\frac{1}{R_e} - \frac{1}{r}\right)$

t = Change in energy

$$\Rightarrow t = \frac{GMm}{2C}\left(\frac{1}{R_e} - \frac{1}{r}\right)$$

Sol 16:

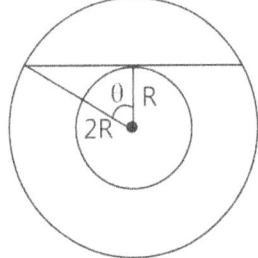

Angle of view = 2θ

$$= 2\cos^{-1}\frac{R}{2R} = 2(60°) = 120° = \frac{2\pi}{3}$$

Angular velocity of earth $\omega_e = \frac{2\pi}{T_0}$

$(T_0 = 24 \text{ Hrs})$

Angular velocity of satellite $\omega = \sqrt{\dfrac{GM}{8R^3}}$

$\left(\omega = \sqrt{\dfrac{GM}{r}}\right)$

Relative angular velocity $\omega_r = \omega - \omega_e$

$= \sqrt{\dfrac{GM}{8R^3}} - \omega_e \; ; \; T = \dfrac{\theta}{\omega} = \dfrac{2\pi}{3\left(\sqrt{\dfrac{Gm}{8R^3}} - \omega_e\right)}$

Sol 17: Let final velocity of launch pad be x

$\Rightarrow 3mv_0 = m(x + v_0) + 2m(x)$

$\Rightarrow x = \dfrac{2v_0}{3}$ ..... (i)

Angular momentum = mvr

$= m \cdot \dfrac{2v_0}{3} \cdot (3R_m)$

$L = 2mv_0 R_m$ ..... (i)

Angular momentum is constant

$\therefore mvx \cdot R_m = L$

$\Rightarrow mv_x \cdot R_m = 2mv_0 R_m$

$v_x = 2v_0$ ...... (ii)

$\Delta KE = \Delta Pt$

$\Rightarrow \dfrac{1}{2}mv^2 - \dfrac{1}{2}mx^2 = GMm\left(\dfrac{1}{R_m} - \dfrac{1}{3R_m}\right)$

$\Rightarrow v^2 = x^2 + 2GM\left(\dfrac{2}{3R_m}\right)$

Now satellite equals

$v = \sqrt{\dfrac{GM}{r}} \; ; \Rightarrow v_0 = \sqrt{\dfrac{GM}{3R_m}}$

$\Rightarrow \dfrac{GM}{3R_m} = v_0^2 \; ; \Rightarrow v^2 = x^2 + 4v_0^2$

$x = \dfrac{2v_0}{3} \; ; \Rightarrow v = \dfrac{2}{3}\sqrt{10}\, v_x$

$\cos\theta = \dfrac{v_x}{v} \; ; \Rightarrow \theta = \cos^{-1}\dfrac{2v_0}{\dfrac{2\sqrt{10}}{3}v_0}$

$\Rightarrow \cos\theta = \dfrac{3}{\sqrt{10}} \Rightarrow \theta = \cos^{-1}\dfrac{3}{\sqrt{10}}$

Sol 18: (a) Orbital velocity $V = \sqrt{\dfrac{GM}{R}}$

(b) At maximum distance, they are mutually perpendicular radially about the winter of planet

$\therefore$ Maximum distance $= \sqrt{2}\, r$

(c) Their relative velocity $= \sqrt{2}\, V = \sqrt{\dfrac{2GM}{R}}$

Sol 19: Gravitational force

$F = \dfrac{GMm}{r^2}$

$\Rightarrow \Delta F = -\dfrac{GMm}{r^3}\Delta r$

$\therefore$ Net force at a height $r + \Delta r$

$= \dfrac{GMm}{r^2}\left(1 - \dfrac{2\Delta r}{r}\right)$

Centrifugal force $f = mrw^2$

$\Delta f = mw^2 \Delta r$

$\therefore$ Net centrifugal force $- f + \Delta f$

$= mrw^2\left(1 + \dfrac{\Delta r}{r}\right)$

$= mrw^2\left(1 + \dfrac{\Delta r}{r}\right) - \dfrac{GMm}{r^2}\left(1 - \dfrac{2\Delta r}{r}\right)$

$= mrw^2\dfrac{\Delta r}{r} + \dfrac{GMm}{r^2}\dfrac{2\Delta r}{r}$

$mrw^2 = \dfrac{GMm}{r^2} = mg \;(\because \text{satellite moon})$

$\therefore T = 3Mg\dfrac{\Delta r}{r} = 3 \times 100 \times 10 \times \dfrac{64 \times 10^{-3}}{6400}$

$T = 3 \times 10^{-2}$ N

Sol 21: (i) P.E $= \left(-\dfrac{GMm}{R}\right)3 + 3\left(-\dfrac{GMm}{\sqrt{3}R}\right)$

$= -\dfrac{3GM}{R}\left(\dfrac{m}{\sqrt{3}} + m\right)$

(ii) Centrifugal force = Force towards center

$\dfrac{mv^2}{R} = \dfrac{GMm}{R^2} + \left(\dfrac{GMm}{(\sqrt{3}R)^2} \cdot \dfrac{\sqrt{3}}{2}\right)R$

$\Rightarrow v = \left(\sqrt{\dfrac{GM}{R}\dfrac{(2\sqrt{3} + R)}{2\sqrt{3}}}\right)$

Sol 22: Surface area of earth $A = 4\pi R^2$

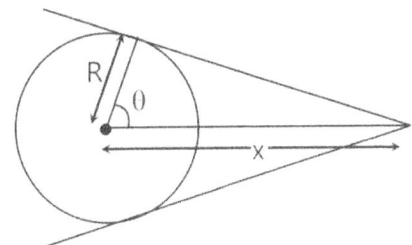

$\cos\theta = \dfrac{R}{x}$

Area covered by the satellite on surface of earth

$A_1 = 2\pi R^2 (1 - \cos\theta)$

where θ is semi-vertical angle

$\therefore$ Area out of reach $= A - A_1$

$= 4\pi R^2 - 2\pi r^2 (1 - \cos\theta)$

$= 2\pi R^2 (1 + \cos\theta)$

$= 2\pi R^2 \left(1 + \dfrac{R}{x}\right)$

Sol 23: (a) $M = 2m$

Center of mass from m $(r_1)\,\dfrac{md}{m+m}$

$= \dfrac{2d}{3}\,;\; \Rightarrow r^2 = \dfrac{d}{3}$

Force between them $(t) = \dfrac{GMm}{d^2}$

$= \dfrac{2GM^2}{d^2}$

Let velocity of mass m be v_1

$\dfrac{mv_1^2}{r_1} = (F)\,;\; \Rightarrow \dfrac{mv_1^2}{\dfrac{2d}{3}} = \dfrac{2GM^2}{d^2}$

$\Rightarrow v_1 = \sqrt{\dfrac{4GM}{3d}}\,;\; T = \dfrac{2\pi}{w} = \dfrac{2\pi r}{v}$

$= \dfrac{2\pi \times \dfrac{2d}{3}}{\sqrt{\dfrac{4GM}{3d}}} = \dfrac{4\pi}{3}\sqrt{\dfrac{3d^3}{4GM}} = \dfrac{2\pi d^{3/2}}{\sqrt{3GM}}$

(b) $\dfrac{v_m}{r_m} = \dfrac{v_M}{r_M}$

$\dfrac{L_m}{L_M} = \dfrac{mv_m r_m}{2mv_m r_m} - \dfrac{1}{2}\left(\dfrac{r_m}{r_M}\right)^2 = \dfrac{1}{2}(2)^2$

$\dfrac{L_m}{L_M} = 2$

(c) $k = \dfrac{1}{2}mv^2$

$\dfrac{k_m}{k_M} = \dfrac{\dfrac{1}{2}mv_m^2}{\dfrac{1}{2}mv_m^2} = \dfrac{m}{2m}\cdot\left(\dfrac{r_m}{r_M}\right)^2 = \dfrac{1}{2}(2)^2$

$\dfrac{k_m}{k_M} = 2$

Sol 24:

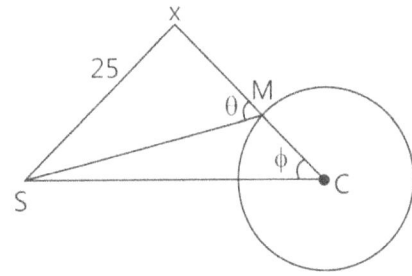

Let S be satellite, M Mumbai, C center of earth.

$\phi = 30°$

let $SC = r$

$MC = R$

$\Rightarrow r\sin\phi = (r\cos\phi - R)\tan q$

$\Rightarrow \theta = \tan^{-1}\left(\dfrac{\sin\phi}{\cos\phi - \dfrac{R}{r}}\right)\,;\; r = \left(\dfrac{T\sqrt{GM}}{2\pi}\right)^{\frac{2}{3}}$

$T = 24\ \mathrm{Hr}$

$\sqrt{GM} = R\sqrt{g}$

$R = $ radius of earth

Upon substitution we get

$\theta = \cot^{-1}\left(\sqrt{3} - \dfrac{32}{105}\right)$

Sol 25: Angular velocity be ω

$mr\omega^2 = mg$

$\omega = \dfrac{2\pi}{T}\,;\; mR.\dfrac{(2\pi)^2}{T^2} = mg$

$T = 2\pi\sqrt{\dfrac{R}{g}}$

$g = \dfrac{GM}{R^2}$

$$= \frac{G}{R^2} \cdot \frac{4}{3} \pi R^3 r$$

$$\therefore T = 2\pi \sqrt{\frac{R}{GR \cdot \frac{4}{3} \pi \rho}}$$

$$T = \sqrt{\frac{3\pi}{G\rho}}$$

Exercise 2

Multiple Correct Choice Type

Sol 1: (A, D)

$$g = \frac{GM}{r^2}; \quad r > R$$

$$\frac{GM}{R^2} r; \qquad r < R$$

Sol 2: (A, D) $f_{12} = f_{21}$

$\Rightarrow m_1 a_1 = m_2 a_2$

$\Rightarrow$ if $m_1 < m_2$

$\therefore a_1 > a_2$

Total energy is constant, by law of conservation of energy centre of mass is in motion in the reference from of the masses.

Sol 3: (B, C, D) Everywhere gravitational field is zero which is same everywhere inside the spherical shell.

Everywhere potential is same as $-\dfrac{dE}{dx} = 0$

i.e., gravitational field is same. Potential inside sphere is equal to that on surface.

Sol 4: (A, C) The satellite will always be in orbital motion at every instant

$|u| = 2|k|$

$|\Delta u| = 2|\Delta k|$

$u = -2k$

$\Rightarrow \Delta u = -2\Delta k$

$\because \Delta u$ is –ve

hence kinetic energy increases

$L = mvr$

$$\Delta L = mvr \left(\frac{\Delta v}{v} + \frac{\Delta r}{r} \right)$$

$$U = -\frac{GMm}{r} ; \Delta u = \frac{GMm}{r^2} \Delta r$$

$$k = \frac{1}{2} mv^2$$

$$\Delta k = mv \, \Delta u$$

$$\Delta u = -2\Delta k$$

$$\frac{GMm}{r^2} \Delta r = -2mv\Delta u$$

$$\frac{GMm}{r^2} \Delta r = 2\left(\frac{1}{2} mv^2 \right)$$

$$\frac{\Delta r}{r} = -\frac{2\Delta v}{v} ; \Rightarrow \Delta L = mvr \left(\frac{\Delta r}{2r} \right)$$

$\Delta r < 0$

$\Rightarrow \Delta L < 0$

Sol 5: (B, D) Communication satellites are geo stationary.

Sol 6: (A, C, D) Only potential energy increase

$$P.E = -\frac{GMm}{r}$$

v decrease; $v \propto \dfrac{1}{\sqrt{r}}$

and hence angular velocity and centripetal acceleration decreases as r increases.

Sol 7: (A, D) Acceleration is always directed towards centre of earth. Centripetal force.

Sol 8: (A, B, C) $v \propto \dfrac{1}{r}$, $T \propto r^{3/2}$

$\therefore$ Speed is maximum and time period is minimum

Potential energy is minimum

$P.E \propto \left(-\dfrac{1}{r} \right)$

Sol 9: (A, D) Satellite has to be above equator at some time

$$T = 2\pi \sqrt{\left(\frac{r}{R} \right)^2 \cdot \frac{r}{g}} ; r = \text{radius of orbit}$$

$$> 2\pi \sqrt{\frac{R}{g}}$$

$$\because r > R, \left(\frac{r}{R}\right)^2 > 1$$

Sol 10: (A, B, D) $T \propto R^{3/2}$

$$\therefore S_1 : S_2 = 1^{3/2} : 4^{3/2} = 1 : 8$$

$$v \propto \frac{1}{\sqrt{r}} \Rightarrow v_1 : v_2 = 4^{1/2} : 1^{1/2} = 2 : 1$$

Angular momentum $L \propto r^{1/2}$

$$L_1 : L_2 = 1 : 4^{1/2} = 1 : 2$$

Let velocities be 2k, k

Relative velocities are 3k, k

i.e. $v_1 : v_2 = 3 : 1$

Relative radii = 4R + R, 4R – R

$\qquad = 5R, 3R$

i.e. $R_1 : R_2 = 4 : 3$.

$$\omega = \frac{v}{R}$$

$$\therefore \omega_1 : \omega_2 = \frac{3}{5} : \frac{1}{3}$$

$$\omega_1 : \omega_2 = 9 : 5$$

Assertion Reasoning Type

Sol 11: (C) There is no such real radial force. It only appears in moon's frame of reference as centrifugal force.

Sol 12: (B) Statement-I is true because there is no net acceleration downward in it.

Sol 13: (D) Geostationary satellites have fixed orbital radius and do have 24 hours of time period of revolution.

Sol 14: (D) Statement-I is only true long distances between them.

Sol 15: (A) For travel, energy required

= maximum P.E – P.E at surface

Comprehension type

Paragraph 1:

Sol 16: (B) Let P.E at $\infty = 0$

Final P.E $= \dfrac{GM_1M_2}{d}$

final distance between center of masses

d = R + 2R = 3R

mass of small sphere is m

mass of smaller sphere

$$= m\left(\frac{R_2}{R}\right)^3 = m\left(\frac{2R}{R}\right)^3$$

$$= 8\,m$$

Final energy $= \dfrac{GM.8m}{3R} = \dfrac{8GM^2}{3R}$

Let initial velocities be v_1, v_2

Let final velocities be v_3, v_4

since centre of mass is at rest

$$\overrightarrow{v}_3 = -e\,\overrightarrow{v}_1$$

(e = Coefficient of restitution)

$$|V_3| = \frac{1}{2}|V_1|$$

$$\Rightarrow \frac{1}{2}mv_3^2 = \frac{1}{2}m\left(\frac{1}{2}v\right)^2$$

$$= \frac{1}{4}\left(\frac{1}{2}mv_1^2\right)$$

Similarly $\dfrac{1}{2}mv_4^2 = \dfrac{1}{4}\left(\dfrac{1}{2}mv_2^2\right)$

$\therefore$ Final energy $= \dfrac{1}{2}mv_3^2 + \dfrac{1}{2}mv_4^2$

$$= \frac{1}{4}\left(\frac{1}{2}mv_1^2 + \frac{1}{2}mv_2^2\right)$$

$$= \frac{1}{4}\left(\frac{8Gm^2}{3R}\right)$$

$$= \frac{2Gm^2}{3R}$$

Sol 17: (A) Change in P.E = kinetic energy

$$\frac{2GM^2}{3R} = P.E - \left(-\frac{8GM^2}{3R}\right)$$

$$\Rightarrow PE = \frac{6GM^2}{3R} = \frac{GM(8M)}{4R}$$

∴ Maximum distance between them is 4R

Note:- Try deriving the result

$\vec{V_3} = -\ e\vec{V_1}$ used in the problem. Here centre of mass is at rest.

Paragraph 2:

Sol 18: (B) Area of ASBC > ASBD

∴ $t_1 > t_2$

Kepler's 2nd law

Sol 19: (C) $|u| > |k|$ always

Because if $|k| \ge |u|$ body escapes from the suns gravitational force

Paragraph 3:

Sol 20: (B) Self energy of a uniform sphere of radius R and mass M is given by

$$E = -\frac{3GM^2}{5R}$$

∴ Change in energy

$$= -\frac{3GM^2}{5}\left(\frac{1}{R/2} - \frac{1}{R}\right)$$

$$= \frac{3GM^2}{5R}$$

Increase in temperature $= \dfrac{energy}{M.S}$

$$= \frac{3GM^2}{5R} \cdot \frac{1}{M.S}$$

$$= \frac{3GM}{5RS}$$

Sol 21: (A) $T_0 = \dfrac{3GM}{5RS}$

$$M = \frac{5SRT_0}{3G}$$

Note: Study the self-energy of objects here is a derivation.

Consider a sphere of density ρ and initial radius r initial mass $m = \dfrac{4}{3}\pi r^3\rho$

Let additional mass added

$dm = 4\pi r^2.dr\rho$

Increase is energy $dE = \dfrac{GMdm}{r}$

$$dE = \frac{G.\frac{4}{3}\pi r^3\rho.4\pi r^2 dr\rho}{r}$$

$$dE = \frac{(4\pi)^2}{3}\rho^2 . G. r^4 . dr$$

$$E = \int_0^R dE = \frac{(4\pi)^2}{3}\rho^2 . G. \frac{r^5}{5}$$

$\rho = \dfrac{M}{\frac{4}{3}\pi R^3}$ (M is final mass)

$$\Rightarrow E = \frac{m^2}{\left(\frac{4\pi}{3}\right)^2 R^6} . \frac{R^5}{5} \times \frac{(4\pi)^2}{3}$$

$$E = \frac{3GM^2}{5R}$$

Paragraph 4:

Sol 22: (D) 1 Bose $= \dfrac{1}{c}$ Newton

Let T be tension in the string.

Let a_1, a_2 be acceleration of m_1, m_2 downward

$$a = \left(\frac{\frac{gm_1^2}{c} - T}{m_1^2}\right)c$$

where $\dfrac{gm_1^2}{c}$ is downward r gravitational force on m_1

Similarly $a_2 = \left(\dfrac{\frac{gm_2^2}{c} - T}{m_2}\right)c$

$a_1 + a_2 = 0$ by constrain equation

$$\left(\frac{gm_1^2 - T}{m_1^2}\right)c + \left(\frac{gm_2^2 - T}{m_2^2}\right) = 0$$

$$\Rightarrow T = 2g\left(\frac{m_1^2 m_2^2}{m_1^2 + m_2^2}\right) \cdot \frac{1}{c}$$

$$= \frac{2 \times 5 \times 2^2 \times 4^2}{2^2 + 4^2} \cdot \frac{1}{c}$$

$$T = \frac{32}{c}$$

T = 32 Bose

Note: If you do not know bose, try guessing what it could be. 1 newton is the force which is produced when an object of mass 1 kg moves with an acceleration of 1 ms^{-2}. Similarly define bose. This is the best assumption you can do with the given amount of information.

Sol 23: (B) Force due to gravity $F = \frac{gm^2}{c} = gm^2$ bose

force along slope $f_1 = f \sin\theta$

$f = f_1$

$= f \sin\theta$

$= gm^2 \sin\theta$ bose

$= 5 \times (2)^2 \cdot \frac{1}{2}$

$= 10$ bose

Previous Years' Questions

Sol 1: (A) Force acting on astronaut is utilized in providing necessary centripetal force, thus he fells weightlessness, as he is in a state of free fall.

Sol 2: (A) The gravitational field is zero at the centre of a solid sphere. The small spheres can be considered as negative mass m located at A and B. The gravitational field due to these masses at O is equal and opposite. Hence, the resultant field at O is zero.

(c and d) → are correct because plane of these circles is y-z, i.e., perpendicular to x-axis i.e., potential at any point on these two circles will be equal due to the positive mass M and negative masses – m and – m.

Sol 3: (B) For $r \leq R$. $F = \frac{GM}{R^3}$, r or $F \propto r$

$$\frac{F_1}{F_2} = \frac{r_1}{r_2} \text{ for } r_1 < R \text{ and } r_2 < R$$

And for $r \leq R$, $F = \frac{GM}{r^2}$ or $F \propto \frac{1}{r^2}$

i.e., $\frac{F_1}{F_2} = \frac{r_2^2}{r_1^2}$ for $r_1 > R$ and $r_2 > R$

Sol 4: $T \propto r^{3/2}$

or $r \propto T^{2/3}$

$$\frac{r_2}{r_1} = \left(\frac{T_2}{T_1}\right)^{2/3}$$

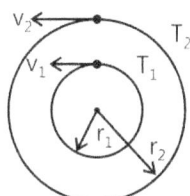

$$r_2 = \left(\frac{T_2}{T_1}\right)^{2/3} \quad r_1 = \left(\frac{8}{1}\right)^{2/3} (10^4) = 4 \times 10^4 \text{ km}$$

Now, $v_1 = \frac{2\pi r_1}{T_1} = \frac{(2\pi)(10^4)}{1} = 2\pi \times 10^4 \text{ km/h}$

$$v_2 = \frac{2\pi r_2}{T_2} = \frac{(2\pi)(4 \times 10^4)}{8} = (\pi \times 10^4) \text{ km/h}$$

(a) Speed of S_2 relative to S_1

$= v_2 - v_1 = -\pi \times 10^4 \text{ km/h}$

(b) Angular speed of S_2 as observed by S_1

$$\omega_r = \frac{|v_2 - v_1|}{|r_2 - r_1|} = \frac{\left(\pi \times 10^4 \times \frac{5}{18} \text{ m/s}\right)}{(3 \times 10^7 \text{ m})}$$

$= 0.3 \times 10^{-3} \text{ rad/s} = 3 \times 10^{-4} \text{ rad/s}$

Sol 5: Centre should be at O and radius r. We can calculate r from figure (b).

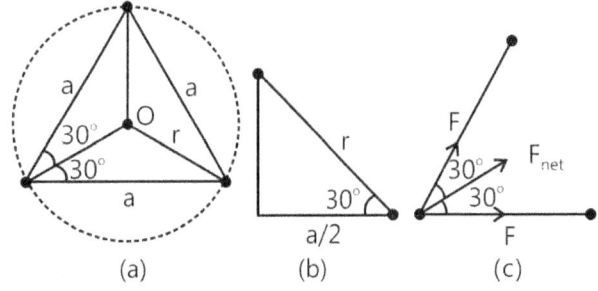

(a)　　　(b)　　　(c)

$$\frac{a/2}{r} = \cos 30° = \frac{\sqrt{3}}{2} \quad \therefore r = \frac{a}{\sqrt{3}}$$

Further net force on any particle towards centre

$F_{net} = 2F \cos 30°$

$$= 2\left(\frac{Gm^2}{a^2}\right)\left(\frac{\sqrt{3}}{2}\right) = \frac{\sqrt{3}\ Gm^2}{a^2}$$

This net force should be equal to $\dfrac{mv^2}{r}$

$$\therefore \frac{\sqrt{3}\ Gm^2}{a^2} = \frac{mv^2}{a/\sqrt{3}} \quad \therefore\ v = \sqrt{\frac{Gm}{a}}$$

Time period of circular motion

$$T = \frac{2\pi r}{v} = \frac{2\pi\,(a/\sqrt{3})}{\sqrt{Gm/a}} = 2\pi\sqrt{\frac{a^3}{3\,Gm}}$$

Sol 6: (a) Orbital speed of a satellite at distance r from centre of earth,

$$v_0 = \sqrt{\frac{GM}{r}} = \sqrt{\frac{GM}{R+h}} \qquad ....(i)$$

Given, $v_0 = \dfrac{v_e}{2} = \dfrac{\sqrt{2GM/R}}{2} = \sqrt{\dfrac{GM}{2R}} \qquad ...(ii)$

From Eqs. (i) and (ii), we get

h = R = 6400 km

(b) Decrease in potential energy=increase in kinetic energy

or $\dfrac{1}{2}mv^2 = \Delta U \quad \therefore\ v = \sqrt{\dfrac{2\,(\Delta U)}{m}}$

$$= \sqrt{\frac{2\left(\dfrac{mgh}{1+h/R}\right)}{m}} = \sqrt{gR}$$

$(h = R) = \sqrt{9.8 \times 6400 \times 10^3} = 7919\ m/s = 7.9\ km/s$

Sol 7: Let there are two stars 1 and 2 as shown below

Let P is a point between C_1 and C_2, where gravitational field strength is zero or at P field strength due to star 1 is equal and opposite to the field strength due to star 2 Hence,

$$\frac{GM}{r_1^2} = \frac{G\,(16M)}{r_2^2} \ \text{or}\ \frac{r_2}{r_1} = 4$$

$r_1 + r_2 = 10\,a \quad \therefore\ r_2 = \left(\dfrac{4}{4+1}\right)(10\,a) = 8a$

and $r_1 = 2\,a$

Now, the body of mass m is projected from the surface of larger star towards the smaller one. Between C_2 and P it is attracted towards 2 and between C_1 and P it will be attracted towards 1

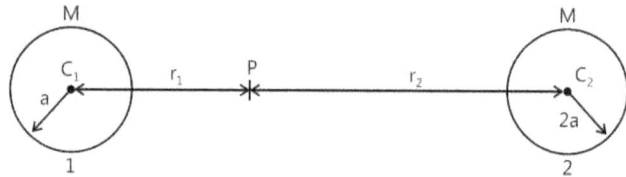

Therefore, the body should be projected to just cross point P because beyond that the particle is attracted towards the smaller star itself.

From conservation of mechanical energy $\dfrac{1}{2}mv^2_{min}$

= Potential energy of the body at P − Potential energy at the surface of larger star.

$$\therefore \frac{1}{2}mv^2_{min} = \left[-\frac{GMm}{r_1} - \frac{16\,GMm}{r_2}\right] - \left[-\frac{GMm}{10a-2a} - \frac{16\,GMm}{2a}\right]$$

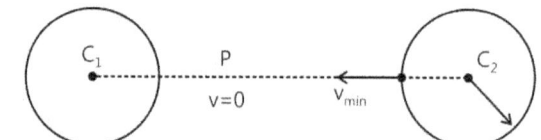

Or $\dfrac{1}{2}mv^2_{min} = \left(\dfrac{48}{8}\right)\dfrac{GMm}{a}$

$$\therefore\ v_{min} = \frac{3\sqrt{5}}{2}\left(\sqrt{\frac{GM}{a}}\right)$$

Sol 8: Speed of particle at A, v_A = escape velocity on the surface of earth

$$= \sqrt{\frac{2GM}{R}}$$

At highest point B, $v_B = 0$

Applying conservation of mechanical energy, decrease in kinetic energy = increase in gravitational potential energy or $\dfrac{1}{2} m v_A^2 = U_B - U_A = m(v_B - v_A)$

or $\dfrac{v_A^2}{2} = v_B - v_A$

$\therefore \dfrac{GM}{R} = -\dfrac{GM}{R+h} - \left[-\dfrac{GM}{R^3}\left(1.5R^2 - 0.5\left(R - \dfrac{R}{100}\right)^2\right) \right]$

or $\dfrac{1}{R} = -\dfrac{1}{R+h} + \dfrac{3}{2R} - \left(\dfrac{1}{2}\right)\left(\dfrac{99}{100}\right)^2 \cdot \dfrac{1}{R}$

Solving this equation, we get

$h = 99.5 \ R$

Sol 9: $g = \dfrac{GM}{R^2} = \dfrac{G\left(\dfrac{4}{3}\pi R^3\right)\rho}{R^2}$ or $g \propto \rho R$

$\quad$ or $R \propto \dfrac{g}{\rho}$

Now escape velocity, $v_e = \sqrt{2gR}$ or $v_e \propto \sqrt{gR}$

or $v_e \propto \sqrt{g \times \dfrac{g}{\rho}} \propto \sqrt{\dfrac{g^2}{\rho}}$

$\therefore (v_e)_{planet} = (11 \ km \ s^{-1})\sqrt{\dfrac{6}{121} \times \dfrac{3}{2}} = 3 \ km \ s^{-1}$

Sol 10: (C) For $r \leq R$

$\dfrac{mv^2}{r} = \dfrac{GmM}{r^2}$ $\quad\quad$ ...... (i)

Here, $M = \left(\dfrac{4}{3}\pi r^3\right)\rho_0$

Substituting in Eq. (i), we get

$v \propto r$

i.e., v-r graph is a straight line passing though origin
For $r > R$

$\dfrac{mv^2}{r} = \dfrac{Gm\left(\dfrac{4}{3}\pi R^3\right)\rho_0}{r^2}$ or $v \propto \dfrac{1}{\sqrt{r}}$

The corresponding v-r graph will be as shown in option (c)

Sol 11: (B, D) $V_{es} = \sqrt{\dfrac{2GM}{R}} = \sqrt{\dfrac{2.G\rho\dfrac{4}{3}\pi R^3}{R}} = \sqrt{\dfrac{4G\rho}{3}}R$

$\quad\quad\quad\quad V_{es} \propto R$

Surface area of $P = A = 4\pi R_P^2$

Surface area of $Q = 4A = 4\pi R_Q^2$

$\Rightarrow R_Q = 2R_P$

Mass R is $M_R = M_P + M_Q$

$\rho\dfrac{4}{3}\pi R_R^3 = \rho\dfrac{4}{3}\pi R_P^3 + \rho\dfrac{4}{3}\pi R_Q^3 \Rightarrow R_R^3 = R_P^3 + R_Q^3 = 9R_P^3$

$R_R = 9^{1/3}R_P \Rightarrow R_R > R_Q > R_P$

Therefore $V_R > V_Q > V_P$

$\dfrac{V_R}{V_P} = 9^{1/3}$ and $\dfrac{V_P}{V_Q} = \dfrac{1}{2}$

Sol 12: (B) Inside planet

$g_i = g_s\dfrac{r}{R} = \dfrac{4}{3}G\pi r\rho$

Force to keep the wire at rest (F)

= Weight of wire

$= \displaystyle\int_{4R/5}^{R} (\lambda dr)\left(\dfrac{4}{3}G\pi r\rho\right) = \left(\dfrac{4}{3}G\pi\rho\right)\left(\dfrac{9\lambda}{50}\right)R^2$

Here, ρ = density of earth $= \dfrac{M_e}{\dfrac{4}{3}\pi R_e^2}$

Also, $R = \dfrac{R_e}{10}$; putting all values, F = 108 N

Sol 13: (A) Measured value of r = (10 ± 1) mm

Δr = 1 mm

Relative error $= \dfrac{\Delta r}{r} = \dfrac{1}{10} = 100\%$

Average value of

$\bar{T} = \dfrac{\displaystyle\sum_{i=1}^{n=5}T_i}{n} = \dfrac{(0.52 + 0.56 + 0.57 + 0.54 + 0.59)}{5} s$

$\Rightarrow \bar{T} = 0.556s \approx 0.56s$

Relative error in time period $\approx \dfrac{0.01}{0.56} = 1.79\%$

Reported value of (R - r) = (50 ± 2) mm

Relative error in (R - r) $= \dfrac{2}{50} = 4\%$

$T = 2\pi\sqrt{\dfrac{7(R-r)}{5g}} \Rightarrow \dfrac{\Delta g}{g} = 2\left(\dfrac{\Delta T}{T}\right) + \dfrac{\Delta(R-r)}{(R-r)}$

$\Rightarrow \dfrac{\Delta g}{g} = 7.57\%$

2. SIMPLE HARMONIC MOTION AND ELASTICITY

SIMPLE HARMONIC MOTION

1. INTRODUCTION

There are so many examples of oscillatory or vibrational motion in our world. E.g. the vibrations of strings in a guitar or a sitar, the vibrations in the speakers of a music system, the to and fro motion of a pendulum, vibration in a suspension bridge as a vehicle passes on it, the oscillations in a tall building during an earthquake etc. Simple harmonic motion (SHM) is a type of oscillatory or vibrational motion. Every kind of oscillation or vibration of a particle or a system is not necessarily simple harmonic. The particle executing SHM like any other oscillatory motion has a variable acceleration, but this variation is different in different kinds of oscillations. The study of SHM is very useful and forms an important tool in understanding the characteristics of sound and light waves and alternating currents. Any oscillatory motion which is not simple harmonic can be expressed as a superposition of several simple harmonic motions of different frequencies.

2. PERIODIC AND OSCILLATORY MOTION

Periodic Motion: A motion which repeats itself after equal intervals of time is called periodic motion.

Oscillatory Motion: A body is said to possess oscillatory or vibratory motion if it moves back and forth repeatedly about a mean position. For an oscillatory motion, a restoring force is required.

Examples of Periodic and Oscillatory motion are revolution of earth around sun and motion of bob of a simple pendulum respectively.

> **NOMORECLASS CONCEPTS**
>
> All Oscillatory motions are periodic but all Periodic motions need not be oscillatory.
>
> A body experiencing force $F = -k(x-a)^n$ is in Oscillatory motion only if n is odd and its mean position is $x=a$. As, if n is odd only then we would have restoring force.

2.1 Periodic Functions

A function is said to be periodic if it repeats itself after time period T i.e. the same function is obtained when the variable t is changed to t + T. Consider the following periodic functions:

$$f(t) = \sin\frac{2\pi}{T}t \quad \text{and} \quad g(t) = \cos\frac{2\pi}{T}t$$

Here T is the time period of the periodic motion. We shall see that if the variable t is changed to t + T, the same function results.

$$f(t+T) = \sin\left[\frac{2\pi}{T}(t+T)\right] = \sin\left[\frac{2\pi t}{T} + 2\pi\right] = \sin\left(\frac{2\pi t}{T}\right) \qquad \therefore \ f(t+T) = f(t)$$

Similarly, $g(t+T) = g(t)$

It can be easily verified that: $f(t+nT) = f(t)$ and $g(t+nT) = g(t)$

where $n = 1, 2, 3, \ldots\ldots$

NOMORECLASS CONCEPTS

These functions could be used to represent periodic motion i.e. Periodic functions represent periodic motion

T is the period of the above function.

To find periodicity of summation of two or more periodic functions the periodicity would be the L.C.M of the periodicities of the each function

Illustration 1: Find the period of the function, $y = \sin\omega t + \sin 2\omega t + \sin 3\omega t$ 　　　　　　　**(JEE MAIN)**

Sol: The function with least angular frequency will have highest time period.

The given function can be written as, $y = y_1 + y_2 + y_3$

Here $y_1 = \sin\omega t$, $T_1 = \frac{2\pi}{\omega}$ $y_2 = \sin 2\omega t$, $T_2 = \frac{2\pi}{2\omega} = \frac{\pi}{\omega}$, and $y_3 = \sin 3\omega t$

$$T_3 = \frac{2\pi}{3\omega} \quad \therefore \ T_1 = 2T_2 \text{ and } T_1 = 3T_3$$

So, the time period of the given function is T_1 or $\frac{2\pi}{\omega}$.

Because in time $T = \frac{2\pi}{\omega}$, first function completes one oscillation, the second function two oscillations and the third, three.

3. SIMPLE HARMONIC MOTION

Simple Harmonic Motion is a periodic motion in which a body moves to and fro about its mean position such that its restoring force or its acceleration is directly proportional to the displacement from its mean position and is directed towards its mean position. It can be expressed mathematically as, $F = m\frac{d^2x}{dt^2} = -kx$. Where m is the mass on which a restoring force F acts to impart an acceleration $\frac{d^2x}{dt^2}$ along x-axis such that the restoring force F or acceleration is directly proportional to the displacement x along x-axis and k is a constant. The negative sign shows that the restoring force or acceleration is directed towards the mean position.

The differential equation of a simple harmonic motion is given by, $\dfrac{d^2x}{dt^2} + \left(\dfrac{K}{m}\right)x = 0$ or $\dfrac{d^2x}{dt^2} + \omega^2 x = 0$

Where $\omega = \sqrt{\dfrac{K}{m}} = \sqrt{\dfrac{\text{acceleration}}{\text{displacement}}}$

The time period T, to complete one complete cycle by a body undergoing simple harmonic motion is given by

$$T = \dfrac{2\pi}{\omega} = 2\pi\sqrt{\dfrac{\text{acceleration}}{\text{displacement}}} = 2\pi\sqrt{\dfrac{m}{K}}$$

3.1 Types of SHM

Two Types of Simple Harmonic Motion

(a) Linear SHM   (b) Angular SHM

Important among all oscillatory motion is the simple harmonic motion. A particle executing linear simple harmonic motion oscillates in straight line periodically in such a way that the acceleration is proportional to its displacement from a fixed point (called equilibrium), and is always directed towards that point.

If a body describes rotational motion in such a way that the direction of its angular velocity changes periodically and the torque acting on is always directed opposite to the angular displacement and magnitude of the torque is directly proportional to the angular displacement, then its motion is called angular SHM.

4. REPRESENTATION OF SIMPLE HARMONIC MOTION

If a point mass m is moving with uniform speed along a circular path of radius a, it's projection on the diameter of the circle along y-axis represents its simple harmonic motion (see Fig. 8.1). $y = a\sin\omega t$

Where ω is the uniform angular velocity of the body of mass m along a circular path of radius a than ωt is angle covered by the radius in time t from the initial position A at $t = 0$ to the position B. As $\angle AOB = \angle OBC = \omega t$, the foot of perpendicular from B to the diameter YOY' gives the projection at the point C such that $y = OC$ is the projection of this body on the diameter and represents the displacement of the body executing SHM along y-axis. If the body does not start its motion from the point A but at a point A' so that $\angle AOA'$ is the phase angle ϕ,

then $y = a\sin(\omega t \pm \phi)$ ... (i)

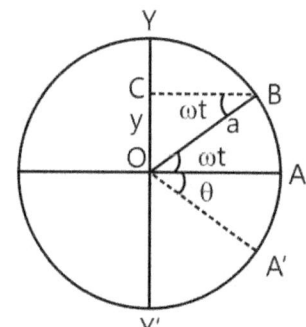

Where ϕ is the phase angle which may be positive or negative. The phase angle represents the fraction of the angle by which the motion of the body is out of step between the initial position of the body and the mean position of simple harmonic motion. The phase difference is the fraction of angle 2π or time period T of SHM by which the body is out of step initially from the mean position of the body.

Figure 8.1: Particle moving in a circle with angular speed ω in X-Y plane

Differentiating equation (i), $\dfrac{dy}{dt} = v = a\omega\cos(\omega t \pm \phi)$

As $\sin(\omega t \pm \phi) = \dfrac{y}{a}$, $\cos(\omega t \pm \phi) = \sqrt{1 - \sin^2(\omega t \pm \phi)} = \sqrt{1 - \dfrac{y^2}{a^2}} = \dfrac{\sqrt{a^2 - y^2}}{a}$

$\therefore v = \omega\sqrt{a^2 - y^2}$ ... (ii)

Differentiating (ii), the acceleration $= \dfrac{d^2y}{dt^2} = -a\omega^2\sin(\omega t \pm \phi)$

$\therefore \dfrac{d^2y}{dt^2} = -\omega^2 y$

2.3

It represents the equation of simple harmonic motion where $\omega = \sqrt{\left(\dfrac{d^2y}{dt^2}\middle/ y\right)} = \sqrt{\dfrac{\text{acceleration}}{\text{displacement}}}$

Time period, $T = 2\pi/\omega = 2\pi\sqrt{\dfrac{\text{displacement}}{\text{acceleration}}}$

4.1 Alternative Method for Finding Velocity and Acceleration in SHM

Let v be the velocity of the reference particle at P. Resolve velocity V into two rectangular components $V\cos\theta$ parallel to YOY' and $V\sin\theta$ perpendicular to YOY' (see Fig. 8.2). The velocity v of the projection N is clearly $V\cos\theta$.

$\therefore\ v = V\cos\theta = A\omega\cos\omega t$ or $v = A\omega\sqrt{1 - \sin^2\omega t}$

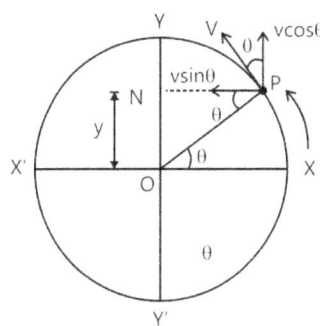

or $v = A\omega\sqrt{1 - \dfrac{y^2}{A^2}}$ or $v = A\omega\sqrt{\dfrac{A^2 - y^2}{A^2}}$ or $v = \omega\sqrt{A^2 - y^2}$

Figure 8.2: Relation between v and ω.

The centripetal acceleration $\dfrac{V^2}{A}$ of the particle at P can be resolved

into two rectangular components $-\dfrac{V^2}{A}\cos\theta$ Perpendicular to YOY'

and $\dfrac{V^2}{A}\sin\theta$ anti-parallel to YOY' Acceleration of $N = -\dfrac{V^2}{A}\sin\theta$

or Acceleration $= -\dfrac{V^2}{A^2}(A\sin\theta) = -\omega^2(A\sin\omega t)$

or Acceleration $= -\omega^2 y$

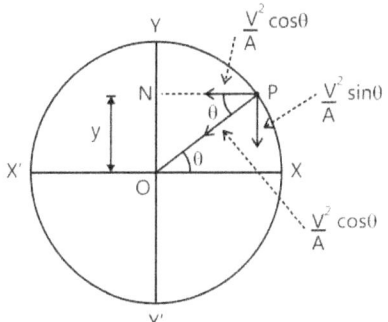

Figure 8.3: Direction of centripetal acceleration of particle

4.2 Time Period or Periodic Time of SHM

It is the smallest interval of time at which the details of motion repeat. It is generally represented by T.

$$'x'\,\text{at}\,(t + T) = A\cos\left(2\pi\frac{t + T}{T} + \phi_0\right) = A\cos\left(\frac{2\pi t}{T} + 2\pi + \phi_0\right) \qquad \text{... (i)}$$

It is clear from here that the details of motion repeat after time T. Time period may also be defined as the time taken by the oscillating particle to complete one oscillation. It is equal to the time taken by the reference particle to complete one revolution. In one revolution, the angle traversed by the reference particle is 2π radian and T is the

time taken. If ω be the uniform angular velocity of the reference particle, then $\omega = \dfrac{2\pi}{T}$ or $T = \dfrac{2\pi}{\omega}$

4.3 Frequency

It is the number of oscillations (or vibrations) completed per unit time. It is denoted by f. In time T second, one vibration is completed.

In 1 second, $\dfrac{1}{T}$ vibrations are completed   or $f = \dfrac{1}{T}$ or $fT = 1$

Also, $\omega = \dfrac{2\pi}{T} = 2\pi \times \dfrac{1}{T} = 2\pi f$ So, equation (i) may also be written as under

$x = A\cos(2\pi ft + \phi_0)$
... (ii)

2.4

The unit of f is s^{-1} or hertz or 'cycles per second' (cps). $\therefore$ $\phi s^{-1} = \phi Hz = \phi cps$.

4.4 Angular Frequency

It is frequency f multiplied by a numerical quantity 2π. It is denoted by ω so that $\omega = 2\pi f = \dfrac{2\pi}{T}$. Equation (vi) may be written as $x = A\cos(\omega t + \phi)$

4.5 Phase

Phase of a vibrating particle at any instant is the state of the vibrating particle regarding its displacement and direction of vibration at that particular instant.

The argument of the cosine in equation $x = A\cos(\omega t + \phi_0)$ gives the phase of oscillation at time t.

It is denoted by ϕ. $\therefore$ $\phi = 2\pi\dfrac{t}{T} + \phi_0$ or $\phi = \omega t + \phi_0$

It is clear that phase ϕ is a function of time t. The phase of a vibrating particle can be expressed in terms of fraction of the time period that has elapsed since the vibrating particle left its initial position in the positive direction. Again,

$\phi - \phi_0 = \omega t = \dfrac{2\pi t}{T}$. So, the phase change in time t is $\dfrac{2\pi t}{T}$. The phase change in T second will be 2π which actually means a 'no change in phase'. Thus, time period may also be defined as the time interval during which the phase of the vibrating particle changes by 2π.

NOMORECLASS CONCEPTS

The phase difference between acceleration and displacement is 180°. In SHM phase difference between velocity and acceleration is $\pi/2$ and velocity and displacement is $\pi/2$.

Fig. 8.4 (a) displacement, (b) velocity and (c) acceleration vs. time in SHM.

$v = \pm\omega\sqrt{A^2 - y^2}$. Graphical variation of v with y is an ellipse.

Max velocity at $y = 0$ i.e. at mean position and

$V_{max} = A\omega$; $a = -\omega^2 y$

Graph between acceleration and displacement of a particle executing SHM is straight line.

Max acceleration at $y = A$ i.e. at extreme position and

$a_{max} = A\omega^2$

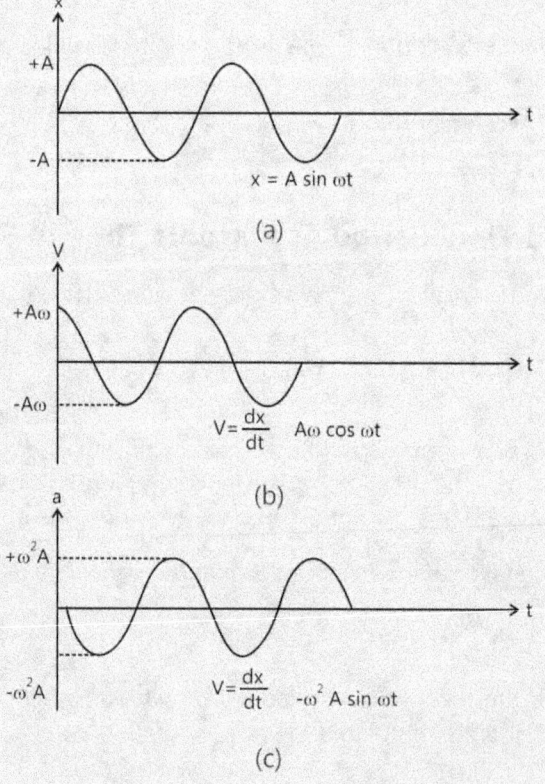

Figure 8.4

2.5

Illustration 2: A particle executes simple harmonic motion about the point $x = 0$. At time $t = 0$ it has displacement $x = 2$ cm and zero velocity. If the frequency of motion is $0.25\,s^{-1}$, find (a) the period, (b) angular frequency, (c) the amplitude, (d) maximum speed, (e) the displacement at $t = 3s$ and (f) the velocity at $t = 3s$. **(JEE MAIN)**

Sol: The standard equation for displacement in SHM is $x = A\sin(\omega t + \phi)$. When velocity is zero, the particle is at maximum displacement.

(a) Period $T = \dfrac{1}{f} = \dfrac{1}{0.25\,s^{-1}} = 4\,s$

(b) Angular frequency $\omega = \dfrac{2\pi}{T} = \dfrac{2\pi}{4} = \dfrac{\pi}{2}\,rad/s = 1.57\,rad/s$

(c) Amplitude is the maximum displacement from mean position. Hence, $A = 2 - 0 = 2\,cm$

(d) Maximum speed $v_{max} = A\omega = 2.\dfrac{\pi}{2} = \pi\,cm/s = 3.14\,cm/s$

(e) The displacement is given by $x = A\sin(\omega t + \phi)$

Initially at t=0; $x = 2cm$, then $2 = 2\sin\phi$ or $\sin\phi = 1 = \sin 90°$ or $\phi = 90°$

Now, at $t = 3s$ $x = 2\sin\left(\dfrac{\pi}{2} \times 3 + \dfrac{\pi}{2}\right) = 0$

(f) Velocity at $x = 0$ is v_{max} i.e., 3.14 cm/s.

Illustration 3: Two particles move parallel to x-axis about the origin with the same amplitude and frequency. At a certain instant, they are found at distance $\dfrac{A}{3}$ from the origin on opposite sides but their velocities are found to be in the same direction. What is the phase difference between the two? **(JEE ADVANCED)**

Sol: The standard equation for displacement in SHM is $x = A\sin(\omega t + \phi)$. Displacement on opposite sides of the mean position has opposite signs. Equation for velocity is $v = A\omega\cos(\omega t + \phi)$. Velocities in same direction have same sign.

Let equations of two SHM be $x_1 = A\sin\omega t$ ... (i)

$x_2 = A\sin(\omega t + \phi)$ ... (ii)

Give that $\dfrac{A}{3} = A\sin\omega t$ and $-\dfrac{A}{3} = A\sin(\omega t + \phi)$ Which gives $\sin\omega t = \dfrac{1}{3}$ ... (iii)

$\sin(\omega t + \phi) = -\dfrac{1}{3}$ ... (iv)

From Eq.(iv), $\sin\omega t\cos\phi + \cos\omega t\sin\phi = -\dfrac{1}{3}$; $\dfrac{1}{3}\cos\phi + \sqrt{1 - \dfrac{1}{9}}\sin\phi = -\dfrac{1}{3}$

Solving this equation, we get or $\cos\phi = -1, \dfrac{7}{9}$; $\phi = \pi$ or $\cos^{-1}\left(\dfrac{7}{9}\right)$

Differentiating Eqs. (i) and (ii), we obtain; $v_1 = A\omega\cos\omega t$ and $v_2 = A\omega\cos(\omega t + \phi)$

If we put $\phi = \pi$, we find v_1 and v_2 are of opposite signs. Hence, $\phi = \pi$ is not acceptable.

$\phi = \cos^{-1}\left(\dfrac{7}{9}\right)$

Illustration 4: With the assumption of no slipping, determine the mass m of the block which must be placed on the top of a 6 kg cart in order that the system period is 0.75s. What is the minimum coefficient of static fraction μ_s for which the block will not slip relative to the cart if the cart is displaced 50mm from the equilibrium position and released? Take $\left(g = 9.8\,m/s^2\right)$. **(JEE ADVANCED)**

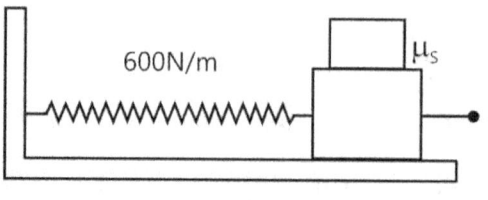

Figure 8.5

Sol: $\omega = \sqrt{\dfrac{k}{M}}$ where M is the total mass attached to the spring. The maximum restoring force on the blocks will be at the extreme position. The limiting friction on mass m should be greater than or equal to the maximum restoring force required for mass m.

(a) $T = 2\pi\sqrt{\dfrac{m+6}{600}}\left(\because T = 2\pi\sqrt{\dfrac{m}{k}}\right)$ $\therefore$ $0.75 = 2\pi\sqrt{\dfrac{m+6}{600}}$; $m = \dfrac{(0.75)^2 \times 600}{(2\pi)^2} - 6 = 2.55\,kg$

(b) Maximum acceleration of SHM is $a_{max} = \omega^2 A$ $\left(A = amplitude\right)$

i.e., maximum force on mass 'm' is $m\omega^2 A$ which is being provided by the force of friction between the mass and the cart. Therefore, $\mu_s mg \geq m\omega^2 A$ or $\mu_s \geq \dfrac{\omega^2 A}{g}$ or $\mu_s \geq \left(\dfrac{2\pi}{T}\right)^2 \dfrac{A}{g}$ or $\mu_s \geq \left(\dfrac{2\pi}{0.75}\right)^2 \left(\dfrac{0.05}{9.8}\right)$ $(A = 50mm)$

or $\mu_s \geq 0.358$. Thus, the minimum value of μ_s should be 0.358.

5. ENERGY IN SHM

The displacement and the velocity of a particle executing a simple harmonic motion are given by

$x = A\sin(\omega t + \delta)$ and $v = A\omega\cos(\omega t + \delta)$. The potential energy at time t is, therefore,

$U = \dfrac{1}{2}kx^2$ and $k = m\omega^2$ Therefore $U = \dfrac{1}{2}m\omega^2 x^2 = \dfrac{1}{2}m\omega^2 A^2 \sin^2(\omega t + \delta)$, and the kinetic energy at time t

is $K = \dfrac{1}{2}mv^2 = \dfrac{1}{2}mA^2\omega^2 \cos^2(\omega t + \delta)$

The total mechanical energy time t is $E = U + K$

$= \dfrac{1}{2}m\omega^2 A^2\left[\sin^2(\omega t + \delta) + \left(\cos^2(\omega t + \delta)\right)\right] = \dfrac{1}{2}m\omega^2 A^2$ Average value of P.E. and K.E

By equation (i) P.E. at distance x is given by

$U = \dfrac{1}{2}m\omega^2 x^2 = \dfrac{1}{2}m\omega^2 A^2 \sin^2(\omega t + \phi)$ $\{since\,at\,time\,t, x = A\sin(\omega t + \phi)\}$

The average value of P.E. of complete vibration is given by

$U_{average} = \dfrac{1}{T}\int_0^T Udt = \dfrac{1}{T}\int_0^T \dfrac{1}{2}m\omega^2 A^2 \sin^2(\omega t + \phi) = \dfrac{m\omega^2 A^2}{4T}\int_0^T 2\sin^2(\omega t + \phi)dt = \dfrac{1}{4}m\omega^2 A^2$

Because the average value of sine square or cosine square function for the complete cycle is 0.

Now, KE at x is given by $K.E. = \dfrac{1}{2}m\left(\dfrac{dx}{dt}\right)^2 = \dfrac{1}{2}m\left[\dfrac{d}{dt}\{A\sin(\omega t + \phi)\}\right]^2 = \dfrac{1}{2}m\omega^2 A^2 \cos^2(\omega t + \phi)$

2.7

The average value of K.E. for complete cycle $K.E._{average} = \dfrac{1}{T}\int_0^T \dfrac{1}{2}m\omega^2 A^2 \cos^2(\omega t + \phi)dt$

$= \dfrac{m\omega^2 A^2}{4T}\int_0^T \{1 + \cos 2(\omega t + \phi)\}dt = \dfrac{m\omega^2 A^2}{4T} \cdot T = \dfrac{1}{4}m\omega^2 A^2$

NOMORECLASS CONCEPTS

Thus average values of K.E. and P.E. of harmonic oscillator are equal to half of the total energy.

The total mechanical energy is constant but the kinetic energy and potential energy of the particle are oscillating

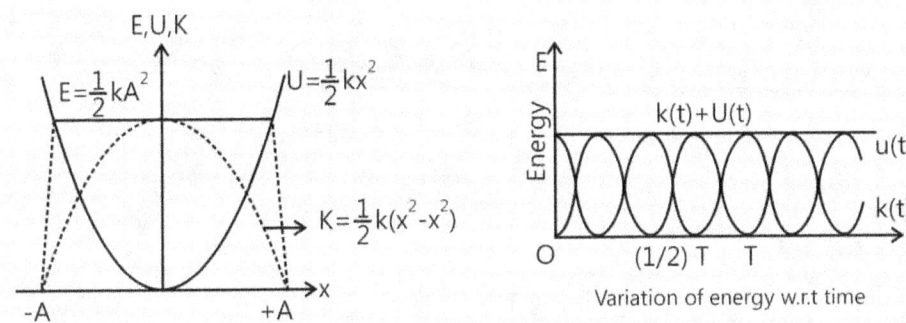

Figure 8.6

Graph for Energy of SHM: Figure 8.6 shows the variation of total energy (E), Potential energy (U) and kinetic energy (K) with Displacement (x).

At a glance

S.No.	Name of the equation	Expression of the Equation	Remarks
1.	Displacement-time	$x = A\cos(\omega t + \phi)$	X varies between $+A$ and $-A$
2.	Velocity-time $\left(V = \dfrac{dx}{dt}\right)$	$v = -A\omega\sin(\omega t + \phi)$	v varies between $+A\omega$ and $-A\omega$
3.	Acceleration-time $\left(a = \dfrac{dv}{dt}\right)$	$a = -A\omega^2\cos(\omega t + \phi)$	a varies between $+A\omega^2$ and $-A\omega^2$
4.	Kinetic energy-time $\left(K = \dfrac{1}{2}mv^2\right)$	$K = \dfrac{1}{2}mA^2\omega^2\sin^2(\omega t + \phi)$	K varies between 0 and $\dfrac{1}{2}mA^2\omega^2$
5.	Potential energy-time $\left(U = \dfrac{1}{2}m\omega^2 x^2\right)$	$K = \dfrac{1}{2}m\omega^2 A^2\cos^2(\omega t + \phi)$	U varies between $\dfrac{1}{2}mA^2\omega^2$ and 0
6.	Total energy-time $\left(E = K + U\right)$	$E = \dfrac{1}{2}m\omega^2 A^2$	E is constant

S.No.	Name of the equation	Expression of the Equation	Remarks
7.	Velocity-displacement	$v = \omega\sqrt{A^2 - X^2}$	$v = 0$ at $x = \pm A$ and at $x = 0$ $v = \pm A\omega$
8.	Acceleration-displacement	$a = -\omega^2 x$	$a = 0$ at $x = 0$ $a = \pm\omega^2 A$ at $x = \mp A$
9.	Kinetic energy-displacement	$K = \dfrac{1}{2}m\omega^2\left(A^2 - X^2\right)$	$K = 0$ at $x = \mp A$ $K = \dfrac{1}{2}m\omega^2 A^2$ at $x = 0$
10.	Potential energy-displacement	$U = \dfrac{1}{2}m\omega^2 x^2$	$U = 0$ at $x = 0$ $U = \dfrac{1}{2}m\omega^2 A^2$ at $x = \pm A$
11.	Total energy-displacement	$E = \dfrac{1}{2}m\omega^2 A^2$	E is constant

NOMORECLASS CONCEPTS

At mean position → K is the maximum and U is the minimum (it may be zero also, but it is not necessarily zero). At extreme positions → K is zero and U is the maximum.

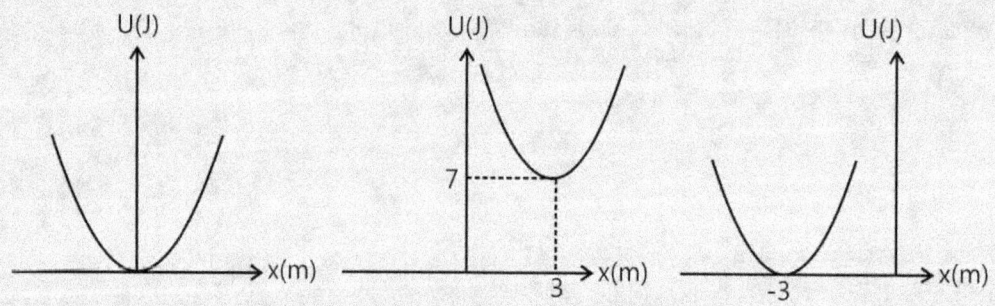

Figure 8.7

Illustration 5: The potential energy of a particle oscillating on x-axis is given as $U = 20 + \left(x - 2\right)^2$

Here, U is in joules and x in meters. Total mechanical energy of the particle is 36J. **(JEE MAIN)**

(a) State whether the motion of the particle is simple harmonic or not.

(b) Find the mean position.

(c) Find the maximum kinetic energy of the particle.

Sol: At the mean position the kinetic energy is the maximum and potential energy is the minimum. The sum of kinetic energy and potential energy is constant throughout the SHM, equal to the total mechanical energy.

(a) $F = -\dfrac{dU}{dx} = -2\left(x - 2\right)$

By assuming $x - 2 = X$, we have $F = -2x$, Since, $F \propto -X$

The motion of the particle is simple harmonic

(b) The mean position of the particle is $X = 0$ or $x - 2 = 0$, which gives $x = 2$ m

(c) Maximum kinetic energy of the particle is, $K_{max} = E - U_{min} = 36 - 20 = 16J$

Note U_{min} is 20J at mean position or at $x = 2m$.

Illustration 6: A block with mass M attached to a horizontal spring with force constant k is moving with simple harmonic motion having amplitude A_1. At the instant when the block passes through its equilibrium position a lump of putty with mass m is dropped vertically on the block from a very small height and sticks to it.

(JEE ADVANCED)

(a) Find the new amplitude and period.

(b) Repeat part (a) for the case in which the putty is dropped on the block when it is at one end of its path.

Sol: Sticking of putty constitutes an inelastic collision. Kinetic energy at equilibrium position converts into potential energy at extreme position, $\frac{1}{2}mv^2 = \frac{1}{2}kA^2$.

(a) Before the lump of putty is dropped the total mechanical energy of the block and spring is $E_1 = \frac{1}{2}kA_1^2$. Since, the block is at the equilibrium position, $U = 0$, and the energy is purely kinetic. Let v_1 be the speed of the block at the equilibrium position, we have $v_1 = \sqrt{\frac{k}{M}}A_1$

During the process momentum of the system in horizontal direction is conserved. Let v_2 be the speed of the combined mass, then $(M+m)v_2 = Mv_1$; $v_2 = \frac{M}{M+m}v_1$

Now, let A_2 be the amplitude afterwards. Then, $E_2 = \frac{1}{2}kA_2^2 = \frac{1}{2}(M+m)v_2^2$

Substituting the proper values, we have $A_2 = A_1\sqrt{\frac{M}{M+m}}$

Note: $E_2 < E_1$, as some energy is lost into heating up the block and putty. Further, $T_2 = 2\pi\sqrt{\frac{M+m}{k}}$

(b) When the putty drops on the block, the block is instantaneously at rest. All the mechanical energy is stored in the spring as potential energy. Again the momentum in horizontal direction is conserved during the process, but now it is zero just before and after putty is dropped. So, in this case, adding the extra mass of the putty has no effect on the mechanical energy, i.e.,

$E_2 = E_1 = \frac{1}{2}kA_1^2$ and the amplitude is still A_1. Thus, $A_2 = A_1$ and $T_2 = 2\pi\sqrt{\frac{M+m}{k}}$

6. ANGULAR SIMPLE HARMONIC MOTION

When a particle executes SHM on a curve path, then it is said to be angular SHM. E.g. - Simple pendulum. In this case, to find out the time period, we find out restoring torque and hence angular acceleration.

i.e. $\tau = -k\theta$ Where k is a constant $\Rightarrow$ $I\alpha = -k\theta$ where I is moment of inertia   ... (i)

$\Rightarrow \alpha = \frac{-k}{I}\theta$ ... (ii)

Also, the equation of SHM for angular SHM, is $\alpha = -\omega^2\theta$. Comparing (i) and (ii), we get ω, hence the time period.

Problem solving strategy

Step 1: Find the stable equilibrium position which usually is also known as the mean position. Net force or torque on the particle in this position is zero. Potential energy is the minimum.

Step 2: Displace the particle from its mean position by a small displacement x (in case of a linear SHM) or θ (in case of an angular SHM).

Step 3: Find net force or torque in this displaced position.

Step 4: Show that this force or torque has a tendency to bring the particle back to its mean position and magnitude of force or torque is proportional to displacement, i.e.,

$F \propto -x$ or $F = -kx$ ...(i); $\tau \propto -\theta$ or $\tau = -k\theta$ ...(ii)

This force or torque is also known as restoring force or restoring torque.

Step 5: Find linear acceleration by dividing Eq.(i) by mass m or angular acceleration by dividing Eq.(ii) by moment of inertia I.

Hence, $\quad a = -\dfrac{k}{m}.x = -\omega^2 x$ or $\alpha = -\dfrac{k}{I}\theta = -\omega^2 \theta$

Step 6: Finally, $\omega = \sqrt{\left|\dfrac{a}{x}\right|}$ or $\sqrt{\left|\dfrac{\alpha}{\theta}\right|}$ or $\dfrac{2\pi}{T} = \sqrt{\left|\dfrac{a}{x}\right|}$ or $\sqrt{\left|\dfrac{\alpha}{\theta}\right|}$

$\therefore \ T = 2\pi\sqrt{\left|\dfrac{x}{a}\right|}$ or $2\pi\sqrt{\left|\dfrac{\theta}{\alpha}\right|}$

Energy Method: Repeat step 1 and step 2 as in method 1. Find the total mechanical energy (E) in the displaced position. Since, mechanical energy in SHM remains constant. $\dfrac{dE}{dt} = 0$ By differentiating the energy equation with respect to time and substituting $\dfrac{dx}{dt} = v$, $\dfrac{d\theta}{dt} = \omega$, $\dfrac{dv}{dt} = a$, and $\dfrac{d\omega}{dt} = \alpha$ we come to step 5. The remaining procedure is same.

Note: (i) E usually consists of following terms:

(a) Gravitational PE (b) Elastic PE (c) Electrostatic PE (d) Rotational KE and (e) Translational KE

(ii) For gravitational PE, choose the reference point (h=0) at mean position.

Illustration 7: Calculate the angular frequency of the system shown in Fig 8.8. Friction is absent everywhere and the threads, spring and pulleys are massless. Given, that $m_A = m_B = m$. **(JEE ADVANCED)**

Figure 8. 8

Sol: This problem can be solved either by restoring force method or by the energy method. The gain in kinetic energy is at the cost of decrease in gravitational and/or elastic potential energy.

Let x_0 be the extension in the spring in equilibrium. Then equilibrium of A and B give $T = kx_0 + mg\sin\theta$ ... (i)

and $2T = mg$ ... (ii)

Here, T is the tension in the string. Now, suppose A is further displaced by a distance x from its mean position and v be its speed at this moment. Then B lowers by $\dfrac{x}{2}$ and speed of B at this instant will be $\dfrac{v}{2}$. Total energy of the system in this position will be,

$$E = \frac{1}{2}k(x + x_0)^2 + \frac{1}{2}m_A v^2 + \frac{1}{2}m_B\left(\frac{V}{2}\right)^2 + m_A g h_A - m_B g h_B$$

or $E = \frac{1}{2}k(x + x_0)^2 + \frac{1}{2}mv^2 + \frac{1}{8}mv^2 + mgx\sin\theta - mg\frac{x}{2}$

or $E = \frac{1}{2}k(x + x_0)^2 + \frac{5}{8}mv^2 + mgx\sin\theta - mg\frac{x}{2}$

Since, E is constant, $\frac{dE}{dt} = 0$ or $0 = k(x + x_0)\frac{dx}{dt} + \frac{5}{4}mv\left(\frac{dv}{dt}\right) + mg(\sin\theta)\left(\frac{dx}{dt}\right) - \frac{mg}{2}\left(\frac{dx}{dt}\right)$

Substituting, $\frac{dx}{dt} = v$; $\frac{dv}{dt} = a$ and $kx_0 + mg\sin\theta = \frac{mg}{2}$ [From Eqs. (i) and (ii)]

We get, $\frac{5}{4}ma = -kx$ Since, $a \propto -x$

Motion is simple harmonic, time period of which is, $T = 2\pi\sqrt{\left|\frac{x}{a}\right|} = 2\pi\sqrt{\frac{5m}{4k}}$ $\therefore$ $\omega = \frac{2\pi}{T} = \sqrt{\frac{4k}{5m}}$

7. SIMPLE PENDULUM

It is an example of angular simple harmonic motion. Let's calculate its time period. Let us suppose that a bob of mass m is executing SHM (see Fig. 8.9). The length of the pendulum is ℓ, which is the distant between the point of oscillation and the center of mass of the bob. Torque acting on the bob about the point O.

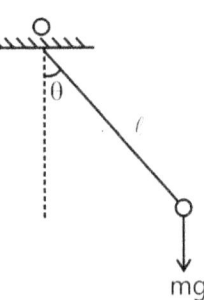

$\Gamma = mg \,\ell \sin\theta$ (And for small θ, $\sin\theta \approx \theta$)

$\Rightarrow \Gamma = mg\ell\theta \Rightarrow I\alpha = -mg\ell\theta \Rightarrow \alpha = -\frac{mg\ell}{I}\theta$

where α is angular acceleration $= -\frac{mg\ell}{m\ell^2}\theta$; $\alpha = -\frac{g}{\ell}\theta$ ... (i)

The equation of SHM is $\alpha = -\omega^2\theta$ ... (ii)

Figure 8.9: Oscillations of simple Pendulum

Comparing (i) and (ii), we get $\omega^2 = \frac{g}{\ell}$; $\omega = \sqrt{\frac{g}{\ell}}$ $\Rightarrow \frac{2\pi}{T} = \sqrt{\frac{g}{\ell}}$; $T = 2\pi\sqrt{\frac{\ell}{g}}$

NOMORECLASS CONCEPTS

The period is independent of the mass of the suspended particle.

Various scenarios: Time period

Pendulum in a lift descending with acceleration "a", $T = 2\pi\sqrt{\frac{\ell}{(g-a)}}$

Pendulum in a lift ascending with acceleration "a", $T = 2\pi\sqrt{\frac{\ell}{(g+a)}}$

Pendulum suspended in a train accelerated with "a" uniformly in horizontal direction $T = 2\pi\sqrt{\frac{\ell}{\left(a^2 + g^2\right)^{\frac{1}{2}}}}$

Pendulum suspended in car taking turn with velocity v in a circular path of radius r, $T = 2\pi \sqrt{\dfrac{\ell}{\left(\left(\dfrac{v^2}{r}\right)^2 + g^2\right)^{\frac{1}{2}}}}$

Note: If the pendulum is suspended in vacuum, then the time period of the pendulum decreases.

Illustration 8: A simple pendulum consists of a small sphere of mass m suspended by a thread of length ℓ. The sphere carries a positive charge q. The pendulum is placed in a uniform electric field of strength E directed vertically upwards. With what period will pendulum oscillate if the electrostatic force acting on the sphere is less than the gravitational force? **(JEE MAIN)**

Sol: The electrostatic force is acting opposite to the weight of the block. So the effective value of acceleration due to gravity will be less than the actual value of g.

The two forces acting on the bob are shown in Fig 8.10.

g_{eff} in this case will be $\dfrac{w - F_e}{m}$ or $g_{eff} = \dfrac{mg - qE}{m} = g - \dfrac{qE}{m}$

$\therefore T = 2\pi \sqrt{\dfrac{\ell}{g_{eff}}} = 2\pi \sqrt{\dfrac{\ell}{g - \dfrac{qE}{m}}}$

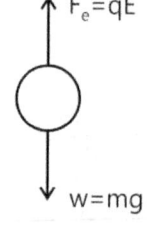

Figure 8.10

NOMORECLASS CONCEPTS

In case of a pendulum clock, time is lost if T increase and gained if T decreases. Time lost or gained in time t is given by.

$\Delta t = \dfrac{\Delta T}{T} t$ e.g., if $T = 2s, T' = 3s$, then $\Delta T = 1s$

$\therefore$ Time lost by the clock in 1 hr. $\Delta t = \dfrac{1}{3} \times 3600 = 1200s$

Second pendulum is a with its time period precisely 2 seconds

Illustration 9: A simple pendulum of length l is suspended from the ceiling of a cart which is sliding without friction on an inclined plane of inclination θ. What will be the time period of the pendulum? **(JEE MAIN)**

Sol: The cart accelerates down the plane with acceleration a = g sinθ.

$g_{eff} = |\vec{g} - \vec{a}| = \sqrt{g^2 + 2g^2 \sin\theta \cos(90° + \theta) + g^2\sin^2\theta} = g\cos\theta$

Here, point of suspension has acceleration. $\vec{a} = g\sin\theta$ (down the Plane). Further, $\vec{g}$ can be resolved into two components g sin θ (along the plane) and g cos θ (perpendicular to plane)

$\therefore \vec{g}_{eff} = \vec{g} - \vec{a} = g\cos\theta$

(perpendicular to plane)

$\therefore T = 2\pi \sqrt{\dfrac{\ell}{|\vec{g}_{eff}|}} = 2\pi \sqrt{\dfrac{\ell}{g\cos\theta}}$

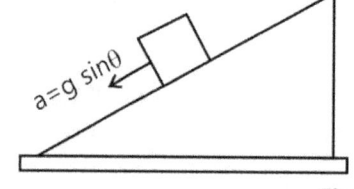
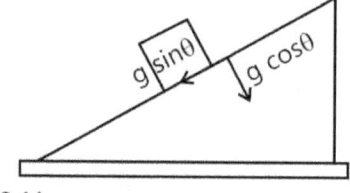

Figure 8.11

2.13

Note: If $\theta = 0^\circ$, $T = 2\pi\sqrt{\dfrac{l}{g}}$ which is quiet obvious.

8. PHYSICAL PENDULUM

Any rigid body mounted so that it can swing in a vertical plane about some axis passing through it is called a physical pendulum (see Fig. 8.12).

A body of irregular shape is pivoted about a horizontal frictionless axis through P and displaced from the equilibrium position by an angle θ. (The equilibrium position is that in which the center of mass C of the body lies vertically below P).

The distance from the pivot to the center of mass is d. The moment of inertia of the body about an axis through the pivot is I and the mass of the body is M. The restoring torque about the point P,

$\tau = Mgd\theta$ (if θ be very small, $\sin\theta = \theta$)

$\tau = Mgd\theta$; $I\alpha = -Mgd\theta$; $\alpha = -\dfrac{Mgd}{I}\theta$ ......(i)

Comparing with the equation of SHM

$\omega^2 = \dfrac{Mgd}{I}$; $\omega = \sqrt{\dfrac{Mgd}{I}}$; $2\pi/T = \sqrt{\dfrac{Mgd}{I}}$; $T = 2\pi\sqrt{\dfrac{I}{Mgd}}$

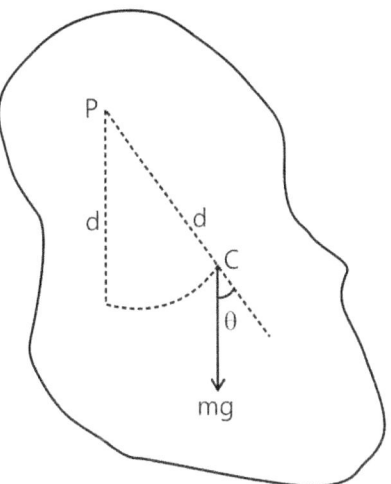

Figure 8.12: Rotation of Physical Pendulum

NOMORECLASS CONCEPTS

It may be necessary to use parallel axis theorem to find Moment of Inertia about the pivoted axis
$I = I_G + ml^2$

9. TORSIONAL PENDULUM

In torsional pendulum, an extended body is suspended by a light thread or a wire (see Fig. 8.13). The body is rotated through an angle about the wire as the axis of rotation. The wire remains vertical during this motion but a twist is produced in the wire. The lower end of the wire is rotated through an angle with the body but the upper end remains fixed with the support. Thus, a twist θ is produced. The twisted wire exerts a restoring torque on the body to bring it back to its original position in which the twist θ in the wire is zero. This torque has a magnitude proportional to the angle of twist which is equal to the angle rotated by the body. The proportionality constant is called the torsional constant of the wire. Thus, if the torsional constant of the wire is κ and the body is rotated through an angle θ, the torque produced is $\Gamma = -\kappa\theta$. If I be the moment of inertia

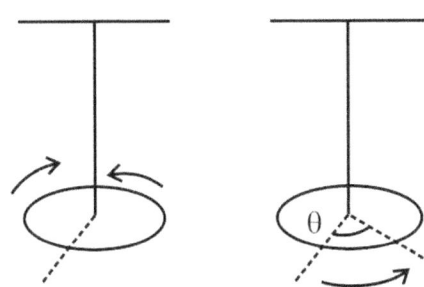

Figure 8.13: Torsional pendulum

of the body about the vertical axis, the angular acceleration is $\alpha = \dfrac{\Gamma}{I} = -\dfrac{\kappa}{I}\theta = -\omega^2\theta$ where $\omega = \sqrt{\dfrac{\kappa}{I}}$

Thus, the motion of the body is simple harmonic and the time period is $T = \dfrac{2\pi}{\omega} = 2\pi\sqrt{\dfrac{I}{\kappa}}$

Illustration 10: A ring of radius r is suspended from a point on its circumference. Determine its angular frequency of small oscillations. **(JEE ADVANCED)**

Sol: This is an example of a physical pendulum. Find moment of inertia about point of suspension and the distance of the point of suspension from the center of mass.

It is physical pendulum, the time period of which is, $T = 2\pi\sqrt{\dfrac{I}{mgl}}$

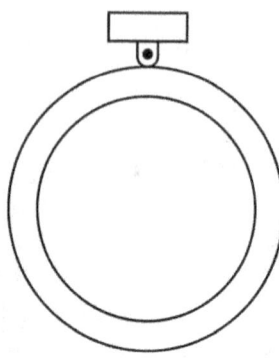

Here, I = moments of inertia of the ring about point of suspension

$= mr^2 + mr^2 = 2mr^2$

and l = distance of point of suspension from centre of mass = r

$\therefore \qquad T = 2\pi\sqrt{\dfrac{2mr^2}{mgr}} = 2\pi\sqrt{\dfrac{2r}{g}}$; $\qquad \therefore$ Angular frequency $\quad \omega = \dfrac{2\pi}{t}$ or $\omega = \sqrt{\dfrac{g}{2r}}$

Figure 8.14

Illustration 11: Find the period of small oscillations of a uniform rod with length l, pivoted at one end. **(JEE MAIN)**

Sol: This is an example of a physical pendulum. Find moment of inertia about point of suspension and the distance of the point of suspension from the center of gravity.

$T = 2\pi\sqrt{\dfrac{I_\circ}{mg(OG)}}$ $\qquad$ Here, $I_\circ = \dfrac{1}{3}ml^2$ and $OG = \dfrac{l}{2}$

$\therefore \qquad T = 2\pi\sqrt{\dfrac{\left(\dfrac{1}{3}ml^2\right)}{(m)(g)\left(\dfrac{l}{2}\right)}}$ $\qquad$ or $T = 2\pi\sqrt{\dfrac{2l}{3g}}$

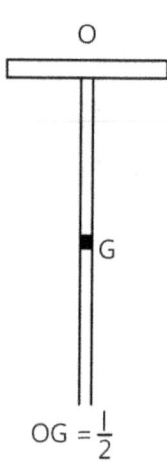

$OG = \dfrac{l}{2}$

Figure 8.15

Illustration 12: A uniform disc of radius 5.0 cm and mass 200 g is fixed at its center to a metal wire, the other end of which is fixed with a clamp. The hanging disc is rotated about the wire through angle and is released. If the disc makes torsional oscillations with time period 0.20 s, find the torsional constant of the wire. **(JEE MAIN)**

Sol: This is an example of a torsional pendulum. Find moment of inertia about the axis passing through the wire.

The Situation is shown in Fig 8.16. The moment of inertia of the disc about the wire is

$I = \dfrac{mr^2}{2} = \dfrac{(0.200kg)(5.0\times10^{-2}m)^2}{2} = 2.5\times10^{-4} kg.m^2$

The time period is given by

$T = 2\pi\sqrt{\dfrac{I}{K}}$; $\qquad K = \dfrac{4\pi^2 I}{T^2} = \dfrac{4\pi^2\left(2.5\times10^{-4}kg - m^2\right)}{(0.20s)^2} = 0.25\dfrac{kg - m^2}{s^2}$

Figure 8.16

10. SPRING - MASS SYSTEM

As shown in the Fig. 8.17 a mass m is attached to a massless spring. It is displaced from its mean position to a distance x. The restoring force is given by

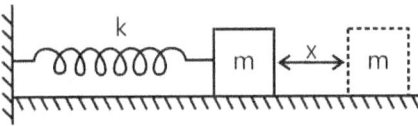

Figure 8.17: Block of mass m attached to spring

$F = -kx$ where k is the force constant.

$$\Rightarrow ma = -kx; \quad a = -x\frac{k}{m} \qquad \ldots(i)$$

$\Rightarrow a \propto -x, \therefore$ Motion is SHM

$$\Rightarrow \omega^2 = \frac{k}{m} \quad \text{or} \quad \omega = \sqrt{\frac{k}{m}}; \qquad T = 2\pi\sqrt{\frac{m}{k}}$$

10.1 Series and Parallel Combination of Springs

10.1.1. Serial Combination of Springs

If springs are connected in series, having force constants k_1, k_2, k_3 then the equivalent force

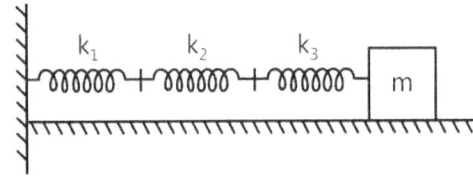

Figure 8.18: Series combination of springs

constant is $\dfrac{1}{k_{eff}} = \dfrac{1}{k_1} + \dfrac{1}{k_2} + \dfrac{1}{k_3}$

10.1.2 Parallel Combination of Springs

If springs are connected in parallel, then the effective force constant is given by
$k_{eff} = k_1 + k_2 + k_3 + \ldots\ldots\ldots\ldots\ldots$

The force constant of a spring is inversely proportional to its length. If a spring of spring constant k is cut into two equal parts, the spring constant of each part becomes 2k. In general, if a spring of spring constant k is divided into n equal parts, the spring constant of each part is nk.

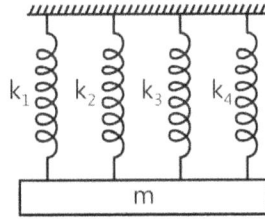

Figure 8.19: Parallel combination of springs

NOMORECLASS CONCEPTS

Time Period for various scenarios at glance

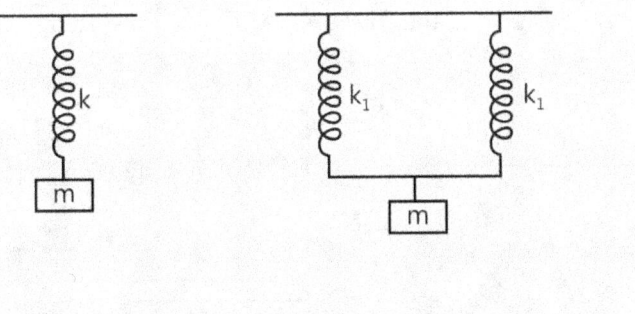

$$T = 2\pi\sqrt{\frac{m}{k}} \qquad T = 2\pi\sqrt{\frac{m}{k_1 + k_2}} \qquad T = 2\pi\sqrt{\frac{m}{k_1 + k_2}}$$

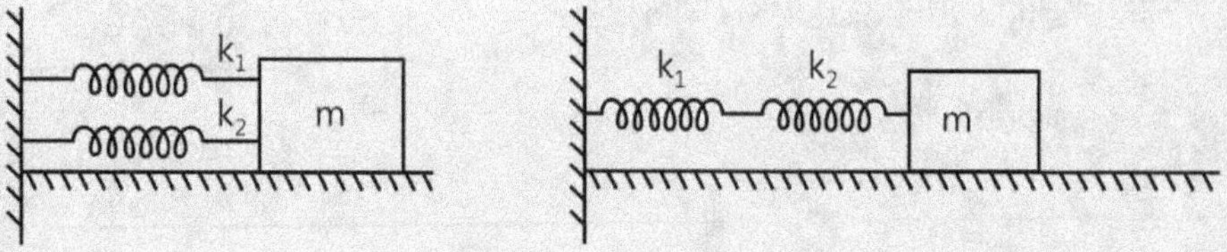

NOMORECLASS CONCEPTS

$$T = 2\pi\sqrt{\frac{m}{k_1 + k_2}} \qquad T = 2\pi\sqrt{\frac{m_1 m_2}{k(m_1 + m_2)}}$$

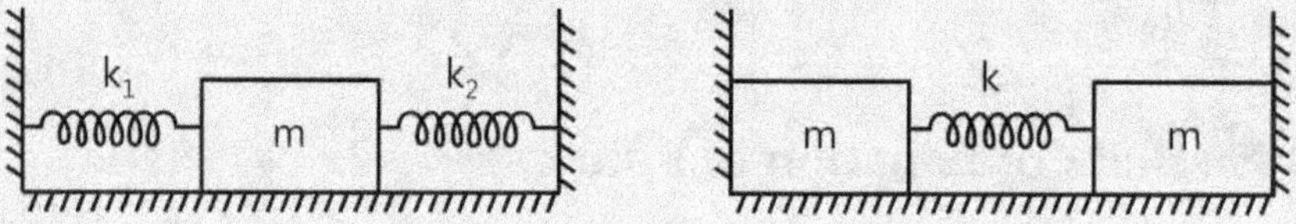

$$T = 2\pi\sqrt{\frac{m}{(k_1 + k_2)}} \qquad T = 2\pi\sqrt{\frac{m(k_1 + k_2)}{k_1 k_2}}$$

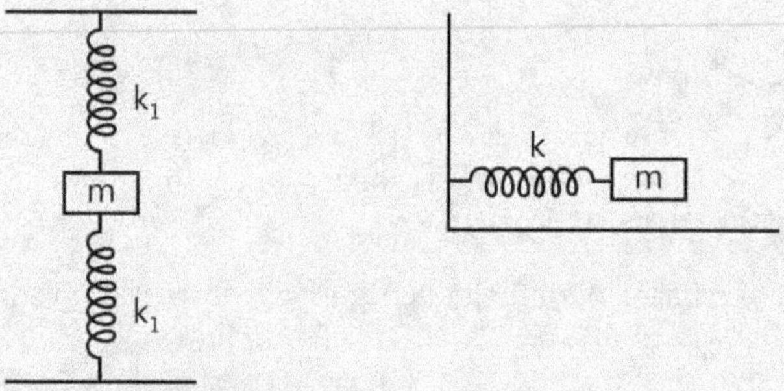

$$T = 2\pi\sqrt{\frac{m(k_1 + k_2)}{k_1 k_2}} \qquad T = 2\pi\sqrt{\frac{m}{k}}$$

Figure 8.20

Illustration 13: For the arrangement shown in Fig 8.21, the spring is initially compressed by 3 cm. When the spring is released the block collides with the wall and rebounds to compress the spring again. **(JEE ADVANCED)**

(a) If the coefficient of restitution is $\frac{1}{\sqrt{2}}$, find the maximum compression in the spring after collision.

Figure 8.21

Sol: Conserve energy to find the velocity of the block. Use equation of restitution for collision of block with the wall.

(a) Velocity of the block just before

collision, $\frac{1}{2}mv_0^2 + \frac{1}{2}kx^2 = \frac{1}{2}kx_0^2$ or $v_0 = \sqrt{\frac{k}{m}\left(x_0^2 - x^2\right)}$

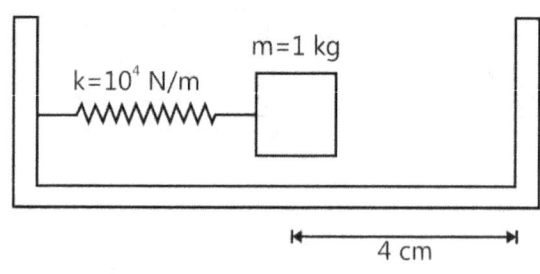

Here, $x_0 = 0.03m$, $\quad\quad x = 0.01m$, $\quad\quad k = 10^4 N/m$, $\quad m = 1kg$ $\quad \therefore \quad v_0 = 2\sqrt{2}\,m/s$

After collision, $\quad v = ev_0 = \dfrac{1}{\sqrt{2}}2\sqrt{2} = 2m/s$

Maximum compression in the spring

$$\frac{1}{2}kx_m^2 = \frac{1}{2}kx^2 + \frac{1}{2}mv^2 \quad\quad \text{or} \quad\quad x_m = \sqrt{x^2 + \frac{m}{\kappa}v^2} = \sqrt{(0.01)^2 + \frac{1(2)^2}{10^4}m} \quad\quad = 2.23cm$$

Illustration 14: Figure 8.22 shows a system consisting of a massless pulley, a spring of force constant k and a block of mass m. If the block is slightly displaced vertically down from its equilibrium position and released, find the period of its vertical oscillation in case (a), (b) and (c). **(JEE ADVANCED)**

Sol: The restoring force on the block will depend on the elongation of the spring. For a small displacement of block find the elongation in the spring.

(a) In equilibrium, $kx_0 = mg$ $\quad\quad\quad\quad\quad$ ...(i)

When further depressed by an amount x, net restoring force (upwards) is,

$F = -\{k(x + x_0) - mg\} = -kx \quad\quad$ (as $kx_0 = mg$)

$a = -\dfrac{k}{m}x \quad \therefore \quad T = 2\pi\sqrt{\left|\dfrac{x}{a}\right|} \quad$ or $\quad\quad T = 2\pi\sqrt{\dfrac{m}{k}}$

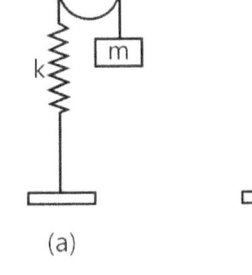
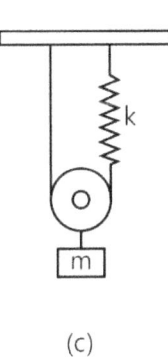

(a)　　　　(b)　　　　(c)

Figure 8.22

(b) In this case if the mass m moves down a distance x from its equilibrium position, then pulley will move down by $\dfrac{x}{2}$. So, the extra force in spring will be $k\dfrac{x}{2}$. Now, as the pulley is massless, this force $\dfrac{kx}{2}$ is equal to extra 2 T or $T = \dfrac{kx}{4}$. This is also the restoring force of the mass. Hence,

$F = -\dfrac{kx}{4}; \quad a = -\dfrac{k}{4m}x \quad$ or $\quad T = 2\pi\sqrt{\left|\dfrac{x}{a}\right|} \quad$ or $\quad T = 2\pi\sqrt{\dfrac{4m}{k}}$

(c) In this situation if the mass m moves down a distance x from its equilibrium position, the pulley will also move by x and so the spring will stretch by 2x. Therefore, the spring force will be 2kx. The restoring force on the block will be 4kx.

Hence, $\quad\quad F = -4kx \quad$ or $\quad\quad a = -\dfrac{4k}{m}, x$

$\therefore \quad T = 2\pi\sqrt{\left|\dfrac{x}{a}\right|} \quad$ or $\quad T = 2\pi\sqrt{\dfrac{m}{4k}}$

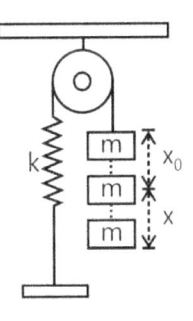

Figure 8.23

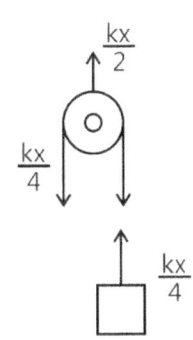

Figure 8.24

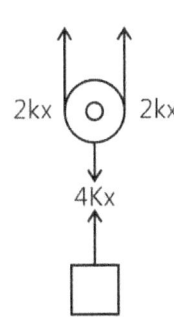

Figure 8.25

Illustration 15: A Spring mass system is hanging from the ceiling of an elevator in equilibrium. The elevator suddenly starts accelerating upwards with acceleration 'a' Find: (a) The frequency and (b) The amplitude of the resulting SHM. **(JEE MAIN)**

2.18

Sol: The time period of spring mass system does not depend on g or acceleration of elevator.

(a) Frequency $= 2\pi\sqrt{\dfrac{m}{k}}$ (Frequency is independent of g in spring)

(b) Extension in spring in equilibrium in initial $= \dfrac{mg}{k}$

Extension in spring in equilibrium in accelerating lift $= \dfrac{m(g+a)}{k}$

∴ Amplitude $= \dfrac{m(g+a)}{k} - \dfrac{mg}{k} = \dfrac{ma}{k}$

Figure 8.26

11. BODY DROPPED IN A TUNNEL ALONG EARTH DIAMETER

Assume earth to be a sphere of radius R and center O. Let a tunnel be dug along the diameter of the earth as shown in Fig. 8.27. If a body of mass m is dropped at one end of the tunnel, the body executes SHM about the center of the earth. Let, at any instant body in the tunnel is at a distance y from the center O of the earth. Only the inner sphere of radius y will exert gravitational force F on the body as the body is inside the earth. The force F serves as the restoring force that tends to bring the body to the equilibrium position O.

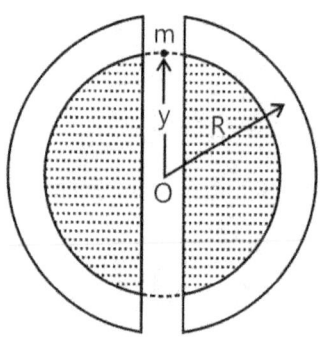

∴ Restoring force, $F = -G\dfrac{\left(4/3\,\pi y^3\rho\right)m}{y^2}$

Figure 8.27: Body moving along diameter of earth

Where ρ is the density of the earth. The negative sign is assigned because the force is of attraction.

Acceleration of the body, $a = \dfrac{F}{m} = -\left(\dfrac{4}{3}\pi G\rho\right)y$ ... (i)

Now the quantity $(4/3)\pi G\rho$ is constant so that: $a \propto -y$

Thus the acceleration of the body is directly proportional to the displacement y and its direction is opposite to the displacement. Therefore, the motion of the body is simple harmonic.

∴ Time period, $T = 2\pi\sqrt{\dfrac{3}{4\pi G\rho}} = \sqrt{\dfrac{3\pi}{G\rho}}$ or $T = \sqrt{\dfrac{3\pi}{G\rho}}$ ... (ii)

12. DAMPED AND UNDAMPED OSCILLATIONS

Damped oscillations is shown in the Fig 8.28 (a) given below. In such a case, during each oscillation, some energy is lost. The amplitude of the oscillation will be reduced to zero as no compensating arrangement for the less is provided. The only parameters that will remain unchanged are the frequency or time period. They will change only according to the circuit parameters.

As shown in Fig 8.28 (b), undamped oscillations have constant have amplitude oscillations.

Damping Force, $F_d = -bV = -b\dfrac{dx}{dt}$ where b is a constant giving the strength of damping. We can write

Newton's law, now including damping force along with the restoring force. For a spring-mass system, we have,

$m\dfrac{d^2x}{dt^2} = -kx - b\dfrac{dx}{dt}$ or $m\dfrac{d^2x}{dt^2} + b\dfrac{dx}{dt} + kx = 0$; $x = ae^{-bt/2m}\cos(\omega t + \phi)$ ... (i) E.q. (i) describes sinusoidal motion whose amplitude (a) decreases exponentially with time. How fast the amplitude drops depends on the damping

constants b and m. The frequency of this damped motion is given by: $f = \dfrac{1}{2\pi}\sqrt{\dfrac{k}{m} - \left(\dfrac{b}{2m}\right)^2}$

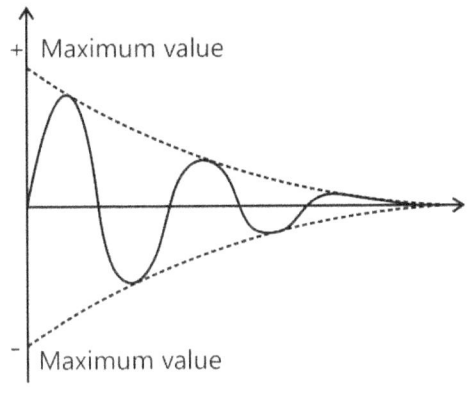

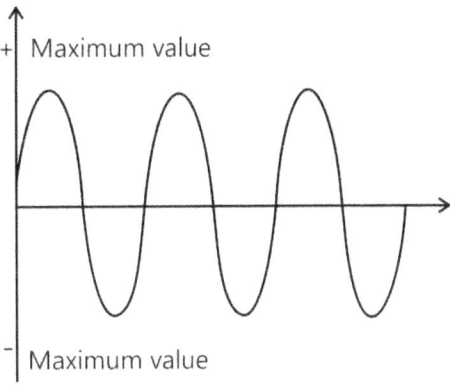

(a) Damped oscillation   (b) Undamped or Sustained osclliation

Figure 8.28: Damped and undamped oscillation

If the frictional forces are absent, b=0 so that: $f = \dfrac{1}{2\pi}\sqrt{\dfrac{k}{m}}$ (undamped oscillations)

13. FREE, FORCED AND RESONANT OSCILLATIONS

(a) **Free oscillations** are executed by an oscillating body that vibrates with its own frequency.

For example, when a simple pendulum is displaced from its mean position and then left free, it executes free oscillations. The natural frequency of the simple pendulum depends upon its length and is given by;

$$f = \frac{1}{2\pi}\sqrt{\frac{g}{l}}$$

(b) **Forced oscillations** - When a body is maintained in a state of oscillations by an external periodic force of frequency other than the natural frequency of the body, it executes forced oscillations.

The frequency of forced oscillations is equal to the frequency of the periodic force. The external applied force on the body is called the driver and the body set into oscillations is called driven oscillator.

Examples. (a) When the stem of a vibrating tuning fork is held in hand, only a feeble sound is heard. However, if the stem is pressed against a table top, the sound becomes louder. It is because the tuning fork forces the table to vibrate with fork's frequency. Since the table has a large vibrating area than the tuning fork, these forced oscillations produce a more intense sound.

Fig 8.29 shown the graph of forced oscillations as a function of ω.

At $\omega = \omega_a$, the value of A_0 is $\left(\dfrac{f_0}{b\omega}\right)$

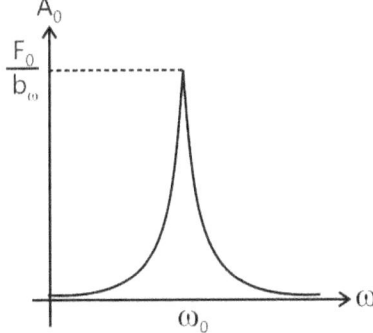

Figure 8.29: Forced oscillation

Notice that amplitude of motion A_0 is directly proportional to the amplitude of driving force.

Mathematical analysis: Most of the oscillations that occur in systems (e.g. machinery) are forced oscillations; oscillations that are produced and sustained by an external force. The simplest driving force is one that oscillates as a sine or a cosine. Suppose such an external force F_{ext} is applied to an oscillator that moves along x axis such as a block connected to a spring. We can represent the external forces as: $F_{ext} = F_0 \cos \omega t$ Where F_0 is the maximum magnitude of the force and $\omega (= 2\pi f)$ is the angular frequency of the force. Then the equation of motion (with damping) is $ma = -kx - bV + F_0 \cos \omega t$. This equation can be written as

$$m\frac{d^2x}{dt^2} = -kx - b\frac{dx}{dt} + F_0 \cos \omega t \qquad \text{or} \qquad m\frac{d^2x}{dt^2} + b\frac{dx}{dt} + kx = F_0 \cos \omega t \qquad \text{... (i)}$$

The solution of eq. (i) is $x = A_0 \cos(\omega t + \phi)$ Where $A_0 = \dfrac{F_0/m}{\sqrt{\left(\omega^2 - \omega_0^2\right)^2 + \left(\dfrac{b\omega}{m}\right)^2}}$ ... (ii)

and $\omega_0 = \sqrt{k/m}$ is the frequency of undamped (b=0) oscillator i.e., natural frequency.

(iii) Resonant oscillations: When a body is maintained in a state of oscillations by a periodic force having the same frequency as the natural frequency of the body, the oscillations are called resonant oscillations. The phenomenon of producing resonant oscillations is called resonance.

(b) The amplitude of motion (A_0) depends on the difference between the applied frequency $\left(\omega\right)$ and natural frequency $\left(\omega_0\right)$. The amplitude is the maximum when the frequency of the driving force equals the natural frequency i.e., when $\omega = \omega_0$. It is because the denominator in eq. (ii) is the minimum when $\omega = \omega_0$. This condition is called resonance. When the frequency of the driving force equals ω_0, the oscillator is said to be in resonance with the driving force.

$$A_0 = \frac{F_0/m}{\sqrt{\left(\omega^2 - \omega_0^2\right) + \left(\dfrac{b\omega}{m}\right)^2}} \qquad \text{At resonance, } \omega = \omega_0 \text{ and } A_0 = \frac{F_0/m}{\sqrt{\left(b\omega/m\right)^2}} = \frac{F_0}{b\omega}$$

PROBLEM-SOLVING TACTICS

To verify SHM see whether force is directly proportional to y or see if $\dfrac{d^2x}{dt^2} + \omega^2 x = 0$ in cases when the equation is directly given compare with general equation to find the time period and other required answers

FORMULAE SHEET

1. **Simple Harmonic Motion (SHM):**

   $F = -kx^n$

   n is even - Motion of particle is not oscillatory

   n is odd - Motion of particle is oscillatory.

   If n = 1, F = -kx or $F \propto -x$. The motion is simple harmonic.

   x = 0 is called the mean position or the equilibrium position.

   Condition for SHM $\dfrac{d^2x}{dt^2} \propto -x$

Acceleration, $a = \dfrac{F}{m} = -\dfrac{k}{m}x = -\omega^2 x$

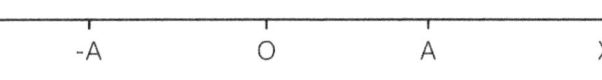

Figure 8.30

Displacement $x = A\cos\underbrace{(\omega t + \phi)}_{\text{phase angle}}$ (A is Amplitude)

Time period of SHM $T = \dfrac{2\pi}{\omega} = 2\pi\sqrt{\dfrac{m}{k}}$

Frequency ν of SHM $\nu = \dfrac{1}{T} = \dfrac{\omega}{2\pi} = \dfrac{1}{2\pi}\sqrt{\dfrac{k}{m}}$

Velocity of particle $v = \dfrac{dx}{dt} = -\omega A\sin(\omega t + \phi)$

Acceleration of particle $a = \dfrac{d^2x}{dt^2} = -\omega^2 A\cos(\omega t + \phi) = -\omega^2 x$

2.   Energy in SHM:

Kinetic energy of particle $= \dfrac{1}{2}m\omega^2\left(A^2 - x^2\right) = \dfrac{1}{2}k\left(A^2 - x^2\right)$

Potential energy $U = \dfrac{1}{2}kx^2 = \dfrac{1}{2}m\omega^2 x^2$

Total energy E = P.E + K.E $= \dfrac{1}{2}m\omega^2 A^2 = \dfrac{1}{2}kA^2$

E is constant throughout the SHM.

3.   Simple pendulum: Time period $T = 2\pi\sqrt{\dfrac{l}{g_{\text{eff}}}}$

Here, l is length of simple pendulum and $\vec{g}_{\text{eff}} = \vec{g} - \vec{a}$ where $\vec{g}$ is acceleration due to gravity and $\vec{a}$ is acceleration of the box or cabin etc. containing the simple pendulum.

4.   Spring-block system: Time period $T = 2\pi\sqrt{\dfrac{m}{k}}$

5.   Physical pendulum: Time period $T = 2\pi\sqrt{\dfrac{I}{mgl}}$

Here I is the moment of inertia about axis of rotation and l is the distance of center of gravity from the point of suspension.

6.   Torsional Pendulum:

$T = 2\pi\sqrt{\dfrac{I}{k}}$

I is the moment of Inertia about axis passing through wire, k is torsional constant of wire.

7. Springs in series and parallel

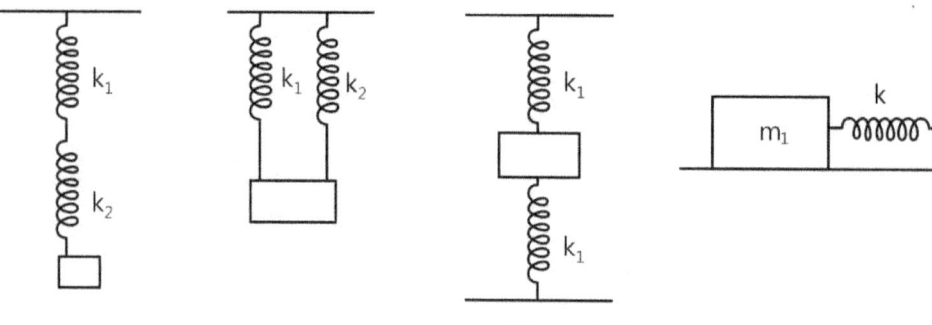

Figure 8.31

Series combination $\dfrac{1}{k}=\dfrac{1}{k_1}+\dfrac{1}{k_2}$

Parallel combination $k=k_1+k_2$

8. For two blocks of masses m_1 and m_2 connected by a spring of constant k:

Time period $\quad T=2\pi\sqrt{\dfrac{\mu}{k}}$

where $\mu=\dfrac{m_1m_2}{m_1+m_2}$ is reduced mass of the two-block system.

Solved Examples

JEE Main/Boards

Example 1: What is the period of pendulum formed by pivoting a meter stick so that it is free to rotate about a horizontal axis passing through 75 cm mark?

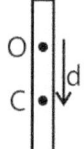

Sol: This is an example of a physical pendulum. Find moment of inertia about point of suspension and the distance of the point of suspension from the center of gravity.

Let m be the mass and ℓ be the length of the stick. $\ell=100$cm The distance of the point of suspension from center of gravity is $d=25$cm

Moment of inertia about a horizontal axis through O is

$$I=I_c+md^2=\dfrac{m\ell^2}{12}+md^2$$

$$T=2\pi\sqrt{\dfrac{I}{mgd}}\,;\quad T=2\pi\sqrt{\dfrac{\dfrac{m\ell^2}{12}+md^2}{mgd}}$$

$$T=2\pi\sqrt{\dfrac{\ell^2+12d^2}{12gd}}=2\pi\sqrt{\dfrac{\ell^2+12(0.25)^2}{12\times9.8\times0.25}}=153\text{ s.}$$

Example 2: A particle executes SHM.

(a) What fraction of total energy is kinetic and what fraction is potential when displacement is one half of the amplitude?

(b) At what value of displacement are the kinetic and potential energies equal?

Sol: The sum of kinetic energy and potential energy is the total mechanical energy which is constant throughout the SHM.

We know that $E_{total}=\dfrac{1}{2}m\omega^2A^2$

$KE=\dfrac{1}{2}m\omega^2\left(A^2-X^2\right)$ and $\quad U=\dfrac{1}{2}m\omega^2x^2$

(a) When $x=\dfrac{A}{2}$, $KE=\dfrac{1}{2}m\omega^2\dfrac{3A^2}{4}\Rightarrow\dfrac{KE}{E_{total}}=\dfrac{3}{4}$

2.23

At $x=\dfrac{A}{2}$, $U=\dfrac{1}{2}m\omega^2\dfrac{A^2}{4}$ $\Rightarrow$ $\dfrac{PE}{E_{total}}=\dfrac{1}{4}$

(b) Since, $K=U$, $\dfrac{1}{2}m\omega^2\left(A^2-x^2\right)=\dfrac{1}{2}m\omega^2x^2$;

$2x^2=A^2$ or $x=\dfrac{A}{\sqrt{2}}=0.707\,A$

Example 3: Show that the period of oscillation of simple pendulum at depth h below earth's surface is inversely proportional to $\sqrt{R-h}$, where R is the radius of earth. Find out the time period of a second pendulum at a depth R / 2 from the earth's surface?

Sol: As we go at a depth below the earth surface, the acceleration due to gravity decreases. The value of g inside the surface of earth is directly proportional to the radial distance from the center of the earth.

At earth's surface the value of time period is given by

$T=2\pi\sqrt{\dfrac{L}{g}}$ or $T\propto\dfrac{1}{\sqrt{g}}$

At a depth h below the surface, $g'=g\left(1-\dfrac{h}{R}\right)$

$\therefore\ \dfrac{T'}{T}=\sqrt{\dfrac{g}{g'}}=\sqrt{\dfrac{1}{\left(1-\dfrac{h}{R}\right)}}=\sqrt{\dfrac{R}{R-h}}$ $\therefore\ T'=T\sqrt{\dfrac{R}{R-h}}$

or $T'\propto\dfrac{1}{\sqrt{R-h}}$ Hence Proved.

Further, $T_{R/2}=2\sqrt{\dfrac{R}{R-R/2}}=2\sqrt{2}\,s$

Example 4: Describe the motion of the mass m shown in figure. The walls and the block are elastic.

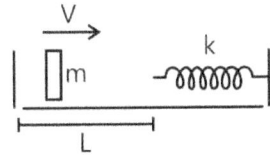

Sol: As the collision of the block with the wall is elastic, there will not be any loss in the kinetic energy and block will execute periodic motion of constant time period.

The block reaches the spring with a speed 'v'. It now compresses the spring. The block is decelerated due to the spring force, comes to rest when $\dfrac{1}{2}mv^2=\dfrac{1}{2}kx^2$ and

return back. It is accelerated due to the spring force till the spring acquires its natural length. The contact of the block with the spring is now broken. At this instant it has regained its speed v (towards left) as the spring is not stretched and no potential energy is stored. This process takes half the period of oscillation, i.e. $\pi\sqrt{m/k}$. The block strikes the left wall after a time L / v and as the collision is elastic, it rebounds with the same speed v. After a time L / v, it again reaches the spring and the process is repeated. The block thus undergoes periodic motion with time period $\pi\sqrt{m/k}+\dfrac{2L}{v}$.

Example 5: A particle is subjected to two simple harmonic motions in the same direction having equal amplitudes and equal frequency. If the resultant amplitude is equal to the amplitude of the individual motions, find the phase difference between the individual motions.

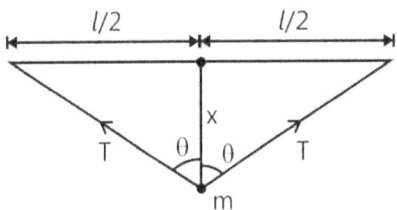

Sol: The amplitude in case of combination of two or more SHMs in same direction and same frequency is obtained by vector addition of the amplitudes of individual SHMs. The angle of each of the individual amplitude with the x-axis is equal to the phase constant of the respective SHM.

Let the amplitudes of the individual motions be A each. The resultant amplitude is also A. If the phase difference between the two motion is δ,

$A=\sqrt{A^2+A^2+2A.A.\cos\delta}$

or $A=A\sqrt{2(1+\cos\delta)}=A\cos\delta/2$

or $\cos\dfrac{\delta}{2}=\dfrac{1}{2}$ or $\delta=2\pi/3$

Example 6: The figure shown below a block collides in-elastically with the right block and sticks to it. Find the amplitude of the resulting simple harmonic motion.

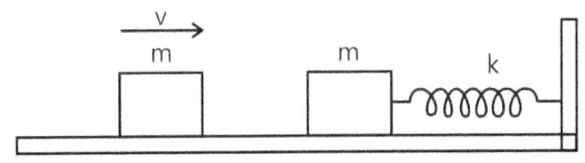

Sol: Conserve momentum before and after collision. The kinetic energy of blocks after collision is converted into elastic potential energy of the spring at the instant of maximum compression. Maximum compression is equal to amplitude of resulting SHM.

Assuming the collision to last for a small interval only, we can apply the principle of conservation of momentum. The common velocity after the collision is $\frac{v}{2}$. The kinetic energy $= \frac{1}{2}(2m)\left(\frac{v}{2}\right)^2 = \frac{1}{4}mv^2$. This is also the total energy of vibration as the spring is unstretched at this moment. If the amplitude is A, the total energy can also be written as $\frac{1}{2}kA^2$.

Thus, $\frac{1}{2}kA^2 = \frac{1}{4}mv^2$, giving $A = \sqrt{\frac{m}{2k}}\,v$.

Example 7: Find the time period of small oscillations in a horizontal plane performed by a ball of mass 40 g fixed at the middle of a horizontally stretched string 1.0 m in length. The tension of the string is assumed to be constant and equal to 10 N.

Sol: Use the restoring force method to find the angular frequency.

Consider a ball of mass m placed at the middle of a string of length l and tension T. The components of tension T towards mean position is $T\cos\theta$.

The force acting on the ball $= 2T\cos\theta$

$$\therefore ma = -\frac{2Tx}{\sqrt{\left(\left(l^2/4\right)+x^2\right)}}$$

$\because T = F$ and $\cos\theta = \dfrac{x}{\sqrt{\left(\left(l^2/4\right)+x^2\right)}}$

As x is small, x_2 can be neglected in the denominator.

$$\therefore a = -\frac{2Tx}{m(l/2)} = -\left(\frac{4T}{ml}\right)x = -\omega^2 x$$

The acceleration is directly proportional to negative displacement x and is directed towards the mean position. Hence the motion is SHM

$$T = \frac{2\pi}{\omega} = \frac{2\pi}{\sqrt{(4T/ml)}} = \pi\sqrt{\frac{ml}{T}}$$

Substituting the given values, we get

$$T = 3.14 \times \sqrt{\left(\frac{\left(4\times10^{-2}\right)(1.0)}{10}\right)} = 0.2\,\text{s}$$

Example 8: If a tunnel is dug through the earth from one side to the other side along a diameter. Show that the motion of a particle dropped into the tunnel is simple harmonic motion. Find the time period. Neglect all the frictional forces and assume that the earth has a uniform density.

$G = 6.67 \times 10^{-11}\,\text{Nm}^2\text{kg}^{-2}$; Density of earth $= 5.51 \times 10^3\,\text{kgm}^{-3}$

Sol: Use the restoring force method to find the angular frequency.

Consider a tunnel dug along the diameter of the earth. A particle of mass m is placed at a distance y from the center of the earth. There will be a gravitational attraction of the earth experienced by this particle due to the mass of matter contained in a sphere of radius y. Force acting on particle at distance y from center

$$F = \frac{GM}{R^3}\cdot y$$

$$\Rightarrow ma = -\frac{GMm}{R^3}\cdot y$$

$$\Rightarrow a = -\frac{GM}{R^3}\cdot y = -\frac{G\times d\times\frac{4}{3}\pi R^3}{R^3}\,y = -\frac{4\pi G}{3}\cdot d.y$$

As the force is directly proportional to the displacement and is directed towards the mean position, the motion is simple harmonic.

$$\Rightarrow \omega^2 = \frac{4}{3}\pi dG. \text{ and } T = 2\pi\sqrt{\left(\frac{3}{4\pi dG}\right)}$$

$$= \sqrt{\left(\frac{3\pi}{dG}\right)} = \sqrt{\left(\frac{3\times3.14}{5.51\times10^3\times6.67\times10^{-11}}\right)}$$

$$= 5062\,\text{s} = 84.4\,\text{min}$$

Example 9: The pulley shown in figure below has a moment of inertia I about its axis and mass m. find the time period of vertical oscillation of its center of mass. The spring has spring constant k and the string does not slip over the pulley.

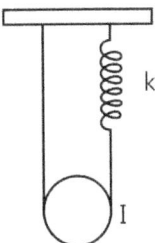

Sol: For a small displacement of the pulley find the extension in the spring. Use the energy method to find the angular frequency.

Let us first find the equilibrium position. For rotational equilibrium of the pulley, the tensions in the two strings should be equal. Only then the torque on the pulley will be zero. Let this tension be T. The extension of the spring will be $y = T/k$, as the tension in the spring will be the same as the tension in the string. For translational equilibrium of the pulley,

$$2T = mg \quad \text{or,} \quad 2ky = mg \quad \text{or,} \quad y = \frac{mg}{2k}.$$

The spring is extended by a distance $\frac{mg}{2k}$ when the pulley is in equilibrium.

Now suppose, the center of the pulley goes down further by a distance x. The total increase in the length of the string plus the spring is 2x (x on the left of the pulley and x on the right). As the string has a constant length, the extension of the spring is 2x. The energy of the system is

$$U = \frac{1}{2}I\omega^2 + \frac{1}{2}mv^2 - mgx + \frac{1}{2}k\left(\frac{mg}{2k} + 2x\right)^2$$

$$= \frac{1}{2}\left(\frac{I}{r^2} + m\right)v^2 + \frac{m^2g^2}{8k} + 2kx^2.$$

As the system is conservative, $\frac{dU}{dt} = 0$,

giving $0 = \left(\frac{I}{r^2} + m\right)v\frac{dv}{dt} + 4kxv$

Or, $\frac{dv}{dt} = -\frac{4kx}{\left(\dfrac{I}{r^2} + m\right)}$

or $a = -\omega^2 x$, where $\omega^2 = \dfrac{4k}{\left(\dfrac{I}{r^2} + m\right)}$

Thus, the center of mass of the pulley executes a simple harmonic motion with time period

$$T = 2\pi\sqrt{\left(\frac{I}{r^2} + m\right) / (4k)}.$$

Example 10: Two light springs of force constant k_1 and k_2 and a block of mass m are in one line AB on a smooth horizontal table such that one end of each spring is fixed on rigid supports and the other end is free as shown in figure.

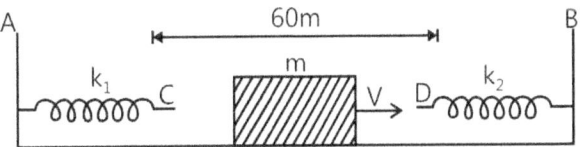

The distance CD between the free ends of the springs 60 cm. If the block moves along AB with a velocity 120 cm/sec in between the springs, calculate the period of oscillation of the block

$$\left(k_1 = 1.8\,N/m, k_2 = 3.2\,N/m, m = 200\,gm\right)$$

If initially block is mid-way of CD.

Sol: As there are no dissipative forces the motion of the block is oscillatory with constant time period. Add the time of motion of different segments to get the time period.

If initially block is mid-way of CD their the time period T is equal to sum of time to travel 30 cm to right, time in contact with spring k_2, time to travel 60 cm to left, time in contact with spring k_1 and time to travel 30 cm to right.

$$\therefore T = \frac{30}{120} + \frac{1}{2}\left[2\pi\sqrt{\frac{m}{k_2}}\right] + \frac{60}{120} + \frac{1}{2}\left[2\pi\sqrt{\frac{m}{k_1}}\right] + \frac{30}{120}$$

$$= 0.25 + \pi\sqrt{\left(\frac{0.2}{3.2}\right)} + 0.5 + \pi\sqrt{\left(\frac{0.2}{1.8}\right)} + 0.25$$

$$= 0.25 + \pi/4 + 0.5 + \pi/3 + 0.25 = 2.83\ \text{s}.$$

Example 11: The moment of inertia of the disc used in torsional pendulum about the suspension wire is $0.2\ kg-m^2$. It oscillates with a period of 2s. Another disc is played over the first one and the time period of the system becomes 2.5 s. Fine the moment of inertia of the second disc about the wire.

Sol: As another disc is placed on the first disc moment of inertia about the axis passing through the wire increases and thus time period increases.

Let the torsional constant of the wire be k. The moment of inertia of the first disc about the wire is $0.2 \, kg-m^2$. hence, the time period is

$$2s = 2\pi\sqrt{\frac{I}{K}} = 2\pi\sqrt{\frac{0.2kg-m^2}{k}} \qquad \text{...(i)}$$

When the second disc having moment of inertia I_1 about. The wire is added, the time period is

$$2.5s = 2\pi\sqrt{\frac{0.2kg-m^2 + I_1}{0.2kg-m^2}} \qquad \text{...(ii)}$$

From (i) and (ii), $\dfrac{6.25}{4} = \dfrac{0.2kgm-m^2 + I_1}{0.2kg-m^2}$.

This gives $I_1 = 0.11 kg-m^2$.

Example 12: A simple pendulum having a bob of mass m undergoes small oscillations with amplitude θ_0. Find the tension in the string as a function of the angle made by the string with the vertical. When is this tension maximum, and when is it minimum?

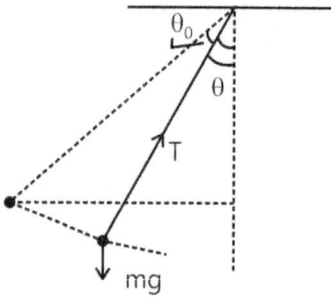

Sol: The forces acting on the bob are tension due to string and weight mg. The bob moves in a circular path. The acceleration of the bob has both radial and tangential component.

Suppose the speed of the bob at angle θ is υ. Using conservation of energy between the extreme position and the position with angle θ,

$$\frac{1}{2}mv^2 = mgl(\cos\theta - \cos\theta_0) \qquad \text{...(i)}$$

As the bob moves in a circular path, the force towards the center should be equal to mv^2 / l. Thus,

$$T - mg\cos\theta = mv^2 / l.$$

Using (i),

$$T - mg\cos\theta = 2mg(\cos\theta - \cos\theta_0)$$

or $T = 3mg\cos\theta - 2mg\cos\theta_0$.

Now $\cos\theta$ is maximum at $\theta = 0$ and decreases as $|\theta|$ increases $(for |\theta| < 90°)$.

Thus, the tension is maximum when $\theta = 0$, i.e., at the mean position and is minimum when $\theta = \pm\theta_0$, i.e., at extreme positions.

JEE Advanced/Boards

Example 1: A simple pendulum is suspended from the ceiling of a car accelerating uniformly on a horizontal road. If the acceleration is a_0 and the length of the pendulum is l, find the time period of small oscillations about the mean position.

Sol: The car accelerates with acceleration a. In the reference frame of car the effective value of acceleration due to gravity is

$$g_{eff} = |\vec{g} - \vec{a}| = \sqrt{g^2 + a^2}$$

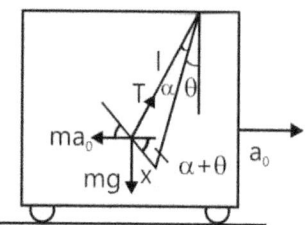

We shall work in the car frame. As it is accelerated with respect to the road, we shall have to apply a pseudo force $m \, a_0$ on the bob of mass m.

For mean position, the acceleration of the bob with respect to the car should be zero. If θ be the angle made by the string with the vertical, then tension, weight and the pseudo force will add to zero in this position.

Suppose at some instant during oscillation, the string is further deflected by an angle α so that the displacement of the bob is x. Taking the components perpendicular to the string, component of T = 0, component of $mg = mg\sin(\alpha + \theta)$ and component of $ma_0 = -ma_0\cos(\alpha + \theta)$. Thus, the resultant component $F = m\left[g\sin(\alpha + \theta) - a_0\cos(\alpha + \theta)\right]$.

Expanding the sine and cosine and putting $\cos\alpha \approx 1$, $\sin\alpha \approx x/l$, we get,

$$F = m\left[g\sin\theta - a_0\cos\theta + (g\cos\theta + a_0\sin\theta)\frac{x}{l}\right]$$

At $x = 0$, the force F on the bob should be zero, as this is the mean position. Thus by (i),

$$0 = m\left[g\sin\theta - a_0\cos\theta\right] \qquad \text{...(ii)}$$

Giving $\tan\theta = \dfrac{a_0}{g}$

Thus, $\sin\theta = \dfrac{a_0}{\sqrt{a_0^2 + g^2}}$ $\qquad$ ...(iii)

$\cos\theta = \dfrac{g}{\sqrt{a_0^2 + g^2}}$ $\qquad$ ...(iv)

Putting (ii), (iii) and (iv) in

(i), $F = m\sqrt{g^2 + a_0^2}\,\dfrac{x}{l}$ or, $F = m\omega^2 x$, where $\omega^2 = \dfrac{\sqrt{g^2 + a_0^2}}{l}$.

This is an equation of simple harmonic motion with

time period $t = \dfrac{2\pi}{\omega} = 2x\dfrac{\sqrt{l}}{(g^2 + a_0^2)^{1/4}}$.

As easy working rule may be found out as follows. In the mean position, the tension, the weight and the pseudo force balance. From figure, the tension is

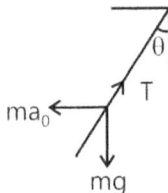

$$T = \sqrt{(ma_0)^2 + (mg)^2}$$

or, $\dfrac{T}{m} = \sqrt{a_0^2 + g^2}$.

This plays the role of effective 'g'. Thus the time period

is $t = 2\pi\sqrt{\dfrac{I}{T/m}} = 2\pi\dfrac{\sqrt{l}}{\left[g^2 + a_0^2\right]^{1/4}}$.

Example 2: A long uniform rod of length L and mass M is free to rotate in a vertical plane about a horizontal axis through its one end 'O'. A spring of force constant k is connected vertically between one end of the rod and ground. When the rod is in equilibrium it is parallel to the ground.

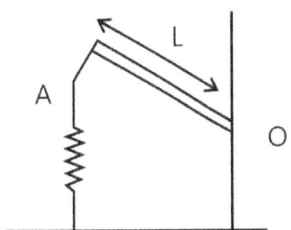

(a) What is the period of small oscillation that result when the rod is rotated slightly and released?

(b) What will be the maximum speed of the displaced end of the rod, if the amplitude of motion is θ_0?

Sol: The rod executes angular SHM. Use restoring torque method to find angular frequency of SHM.

(a) Restoring torque about 'O' due to elastic force of the spring

$\tau = -FL = -kyL$ $\qquad$ (F = ky)

$\tau = -kL^2\theta$ $\qquad$ (as y = Lθ)

$\tau = I\alpha = \dfrac{1}{3}ML^2\dfrac{d^2\theta}{dt^2}$

$\dfrac{1}{3}ML^2\dfrac{d^2\theta}{dt^2} = -kL^2\theta;$ $\quad \dfrac{d^2\theta}{dt^2} = -\dfrac{3k}{M}\theta$

$\omega = \sqrt{\dfrac{3k}{M}} \Rightarrow T = 2\pi\sqrt{\dfrac{M}{3k}}$

(b) In angular SHM, maximum angular velocity

$\left(\dfrac{d\theta}{dt}\right)_{max} = \theta_0,$ $\quad \omega = \theta_0\sqrt{\dfrac{3k}{M}},$ $\quad v = r\left(\dfrac{d\theta}{dt}\right)$

So, $v_{max} = L\left(\dfrac{d\theta}{dt}\right)_{max} = L\theta_0\sqrt{\dfrac{3k}{M}}$

Example 3: A block with mass of 2 kg hangs without vibrating at the end of a spring of spring constant 500 N/m, which is attached to the ceiling of an elevator. The elevator is moving upwards with acceleration $\dfrac{g}{3}$. At time t=0, the acceleration suddenly ceases.

(a) What is the angular frequency of oscillation of the block after the acceleration ceases?

(b) By what amount is the spring stretched during the time when the elevator is accelerating?

(c) What is the amplitude of oscillation and initial phase angle observed by a rider in the elevator? Take the upward direction to be positive. Take $g = 10.0 \, m/s^2$.

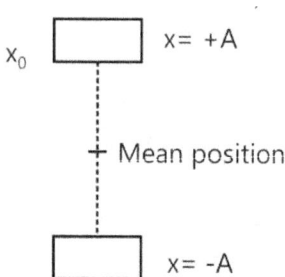

Sol: The angular frequency of the spring block system in vertical oscillations does not depend on the acceleration due to gravity or the acceleration of the elevator. The equilibrium position depends on the acceleration due to gravity and the elevator. When the acceleration of the elevator ceases the block moves to the new equilibrium position.

(a) Angular frequency

$$\omega = \sqrt{\frac{k}{m}} \text{ or } \omega = \sqrt{\frac{500}{2}}$$

or $\omega = 15.81 \, rad/s$

(b) Equation of motion of the block (while elevator is accelerating) is,

$$kx - mg = ma = m\frac{g}{3}$$

$$\therefore x = \frac{4mg}{3k} = \frac{(4)(2)(10)}{(3)(500)} = 0.053 \, m$$

or $x = 5.3 \, cm$

(c) (i) In equilibrium when the elevator has zero acceleration, the equation of motion is

$$kx_0 = mg \text{ or } \qquad x_0 = \frac{mg}{k} = \frac{(2)(10)}{500} = 0.04 \, m$$

$= 4 \, cm$

$\therefore$ Amplitude $A = x - x_0$

$= 5.3 - 4.0$

$= 1.3 \, cm$

(ii) At time $t = 0$, block is at $x = -A$. Therefore, substituting $x = -A$ and $t = 0$ in equation,

$x = A \sin(\omega t + \phi)$ We get initial phase $\phi = \dfrac{3\pi}{2}$

Example 4: A solid sphere (radius = R) rolls without slipping in a cylindrical through (radius = 5R). Find the time period of small oscillations.

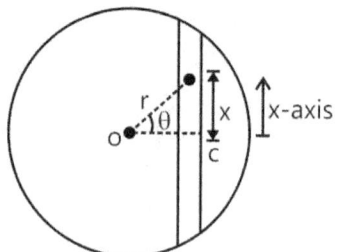

Sol: The sphere executes pure rolling in the cylinder. The mean position is at the lowest point in the cylinder. Find the acceleration for small displacement from the mean position and compare with standard equation of SHM to find angular frequency.

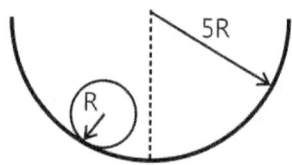

For pure rolling to take place, $v = R\omega$

ω' = Angular velocity of COM of sphere C about O

$$= \frac{v}{4R} = \frac{R\omega}{4R} = \frac{\omega}{4}$$

$$\therefore \frac{d\omega'}{dt} = \frac{1}{4}\frac{d\omega}{dt} \text{ or } \alpha' = \frac{\alpha}{4}$$

$\alpha = \dfrac{a}{R}$ for pure rolling;

Where, $a = \dfrac{g\sin\theta}{I + mR^2} = \dfrac{5g\sin\theta}{7}$

As, $I = \dfrac{2}{5}mR^2 \quad \therefore \quad \alpha' = \dfrac{5g\sin\theta}{28R}$

For small θ, $\sin\theta = \theta$, being restoring in nature,

$$\alpha' = -\frac{5g}{28R}\theta \qquad \therefore T = 2\pi\sqrt{\left|\frac{\theta}{\alpha'}\right|} = 2\pi\sqrt{\frac{28R}{5g}}$$

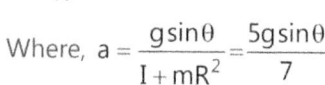

Example 5: Consider the earth as a uniform sphere of mass M and radius R. Imagine a straight smooth tunnel made through the earth which connects any two points on its surface. Show that the motion of a particle of

mass in along this tunnel under the action of gravitation would be simple harmonic. Hence, determine the time that a particle would take to go from one end to the other through the tunnel.

Sol: Use the restoring force method to find the angular frequency.

Suppose at some instant the particle is at radial distance r from center of earth O. Since, the particle is constrained to move along the tunnel, we define its position as distance x from C. Hence, equation of motion of the particle is, $ma_x = F_x$

The gravitational force on mass m at distance r is,

$$F = \frac{GMmr}{R^3} \text{ (Towards O)}$$

Therefore, $F_x = -F\sin\theta = -\frac{GMmr}{R^3}\left(\frac{x}{r}\right)$

Since $F_x \propto -x$, motion is simple harmonic in nature. Further,

$$ma_x = -\frac{GMm}{R^3}.x \quad \text{or} \quad a_x = -\frac{GM}{R^3}.x$$

∴ Time period of oscillation is,

$$T = 2\pi\sqrt{\frac{x}{a_x}} = 2\pi\sqrt{\frac{R^3}{GM}}$$

The time taken by particle to go from one end to the other is $\frac{T}{2}$

$$\therefore \quad t = \frac{T}{2} = \pi\sqrt{\frac{R^3}{GM}}$$

Example 6: Two identical balls A and B, each of mass 0.1 kg are attached to two identical massless springs. The spring mass system is constrained to move inside a rigid smooth pipe bent in the form of a circle as shown in figure. The pipe is fixed in a horizontal plane. The centers of the balls can move in a circle of radius 0.06 m. Each spring has a natural length 0.06 π m and spring constant 0.1 N/m. Initially both the balls are displaced by angle π/6 radian with respect to the diameter PQ of the circle and released from rest.

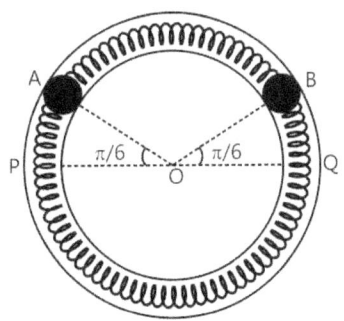

(a) Calculate the frequency of oscillation of ball B.

(b) Find the speed of the ball A when A and B are at the two ends of diameter PQ

(c) What is the total energy of the system

Sol: Here the two balls connected by the springs are free to oscillate along the length of the springs, so the time period will depend on the reduced mass of the two-ball system.

(a) Restoring force on A or B $= k\Delta x + k\Delta x = 2k\Delta x$.

Where Δx is compression in the spring at one end? Effective force constant $= 2k$

$$\text{Frequency } v = \frac{1}{2\pi}\sqrt{\frac{2k}{\mu}}$$

Where μ is reduced mass of system.

$$\text{reduced mass. } \mu = \frac{mm}{m+m} = \frac{m}{2}$$

$$v = \frac{1}{2\pi}\sqrt{\frac{2k}{m/2}} = \frac{1}{3.14}\sqrt{\frac{0.1}{0.1}} = \frac{1}{3.14}s$$

(b) P and Q are equilibrium position. Balls A and B at P and Q have only kinetic energy and it is equal the potential energy at extreme positions.

Potential energy at extreme position

$$= \frac{1}{2}k(2\Delta x)^2 + \frac{1}{2}k(2\Delta x)^2 = 4k(\Delta x)^2$$

Where $\Delta x = Rx\frac{\pi}{6}$

$$\Rightarrow \text{P.E.} = \frac{\pi^2 kR^2}{36} = \frac{(3.14)^2 x0.1x(0.06)^2}{36} \approx 3.94x10^{-4}\,J$$

When the balls A and B are at points P and Q respectively.

$$KE_{(A)} + KE_{(B)} = \text{P.E.} \; ; \; 2KE_{(A)} = \text{P.E.}$$

$$2x\frac{1}{2}mv^2 = 3.94 \times 10^{-4}$$

$$\Rightarrow v = \left(\frac{3.94}{0.1}\right)^{\frac{1}{2}} x10^{-2} = 6.28x10^{-2} = 0.0628\,ms^{-1}$$

(c) Total potential and kinetic energy of the system is equal to total potential energy at the extreme position $=3.94x10^{-4}$J.

Exercise 1

Q.1 A simple harmonic motion is represented by y(t)=10 sin (20t+0.5). Write down its amplitude, angular frequency, time period and initial phase, if displacement is measured in meters and time in seconds.

Q.2 A particle executing SHM along a straight line has a velocity of 4 ms⁻¹, when at a distance of 3 m from its mean position and 3 ms⁻¹, when at a distance of 4 m from it. Find the time it takes to travel 2.5 m from the positive extremity of its oscillation.

Q.3 A simple harmonic oscillation is represented by the equation.

Y=0.4sin (440t+0.61)

Here y and t are in m and s respectively. What are the values of (i) amplitude (ii) angular frequency (iii) frequency of oscillation (iv) time period of oscillation and (v) initial phase?

Q.4 A particle executing SHM of amplitude 25 cm and time period 3 s. What is the minimum time required for the particle to move between two points 12.5 cm on either side of the mean position?

Q.5 A particle executes SHM of amplitude a. At what distance from the mean position is its K.E. equal to its P.E?

Q.6 An 8 kg body performs SHM of amplitude a. At what distance from the mean position is its K.E. equal to its P.E?

Q.7 A spring of force constant 1200 Nm⁻¹ is mounted on a horizontal table as shown in figure. A mass of 3.0 kg is attached to the free end of the spring. Pulled sideways to a distance of 2cm and released, what is

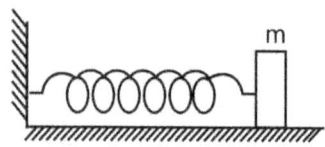

(a) The speed of the mass when the spring is compressed by 1.0 cm?

(b) Potential energy of the oscillating mass.

Q.8 A trolley of mass 3.0 kg is connected to two identical springs each of force constant 600 Nm⁻¹ as shown in figure. If the trolley is displaced from its equilibrium position by 5.0 cm and released, what is the total energy stored?

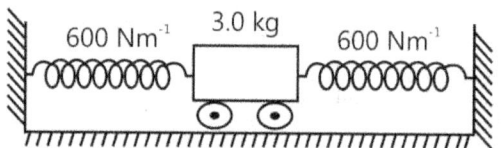

Q.9 A pendulum clock normally shows correct time. On an extremely cold day, its length decreases by 0.2%. Compute the error in time per day.

Q.10 Two particles execute SHM of same amplitude and frequency on parallel lines. They pass one another when moving in opposite directions and at that time their displacement is one third their amplitude. What is the phase difference between them?

Q.11 What is the frequency of a second pendulum in an elevator moving up with an accelerating of g/2?

Q.12 Explain periodic motion and oscillatory motion with illustration.

Q.13 What is a simple pendulum? Find an expression for the time period and frequency of a simple pendulum.

Q.14 Explain the oscillations of a loaded spring and find the relations for the time period and frequency in case of (i) horizontal spring (ii) vertical spring

Q.15 What is a spring factor? Find its value in case of two springs connected in (i) series and (ii) parallel.

Q.16 Explain phase and phase difference, angular frequency, displacement in periodic motion with illustrations.

Q.17 Explain displacement, velocity, acceleration and time period in SHMs. Find the relation between them.

Q.18 From the figure (a) and (b). Obtain the equation of simple harmonic motion of the y-projection of the radius vector of the revolving particle P in each case.

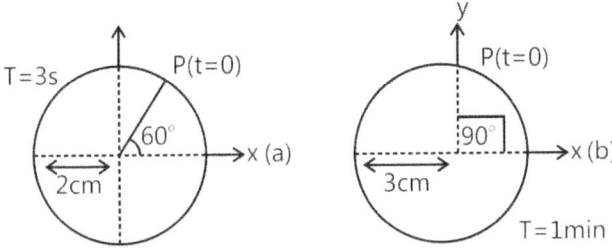

Q.19 Two particles execute SHM of the same amplitude and frequency along does parallel lines. They pass each other moving in opposite directions, each time their displacement in half their amplitude. What is their phase difference?

Q.20 A body oscillates with SHM according to the equation, $X = 6 \cos (3\pi t + \pi/3)$ metres. What is (a) amplitude and (b) the velocity at $t = 2s$.

Q.21 A bob of simple pendulum executes SHM of period 20 s. Its velocity is 5 ms^{-1}, two seconds after it has passed through its mean position. Determine the amplitude of SHM.

Q.22 A particle is moving in a straight line with SHM Its velocity has values 3 ms^{-1} and 2 ms^{-1} when its distance from the mean positions are 1 m and 2 m respectively. Find the period of its motion and length of its path.

Q.23 A particle executes SHM with an amplitude 4 cm. Locate the position of point where its speed is half its speed is half its maximum speed. At what displacement is potential energy equal to kinetic energy?

Q.24 A block whose mass is 1 kg is fastened to a spring. The spring has a spring constant 50 N m^{-1}. The block is pulled to a distance x=10 cm from its equilibrium position at x=0 on a frictionless surface at t=0. Calculate the kinetic, potential and total energies of the block when it is 5 cm away from the mean position.

Q.25 Two point masses of 3.0 kg and 1.0 kg are attached to opposite ends of a horizontal spring whose spring constant in 300 Nm^{-1} as shown in figure. Find the natural frequency of vibration of the system.

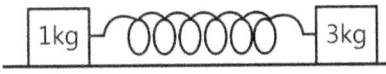

Q.26 A system of springs with their spring constants are as shown in figure . What is the frequency of oscillations of the mass m?

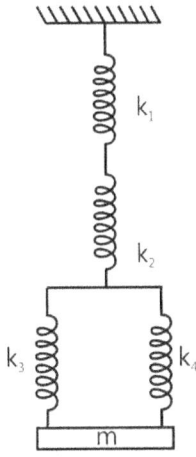

Exercise 2

Single Correct Choice Type

Q.1 A simple harmonic motion having an amplitude A and time period T is represented by the equation:
$y = 5 \sin \pi (t + 4)m$

Then the values of A (in m) and T (in s) are:

(A) A = 5; T = 2   (B) A = 10; T = 1

(C) A = 5; T = 1   (D) A = 10; T = 2

Q.2 The maximum acceleration of a particle in SHM is made two times keeping the maximum speed to be constant. It is possible when

(A) Amplitude of oscillation is doubled while frequency remains constant

(B) Amplitude is doubled while frequency is halved

(C) Frequency is doubled while amplitude is halved

(D) Frequency is doubled while amplitude remains constant

Q.3 A stone is swinging in a horizontal circle 0.8 m in diameter at 30 rev/min. A distant horizontal light beam causes a shadow of the stone to be formed on a nearly vertical wall. The amplitude and period of the simple harmonic motion for the shadow of the stone are

(A) 0.4 m, 4 s   (B) 0.2 m, 2 s

(C) 0.4 m, 2 s   (D) 0.8 m, 2 s

Q.4 A small mass executes linear SHM about O with amplitude a and period T. Its displacement from O at time T/8 after passing through O is:

(A) a/8   (B) $a/2\sqrt{2}$ (C) a/2   (D) $a/\sqrt{2}$

Q.5 The displacement of a body executing SHM is given by $x = A\sin(2\pi t + \pi/3)$. The first time from t=0 when the velocity is maximum is

(A) 0.33 s   (B) 0.16 s   (C) 0.25 s   (D) 0.5 s

Q.6 A particle executes SHM of period 1.2 s. and amplitude 8 cm. Find the time it takes to travel 3cm from the positive extremely of its oscillation.

(A) 0.28 s   (B) 0.32 s   (C) 0.17 s   (D) 0.42 s

Q.7 A particle moves along the x-axis according to :$x=A[1+\sin\omega t]$. What distance does it travel between? $t = 0$ and $t = 2.5\pi/\omega$?

(A) 4A   (B) 6A   (C) 5A   (D) None

Q.8 Find the ratio of time periods of two identical springs if they are first joined in series & then in parallel & a mass m is suspended from them:

(A) 4   (B) 2   (C) 1   (D) 3

Q.9 The amplitude of the vibrating particle due to superposition of two SHMs,

$$y_1 = \sin\left(\omega t + \frac{\pi}{3}\right) \text{ and } y_2 = \sin\omega t \text{ is:}$$

(A) 1   (B) $\sqrt{2}$ (C) $\sqrt{3}$ (D) 2

Q.10 Two simple harmonic motions $y_1 = A \sin \omega t$ are superimposed on a particle of mass m. The total mechanical energy of the particle is:

(A) $\frac{1}{2}m\omega_2 A_2$ (B) $m\omega_2 A_2$

(C) $\frac{1}{4}m\omega_2 A_2$ (D) Zero

Q.11 A block of mass 'm' is attached to a spring in natural length of spring constant 'k'. The other end A of the spring is moved with a constant velocity v away from the block. Find the maximum extension in the spring.

(A) $\frac{1}{4}\sqrt{\frac{mv^2}{k}}$ (B) $\sqrt{\frac{mv^2}{k}}$

(C) $\frac{1}{2}\sqrt{\frac{mv^2}{k}}$ (D) $2\sqrt{\frac{mv^2}{k}}$

Q.12 In the above question, the find amplitude of oscillation of the block in the reference frame of point A of the spring.

(A) $\frac{1}{4}\sqrt{\frac{mv^2}{k}}$ (B) $\frac{1}{2}\sqrt{\frac{mv^2}{k}}$

(C) $\sqrt{\frac{mv^2}{k}}$ (D) $2\sqrt{\frac{mv^2}{k}}$

Q.13 For a particle acceleration is defined as

$$\vec{a} = \frac{-5x\vec{i}}{|x|} \text{ for } x \neq 0 \text{ and } \vec{a} = 0 \text{ for } x = 0.$$

If the particle is initially at rest (a, 0) what is period of motion of the particle.

(A) $4\sqrt{2a/5}$ sec.   (B) $8\sqrt{2a/5}$ sec.

(C) $2\sqrt{2a/5}$ sec.   (D) Cannot be determined

Q.14 A mass m, which is attached to a spring with spring constant k, oscillates on a horizontal table, with amplitude A. At an instant when the spring is stretched by $\sqrt{3}A/2$, a second mass m is dropped vertically onto the original mass and immediately sticks to it. What is the amplitude of the resulting motion?

(A) $\frac{\sqrt{3}}{2}A$ (B) $\sqrt{\frac{7}{8}}A$

(C) $\sqrt{\frac{13}{16}}A$ (D) $\sqrt{\frac{2}{3}}A$

Q.15 A particle is executing SHM of amplitude A, about the mean position x=0. Which of the following cannot be a possible phase difference between the positions of the particle at x=+ A/2 and $x = -A/\sqrt{2}$

(A) 75°   (B) 165°   (C) 135°   (D) 195°

Previous Years' Questions

Q.1 A particle executes simple harmonic motion with a frequency f. The frequency with which its kinetic energy oscillates is *(1987)*

(A) $f/2$ (B) f (C) $2f$ (D) $4f$

Q.2 Two bodies M and N of equal masses are suspended from two separate massless springs of spring constants k_1 and k_2 respectively. If the two bodies oscillate vertically such that their maximum velocities are equal, the ratio of the one amplitude of vibration of M to that of N is *(1988)*

(A) k_1/k_2 (B) $\sqrt{k_2/k_1}$

(C) k_2/k_1 (D) $\sqrt{k_1/k_2}$

Q.3 A highly rigid cubical block A of small mass M and side L is fixed rigidly on to another cubical block B of the same dimensions and of low modulus of rigidity η such that the lower face of A completely covers the upper face of B. The lower face of B is rigidly held on a horizontal surface. A small force F is applied perpendicular to one of the side faces of A. After the force is withdrawn. Block A executes small oscillations. The time period of which is given by *(1992)*

(A) $2\pi\sqrt{M\eta L}$ (B) $2\pi\sqrt{\dfrac{M\eta}{L}}$

(C) $2\pi\sqrt{\dfrac{ML}{\eta}}$ (D) $2\pi\sqrt{\dfrac{M}{\eta L}}$

Q.4 One end of a long metallic wire of length L is tied to the ceiling. The other end is tied to a massless spring of spring constant k. A mass m hangs freely from the free end of the spring. The area of cross-section and the Young's modulus of the wire are A and Y respectively. If the mass is slightly pulled down and released, it will oscillate with a time period T equal to *(1993)*

(A) $2\pi(m/k)^{1/2}$ (B) $2\pi\sqrt{\dfrac{m(YA+kL)}{YAk}}$

(C) $2\pi[(mYA/kL)^{1/2}$ (D) $2\pi[(mL/YA)^{1/2}$

Q.5 A particle of mass m is executing oscillation about the origin on the x-axis. Its potential energy is $U(x) = k|x|^3$, Where k is a positive constant. If the amplitude of oscillation is a then its time period T is *(1998)*

(A) Proportional to $1/\sqrt{a}$

(B) Independent of a

(C) Proportional to $\sqrt{a}$

(D) Proportional to $a^{3/2}$

Q.6 A spring of force constant k is cut into two pieces such that one piece is double the length of the other the long piece will have a force constant of *(1999)*

(A) 2/3 k   (B) 3/2 k   (C) 3k   (D) 6k

Q.7 A particle free to move along the x−axis has potential energy by $U(x) = k[1 - \exp(-x^2)]$ for $-\infty \le x \le +\infty$ Where k is a positive constant of appropriate dimensions. Then *(1999)*

(A) At points away from the origin, the particle is in unstable equilibrium

(B) For any finite non-zero value of x, there is a force directed away from the origin

(C) If its total mechanical energy is k/2, it has its minimum kinetic energy at the origin

(D) For small displacements from x=0, the motion is simple harmonic

Q.8 The period of oscillation of simple pendulum of length L suspended from the roof of the vehicle which moves without friction, down an inclined plane of inclination α, is given by *(2000)*

(A) $2\pi\sqrt{\dfrac{L}{g\cos\alpha}}$ (B) $2\pi\sqrt{\dfrac{L}{g\sin\alpha}}$

(C) $2\pi\sqrt{\dfrac{L}{g}}$ (D) $2\pi\sqrt{\dfrac{L}{g\tan\alpha}}$

Q.9 A particle executes simple harmonic motion between x= -A and x= + A. The time taken for it to go from O to A/2 is T_1 and to go from A/2 to A is T_2, then *(2001)*

(A) $T_1 < T_2$ (B) $T_1 > T_2$

(C) $T_1 = T_2$ (D) $T_1 = 2T_2$

Q.10 For a particle executing SHM the displacement x is given by x=A $\cos\omega t$. Identify the graph which represents the variation of potential energy (PE) as a function of time t and displacement. *(2003)*

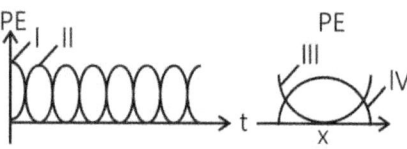

(A) I, III   (B) II, IV   (C) II, III   (D) I, IV

Q.11 A block P of mass m is placed on a horizontal frictionless plane. A second block of same mass m is placed on it and is connected to a spring of spring constant k, the two blocks are pulled by a distance A. Block Q oscillates without slipping. What is the maximum value of frictional force between the two blocks? *(2004)*

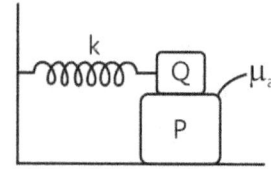

(A) kA

(B) kA

(C) $\mu_s mg$

(D) Zero

Q.12 A simple pendulum has time period T_1. The point of suspension is now moved upward according to the relation $y = kt^2, (k = 1m/s^2)$ where y is the vertical displacement.

The time period now beomes T_2.

The ratio of $\dfrac{T_1^2}{T_2^2}$ is   (Take g $=10m/s^2$)   *(2005)*

(A) 6/5   (B) 5/6   (C) 1   (D) 4/5

Q.13 The x-t graph of a particle undergoing simple harmonic motion is shown below. The acceleration of the particle at t=4/3 s is *(2009)*

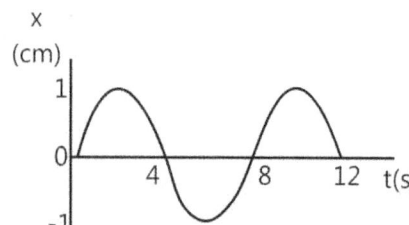

(A) $\dfrac{\sqrt{3}}{32}\pi^2$ cms^{-2} (B) $\dfrac{-\pi^2}{32}$ cms^{-2}

(C) $\dfrac{\pi^2}{32}$ cms^{-2} (D) $-\dfrac{\sqrt{3}}{32}\pi^2$ cms^{-2}

Q.14 A uniform rod of length L and mass M is pivoted at the center. Its two ends are attached to two springs of equal spring constants k. The spring are fixed to rigid supports as shown in the Fig, and rod is free to oscillate in the horizontal plane. The rod is gently pushed through a small angle θ in one direction and released. The frequency of oscillation is *(2009)*

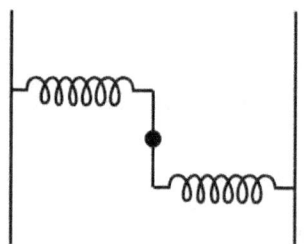

(A) $\dfrac{1}{2\pi}\sqrt{\dfrac{2k}{M}}$ (B) $\dfrac{1}{2\pi}\sqrt{\dfrac{k}{M}}$

(C) $\dfrac{1}{2\pi}\sqrt{\dfrac{6k}{M}}$ (D) $\dfrac{1}{2\pi}\sqrt{\dfrac{24k}{M}}$

Q.15 The mass M shown in the figure oscillates in simple harmonic motion with amplitude A. The amplitude of the point P is *(2009)*

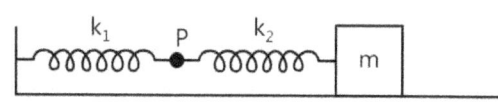

(A) $\dfrac{k_1 A}{k_2}$ (B) $\dfrac{k_2 A}{k_1}$

(C) $\dfrac{k_1 A}{k_1 + k_2}$ (D) $\dfrac{k_2 A}{k_1 + k_2}$

Q.16 A wooden block performs SHM on a frictionless surface with frequency v_0. The block carries a charge +Q on its surface. If now a uniform electric field $\vec{E}$ is switched-on as shown, then the SHM of the block will be *(2011)*

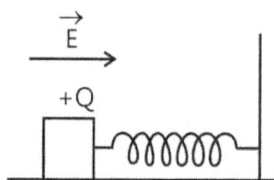

(A) Of the same frequency and with shifted mean position

(B) Of the same frequency and with the same mean position

(C) Of changed frequency and with shifted mean position

(D) Of changed frequency and with the same mean position

Q.17 A point mass is subjected to two simultaneous sinusoidal $\sqrt{2A,\dfrac{3\pi}{4}}$ displacements in x-direction

$x_1(t) = A \sin \omega t$ and $x_2(t) = A \sin\left(\omega t + \dfrac{2\pi}{3}\right)$.

Adding a third sinusoidal displacement

$x_3(t) = B \sin(\omega t + \phi)$ brings the mass to a complete rest. The values of B and ϕ are **(2011)**

(A) $A, \dfrac{6\pi}{3}$

(B) $A, \dfrac{4\pi}{3}$

(C) $\sqrt{3A, \dfrac{5\pi}{6}}$

(D) $A, \dfrac{\pi}{3}$

Q.18 If a simple pendulum has significant amplitude (up to a factor of 1/e of original) only in the period between t = 0s to t = τs, then τ may be called the average life of the pendulum. When the spherical bob of the pendulum suffers a retardation (due to viscous drag) proportional to its velocity, with ' b' as the constant of proportionality, the average life time of the pendulum is (assuming damping is small) in seconds: **(2012)**

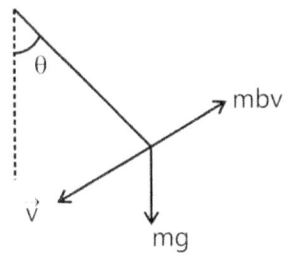

(A) $\dfrac{0.693}{b}$

(B) b

(C) $\dfrac{1}{b}$

(D) $\dfrac{2}{b}$

Q.19 The amplitude of a damped oscillator decreases to0.9 times its original magnitude is 5s. In another 10s it will decrease to α times its original magnitude, where α equals. **(2013)**

(A) 0.81   (B) 0.729   (C) 0.6   (D) 0.7

Q.20 For a simple pendulum, a graph is plotted between its kinetic energy (KE) and potential energy (PE) against its displacement d. Which one of the following represents these correctly?

(Graphs are schematic and not drawn to scale) **(2015)**

(A)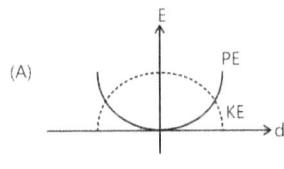

(B)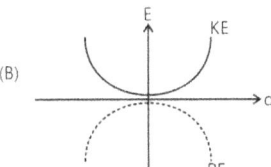

(C)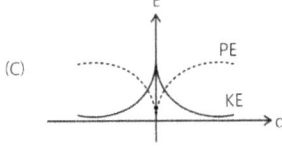

(D)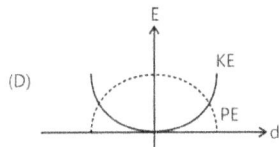

Q.21 A particle performs simple harmonic motion with amplitude A. Its speed is trebled at the instant that it is at a distance $\dfrac{2A}{3}$ from equilibrium position. The new amplitude of the motion is: **(2016)**

(A) 3 A

(B) $A\sqrt{3}$

(C) $\dfrac{7A}{3}$

(D) $\dfrac{A}{3}\sqrt{41}$

Exercise 1

Q.1 A body is in SHM with period T when oscillated from a freely suspended spring. If this spring is cut in two parts of length ratio 1:3 & again oscillated from the two parts separately, then the periods are T_1 & T_2 then find T_1/T_2.

Q.2 A body undergoing SHM about the origin has its equation is given by $x = 0.2\cos 5\pi t$. Find its average speed from $t = 0$ to $t = 0.7$ sec.

Q.3 Two particles A and B execute SHM along the same line with the same amplitude a, same frequency and same equilibrium position O. If the phase difference between them is $\phi = 2\sin^{-1}(0.9)$, then find the maximum distance between the two.

Q.4 The acceleration-displacement $(a - x)$ graph of a particle executing simple harmonic motion is shown in the figure. Find the frequency of oscillation.

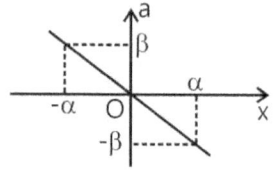

Q.5 A point particle of mass 0.1kg is executing SHM with amplitude of 0.1m. When the particle passes through the mean position, its K.E. is 8×10^{-3} J. Obtain the equation of motion of this particle if the initial phase of oscillation is $45°$.

Q.6 One end of an ideal spring is fixed to a wall at origin O and the axis of spring is parallel to x-axis. A block of mass m=1 kg is attached to free end of the spring and it is performing SHM. Equation of position of block in coordinate system shown is $x = 10 + 3\sin 10t$, is in second and x in cm. Another block of mass M=3kg, moving towards the origin with velocity 30cm/s collides with the block performing SHM at t=0 and gets stuck to it, calculate:

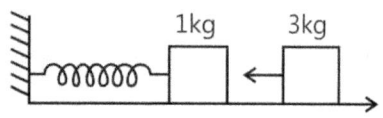

(i) New amplitude of oscillations.

(ii) New equation for position of the combined body.

(iii) Loss of energy during collision. Neglect friction.

Q.7 A mass M is in static equilibrium on a massless vertical spring as shown in the figure. A ball of mass m dropped from certain height sticks to the mass M after colliding with it. The oscillations they perform reach to height 'a' above the original level of scales & depth 'b' below it.

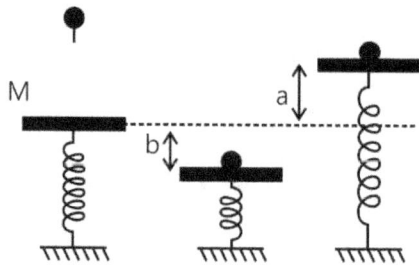

(a) Find the constant of force of the spring;

(b) Find the oscillation frequency.

(c) What is the height above the initial level from which the mass m was dropped?

Q.8 Two identical balls A and B each of mass 0.1 kg are attached to two identical massless springs. The spring mass system is constrained to move inside a rigid smooth pipe in the form of a circle as in figure. The pipe is fixed in a horizontal plane. The centers of the ball can move in a circle of radius 0.06m. Each spring has a natural length 0.06πm and force constant 0.1N/m. Initially both the balls are displaced by an angle of $\theta = \pi / 6$ radian with respect to diameter PQ of the circle and released from rest

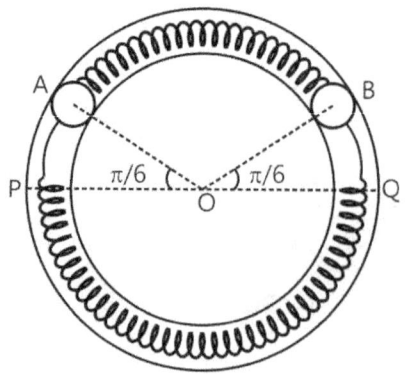

(a) Calculate the frequency of oscillation of the ball B.

(b) What is the total energy of the system?

(c) Find the speed of the ball A when A and B are at the two ends of the diameter PQ.

Q.9 Two blocks A(2kg) and B(3kg) rest up on a smooth horizontal surface are connected by a spring of stiffness 120 N/m. Initially the spring is unreformed. A is imparted a velocity of 2m/s along the line of the spring away from B. Find the displacement of A, t seconds later.

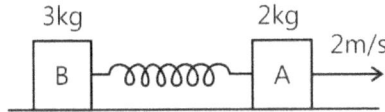

Q.10 A force $F = 10x + 2$ acts on a particle of mass 0.1 kg, where 'k' is in m and F in newton. If it is released from rest at $x = 0.2m$, find :

(a) Amplitude; (b) time period; (c) equation of motion.

Q.11 Potential Energy (U) of a body of unit mass moving in one-dimension conservative force field is given by, $U = (x^2 - 4x + 3)$. All units are in S.I.

(i) Find the equilibrium position of the body.

(ii) Show that oscillations of the body about this equilibrium position are simple harmonic motion & find its time period.

(iii) Find the amplitude of oscillations if speed of the body at equilibrium position is $2\sqrt{6}$ m/s.

Q.12 A body is executing SHM under the action of force whose maximum magnitude is 50N. Find the magnitude of force acting on the particle at the time when its energy is half kinetic and half potential.

Q.13 The system shown in the figure can move on a smooth surface. The spring is initially compressed by 6cm and then released. Find

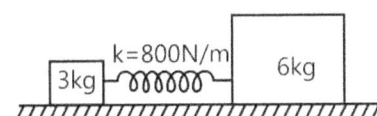

(a) Time period

(b) Amplitude of 3kg block

(c) Maximum momentum of 6kg block

Q.14 The resulting amplitude A' and the phase of the vibrations δ

$$S = A\cos(\omega t) + \frac{A}{2}\cos\left(\omega t + \frac{\pi}{2}\right) + \frac{A}{4}\cos(\omega t + \pi)$$
$$+ \frac{A}{8}\cos\left(\omega t + \frac{3\pi}{2}\right) = A'\cos(\omega t + \delta)$$

are _____ and _____ respectively.

Q.15 A spring block (force constant k=1000N/m and mass m=4kg) system is suspended from the ceiling of an elevator such that block is initially at rest. The elevator begins to move upwards at t=0. Acceleration time graph of the elevator is shown in the figure. Draw the displacement x (from its initial position taking upwards as positive) vs time graph of the block with respect to the elevator starting from t=0 to t=1 sec. Take $\pi^2 = 10$.

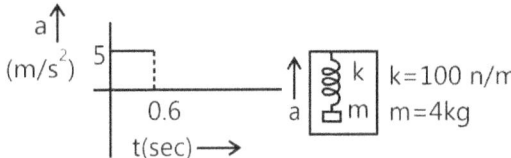

Q.16 A particle of mass m moves in the potential energy U shown below. Find the period of the motion when the particle has total energy E.

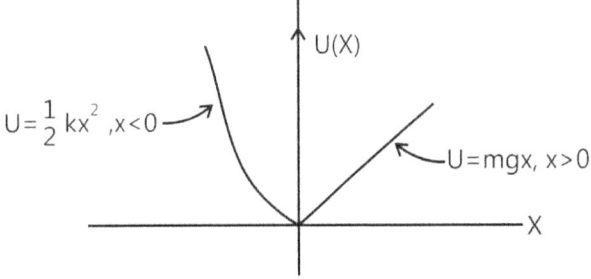

Q.17 The motion of a particle is described by $x = 30 \sin(\pi t + \pi/6)$, where x is in cm and t in sec. Potential energy of the particle is twice of kinetic energy for the first time after t=0 when the particle is at position _____ after _____ time.

Q.18 Two blocks A (5kg) and B (2kg) attached to the ends of a spring constant 1120N/m are placed on a smooth horizontal plane with the spring undeformed. Simultaneously velocities of 3m/s and 10m/s along the line of the spring in the same direction are imparted to A and B then

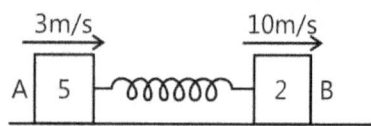

(a) Find the maximum extension of the spring.

(b) When does the first maximum compression occurs after start.

Q.19 Two identical rods each of mass m and length L, are rigidly joined and then suspended in a vertical plane so as to oscillate freely about an axis normal to the plane of paper passing through 'S' (point of suspension). Find the time period of such small oscillations.

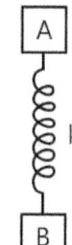

Q.20 (a) Find the time period of oscillations of a torsional pendulum, if the torsional constant of the wire is $K = 10\pi^2$ J/rad. The moment of inertia of rigid body is 10kg-m^2 about the axis of rotation.

(b) A simple pendulum of length $I = 0.5$m is hanging from ceiling of a car. The car is kept on a horizontal plane The car starts accelerating on the horizontal road with acceleration of 5m/s^2. Find the time period of oscillations of the pendulum for small amplitudes about the mean position.

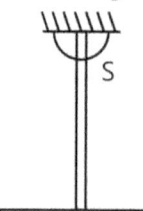

Q.21 An object of mass 0.2kg executes SHM along the x-axis with frequency of $(25/\pi)$Hz. At the point $x = 0.04$m the object has KE 0.5 J and PE 0.4 J. The amplitude of oscillation is _____.

Q.22 A body of mass 1kg is suspended from a weightless spring having force constant 600N/m. Another body of mass 0.5 kg moving vertically upwards hits the suspended body with a velocity of 3.0m/s and get embedded in it. Find the frequency of oscillations and amplitude of motion.

Q.23 A body A of mass $m_1 = 1$kg and a body B of mass $m_2 = 4$kg are attached to the ends of a spring. The body a performs vertical simple harmonic oscillations of amplitude a=1.6 cm and angular frequency $\omega = 25$ rad/s. Neglecting the mass of the spring determine the maximum and minimum values of force the system exerts on the surface on which it rests. [Take $g = 10$m/s^2]

Q.24 A spring mass system is hanging from the ceiling of an elevator in equilibrium Elongation of spring is l. The elevator suddenly starts accelerating downwards with accelerating g/3 find

(a) The frequency and

(b) The amplitude of the resulting SHM.

Exercise 2

Single Correct Choice Type

Q.1 A particle executes SHM on a straight line path. The amplitude of oscillation is 2 cm. When the displacement of the particle from the mean position is 1 cm, the numerical value of magnitude of acceleration is equal to the numerical value of magnitude of velocity. The frequency of SHM (in second^{-1}) is:

(A) $2\pi\sqrt{3}$ (B) $\dfrac{2\pi}{\sqrt{3}}$ (C) $\dfrac{\sqrt{3}}{2\pi}$ (D) $\dfrac{1}{2\pi\sqrt{3}}$

Q.2 A particle executed SHM with time period T and amplitude A. The maximum possible average velocity in time $\dfrac{T}{4}$ is

(A) $\dfrac{2A}{T}$ (B) $\dfrac{4A}{T}$ (C) $\dfrac{8A}{T}$ (D) $\dfrac{4\sqrt{2}A}{T}$

Q.3 A particle performs SHM with a period T and amplitude a. The mean velocity of the particle over the time interval during which it travels a distance a/2 from the extreme position is

(A) a/T   (B) 2a/T   (C) 3a/T   (D) a/2T

Q.4 Two particles are in SHM on same straight line with amplitude A and 2A and with same angular frequency ω. It is observed that when first particle is at a distance $A/\sqrt{2}$ from origin and going toward mean position, other particle is at extreme position on other side of mean position. Find phase difference between the two particles

(A) 45°   (B) 90° (C) 135°   (D) 180°

Q.5 A body performs simple harmonic oscillations along the straight line ABCDE with C as the midpoint of AE. Its kinetic energies at B and D are each one fourth of its maximum value. If AE=2R, the distance between B and D is

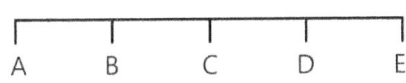

(A) $\dfrac{\sqrt{3}R}{2}$ (B) $\dfrac{R}{\sqrt{2}}$

(C) $\sqrt{3}R$ (D) $\sqrt{2}R$

Q.6 In an elevator, a spring clock of time period T_s (mass attached to a spring) and a pendulum clock of time period T_p are kept. If the elevator accelerates upwards

(A) T_s well as T_p increases

(B) T_s remain same, T_p increases

(C) T_s remains same, T_p decreases

(D) T_s as well as T_p decreases

Q.7 Two bodies P & Q of equal mass are suspended from two separate massless springs of force constants k_1 and k_2 respectively. If the maximum velocities of them are equal during their motion, the ratio of amplitude of P to Q is:

(A) $\dfrac{k_1}{k_2}$ (B) $\sqrt{\dfrac{k_2}{k_1}}$

(C) $\dfrac{k_2}{k_1}$ (D) $\sqrt{\dfrac{k_1}{k_2}}$

Q.8 The spring in figure. A and B are identical but length in A is three times each of that in B. the ratio of period T_A/T_B is

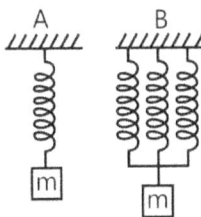

(A)$\sqrt{3}$ (B) 1/3   (C) 3   (D) 1/$\sqrt{3}$

Q.9 In the figure the block of mass m, attached to the spring of stiffness k is in contact with the completely elastic wall, and the compression in the spring is 'e'. The spring is compressed further by 'e' by displacing the block towards left and is then released. If the collision between the block and the wall is completely elastic then the time period of oscillations of the block will be:

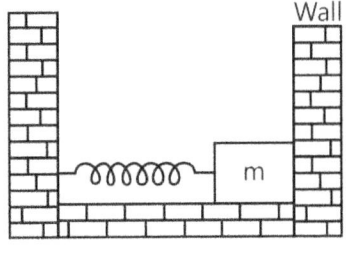

(A) $\dfrac{2\pi}{3}\sqrt{\dfrac{m}{k}}$ (B) $2\pi\sqrt{\dfrac{m}{k}}$

(C) $\dfrac{\pi}{3}\sqrt{\dfrac{m}{k}}$ (D) $\dfrac{\pi}{6}\sqrt{\dfrac{m}{k}}$

Q.10 A 2 kg block moving with 10 m/s strikes a spring of constant $\pi2$ N/m attached to 2 Kg block at rest kept on a smooth floor. The time for which rear moving block remain in contact with spring will be

(A) $\sqrt{2}$ sec   (B) $\dfrac{1}{\sqrt{2}}$ sec

(C) 1sec   (D) $\dfrac{1}{2}$ sec

Q.11 In the above question, the velocity of the rear 2 kg block after it separates from the spring will be:

(A) 0 m/s   (B) 5 m/s

(C) 10 m/s   (D) 7.5 m/s

Q.12 A rod whose ends are A & B of length 25 cm is hanged in vertical plane. When hanged from point A and point B the time periods calculated are 3 sec & 4 sec respectively. Given the moment of inertia of rod about axis perpendicular to the rod is in ratio 9:4 at points A and B. Find the distance of the center of mass from point A.

(A) 9 cm   (B) 5 cm   (C) 25 cm   (D) 20 cm

Q. 13 A circular disc has a tiny hole in it, at a distance z from its center. Its mass is M and radius R (R > z). A horizontal shaft is passed through the hole and held fixed so that the disc can freely swing in the vertical plane. For small disturbance, the disc performs SHM whose time period is the minimum for z =

(A) R/2   (B) R/3

(C) R /$\sqrt{2}$ (D) R/$\sqrt{3}$

Multiple Correct Choice Type

Q.14 The displacement-time graph of a particle executing SHM is shown which of the following statement is/are true?

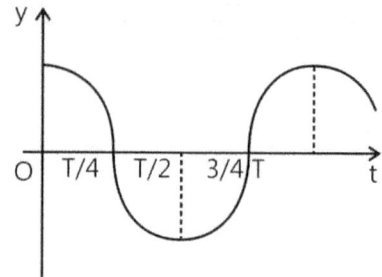

(A) The velocity is maximum at t=T/2

(B) The acceleration is maximum at t=T

(C) The force is zero at t= 3T/4

(D) The potential energy equals the oscillation energy at t=T/2.

Q.15 The amplitude of a particle executing SHM about O is 10 cm. Then:

(A) When the K.E. is 0.64 of its max. K.E. its displacement is 6cm from O.

(B) When the displacement is 5cm from O its K.E.is 0.75 of its max. P.E.

(C) Its total energy at any point is equal to its maximum K.E.

(D) Its velocity is half the maximum velocity when its displacement is half the maximum displacement.

Q.16 A particle of mass m performs SHM along a straight line with frequency f and amplitude A.

(A) The average kinetic energy of the particle is zero.

(B) The average potential energy is $m\pi 2f2A^2$.

(C) The frequency of oscillation of kinetic energy is 2f.

(D) Velocity function leads acceleration by $\pi / 2$

Q.17 A system is oscillating with undamped simple harmonic motion. Then the

(A) Average total energy per cycle of the motion is its maximum kinetic energy.

(B) Average total energy per cycle of the motion is $\dfrac{1}{\sqrt{2}}$ times its maximum kinetic energy.

(C) Root means square velocity $\dfrac{1}{\sqrt{2}}$ times its maximum velocity.

(D) Mean velocity is $\dfrac{1}{2}$ of maximum velocity.

Q.18 A spring has natural length 40 cm and spring constant 500 N/m. A block of mass 1 kg is attached at one end of the spring and other end of the spring is attached to ceiling. The block released from the position, where the spring has length 45cm.

(A) The block will performs SHM of amplitude 5 cm.

(B) The block will have maximum velocity $30\sqrt{5}$ cm / sec.

(C) The block will have maximum acceleration $15 m / s^2$

(D) The minimum potential energy of the spring will be zero.

Q.19 The figure shows a graph between velocity and displacement (from mean position) of a particle performing SHM:

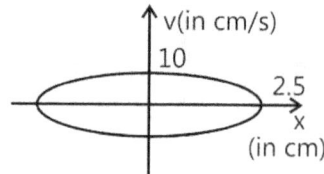

(A) The time period of the particle is 1.57s

(B) The maximum acceleration will be $40 cm / s^2$

(C) The velocity of particle is $2\sqrt{21}$ cm / s when it is at a distance 1 cm from the mean position.

(D) None of these

Q.20 Two blocks of masses 3 kg and 6 kg rest on a horizontal smooth surface. The 3 kg block is attached to A Spring with a force constant

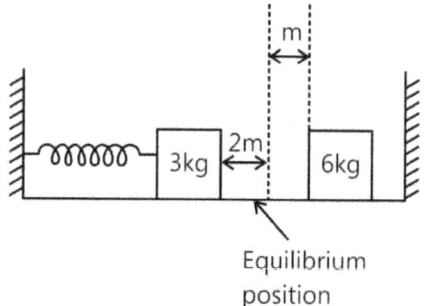

Equilibrium position

$k = 900 Nm^{-1}$ Which is compressed 2 m from beyond the equilibrium position. The 6 kg mass is at rest at 1m from mean position 3kg mass strikes the 6kg mass and the two stick together.

(A) Velocity of the combined masses immediately after the collision is $10 ms^{-1}$

(B) Velocity of the combined masses immediately after the collision is 5ms^{-1}

(C) Amplitude of the resulting oscillations is $\sqrt{2}\,m$

(D) Amplitude of the resulting oscillation is $\sqrt{\dfrac{5}{2}}\,m$.

Q.21 A particle is executing SHM with amplitude A, time period T, maximum acceleration a_0 and maximum velocity v_0. Its starts from mean position at t-0 and at time t, it has the displacement A/2, acceleration a and velocity v then

(A) t=T/12

(B) $a = a_0 / 2$

(C) $v = v_0 / 2$

(D) t=T/8

Q.22 For a particle executing SHM, x=displacement from equilibrium position, v= velocity at any instant and a = acceleration at any instant, then

(A) v-x graph is a circle

(B) v-x graph is an ellipse

(C) a-x graph is a straight line

(D) a-v graph is an ellipse

Q.23 A particle starts from a point P at a distance of A/2 from the mean position O & travels towards left as shown in the figure. If the time period of SHM, executed about O is T and amplitude A then the equation of motion of particle is:

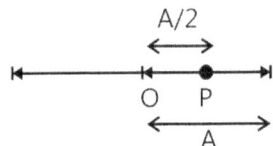

(A) $x = A\sin\left(\dfrac{2\pi}{T}t + \dfrac{\pi}{6}\right)$

(B) $x = A\sin\left(\dfrac{2\pi}{T}t + \dfrac{5\pi}{6}\right)$

(C) $x = A\cos\left(\dfrac{2\pi}{T}t + \dfrac{\pi}{6}\right)$

(D) $x = A\cos\left(\dfrac{2\pi}{T}t + \dfrac{\pi}{3}\right)$

Q.24 Two particles execute SHM with amplitude A and 2A and angular frequency ω and 2ω respectively. At t=0 they starts with some initial phase difference. At, t= difference is: $\dfrac{2\pi}{3\omega}$. They are in same phase. Their initial phase

(A) $\dfrac{\pi}{3}$

(B) $\dfrac{2\pi}{3}$

(C) $\dfrac{4\pi}{3}$

(D) π

Q.25 A mass of 0.2 kg is attached to the lower end of a massless spring of force-constant 200 N/m, the upper end of which is fixed to a rigid support. Which of the following statements is/are true?

(A) In equilibrium, the spring will be stretched by 1cm.

(B) If the mass is raised till the spring is in not stretched state and then released, it will go down by 2 cm before moving upwards.

(C) The frequency of oscillation will be nearly 5 Hz.

(D) If the system is taken to moon, the frequency of oscillation will be the same as on the earth.

Q.26 The potential energy of particle of mass 0.1kg, moving along x-axis, is given by U=5x(x-4)J where x is in meters. It can be concluded that

(A) The particle is acted upon by a constant force.

(B) The speed of the particle is maximum at x=2m

(C) The particle executes simple harmonic motion

(D) The period of oscillation of the particle is π /5 s

Q.27 The displacement of a particle varies according to the relation x=3 sin 100t + $\cos^2$ 50t. Which of the following is/are correct about this motion.

(A) The motion of the particle is not SHM

(B) The amplitude of the SHM of the particle is 5 units

(C) The amplitude of the resultant SHM is $\sqrt{73}$ units.

(D) The maximum displacement of the particle from the origin is 9 units.

Q.28 The equation of motion for an oscillating particle is given by x=3sin (4πt) + 4cos (4πt), where x is in mm and t is in second

(A) The motion is simple harmonic

(B) The period of oscillation is 0.5 s

(C) The amplitude of oscillation is 5 mm

(D) The particle starts its motion from the equilibrium

Q.29 A linear harmonic oscillator of force constant $2\times10^6 \text{Nm}^{-1}$ and amplitude 0.01 m has a total mechanical energy of 160 J. Its

(A) Maximum potential energy is 100 J

(B) Maximum kinetic energy is 100 J

(C) Maximum potential energy is 160

(D) Minimum potential energy is zero.

Q.30 The two blocks shown here rest on a frictionless surface. If they are pulled apart by a small distance and released at t=0, the time when

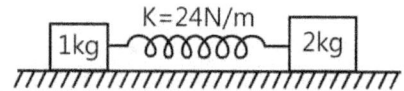

1 kg block comes to rest can be

(A) $\frac{2\pi}{3}$ sec

(B) π sec.

(C) $\frac{\pi}{2}$ sec

(D) $\frac{\pi}{9}$ sec

Assertion Reasoning Type

Q.31 Statement-I: A particle is moving along x-axis. The resultant force F acting on it at position x is given by F=-ax-b. Where a and b are both positive constants. The motion of this particle is not SHM.

Statement-II: In SHM restoring force must be proportional to the displacement from mean position.

(A) Statement-I is true, statement-II is true and statement-II is correct explanation for statement-I

(B) Statement-I is true, statement-II is true and statement-II is NOT the correct explanation for statement-I

(C) Statement-I is true, statement-II is false.

(D) Statement-I is false, statement-II is true.

Q.32 Statement-I: For a particle performing SHM, its speed decreases as it goes away from the mean position.

Statement-II: In SHM, the acceleration is always opposite to the velocity of the particle.

(A) Statement-I is true, statement-II is true and statement-II is correct explanation for statement-I.

(B) Statement-I is true, statement-II is true and Statement-II is NOT the correct explanation for statement-I

(C) Statement-I is true, statement-II is false.

(D) Statement-I is false, statement-II is true.

Q.33 Statement-I: Motion of a ball bouncing elastically in vertical direction on a smooth horizontal floor is a periodic motion but not an SHM.

Statement-II: Motion is SHM when restoring force is proportional to displacement from mean position.

(A) Statement-I is true, statement-II is true and statement-II is correct explanation for statement-I

(B) Statement-I is true, statement-II is true and statement-II is NOT the correct explanation for statement-I

(C) Statement-I is true, statement-II is false.

(D) Statement-I is false, statement-II is true

Q.34 Statement-I: A particle, simultaneously subjected to two simple harmonic motions of same frequency and same amplitude, will perform SHM only if two SHM's are in the same direction

Statement-II: A particle, simultaneously subjected to two simple harmonic motions of same frequency and same amplitude, perpendicular to each other the particle can be in uniform circular motion.

(A) Statement-I is true, statement-II is true and statement-II is correct explanation for statement-I

(B) Statement-I is true, statement-II is true and statement-II is NOT the correct explanation for statement-I.

(C) Statement-I is true, statement-II is false.

(D) Statement-I is false, statement-II is true.

Q.35 Statement-I: In case of oscillatory motion the average speed for any time interval is always greater than or equal to its average velocity.

Statement-II: Distance travelled by a particle cannot be less than its displacement.

(A) Statement-I is true, statement-II is true and statement-II is correct explanation for statement-I

(B) Statement-I is true. statement-II is true and statement-II is NOT the correct explanation for statement-I.

(C) Statement-I is true, statement-II is false.

(D) Statement-I is false, statement-II is true.

Comprehension Type

Paragraph 1: When force acting on the particle is of nature $F = -kx$, motion of particle is SHM, Velocity at extreme is zero while at mean position it is maximum. In case of acceleration situation is just reverse. Maximum displacement of particle from mean position on both sides is same and is known as amplitude. Refer to figure One kg block performs vertical harmonic oscillations with amplitude 1.6 cm and frequency 25 rad s^{-1}.

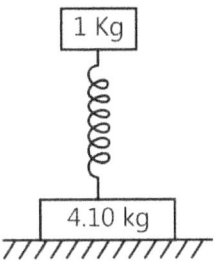

Q.36 The maximum value of the force that the system exerts on the surface is

(A) 20 N   (B) 30 N   (C) 40 N   D) 60 N

Q.37 The minimum force is

(A) 20 N   (B) 30 N   (C) 40 N   (D) 60 N

Paragraph 2: The graphs in figure show that a quantity y varies with displacement d in a system undergoing simple harmonic motion.

(A) 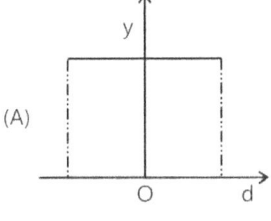   (B)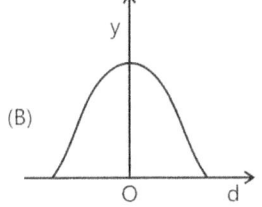

(C) 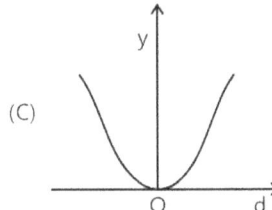   (D)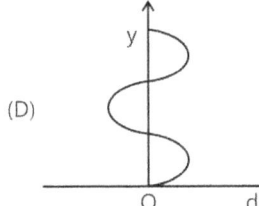

Which graphs best represents the relationship obtained when Y is

Q. 38 The total energy of the system

(A) I   (B) II   (C) III   (D) IV

Q.39 The time

(A) I   (B) II   (C) III   (D) IV

Q.40 The unbalanced force acting on the system

(A) I   (B) II   (C) III   (D) None

Match the Columns

Q.41 The graph plotted between phase angle (ϕ) and displacement of a particle from equilibrium position (y) is a sinusoidal curve as shown below. Then the best matching is

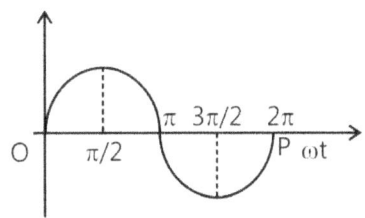

Column A	Column B
(a) K.E. versus phase angle curve	(i)
(b) P.E. versus phase angle curve	(ii)
(c) T.E. versus phase angle curve	(iii)
(d) Velocity versus phase angle curve	(iv)

(A) (a)-(i), (b)-(ii), (c)-(iii) & (d)-(iv)

(B) (a)-(ii), (b)-(i), (c)-(iii) & (d)-(iv)

(C) (a)-(ii), (b)-(i), (c)-(iv) & (d) – (iii)

(D) (a)-(ii), (b)-(iii), (c)-(iv) & (d)-(i)

Q.42 Column I is a list of possible set of parameters measured in some experiments. The variations of the parameters in the form of graphs are shown in Column II. Match the set of parameters given in Column I with the graphs given in Column II. Indicate your answer by darkening the appropriate bubbles of the 4 x 4 matrix given in the ORS.

Column I		Column II	
(A) Potential energy of a simple pendulum (y axis) as a function of displacement (x axis)		(p)	![parabola graph]
(B) Displacement (y axis) as a function of time (x axis) for a one dimensional motion at zero or constant acceleration when the body is moving along the positive x-direction.		(q)	![line through origin graph]
(C) Range of projectile (y axis) as a function of its velocity (x axis) when projected at a fixed angle.		(r)	![line with positive intercept graph]
(D) The square of the time period (y axis) of a simple pendulum as a function of its length (x axis)		(s)	![increasing curve graph]

Previous Years' Questions

Paragraph 1: When a particle of mass m moves on the x-axis in a potential of the form $V(x) = kx^2$, it performs simple harmonic motion. The corresponding time period is proportional to $\sqrt{\dfrac{m}{k}}$, as can be seen easily using dimensional analysis. However, the motion of a particle can be periodic even when its potential energy increases on both sides of $x = 0$ in a way different from kx^2 and its total energy is such that the particle does not escape to infinity. Consider a particle of mass m moving on the x-axis. Its potential energy is $v(x) = \alpha x^2 (\alpha > 0)$ for $|x|$ near the origin and becomes a constant equal to V_0 for $|x| \geq X_0$ (see figure below) **(2010)**

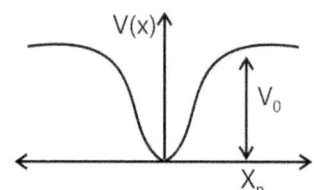

Q.1 If the total energy of the particle is E, it will perform periodic motion only if

(A) $E < 0$ (B) $E > 0$

(C) $V_0 > E > 0$ (D) $E > V_0$

Q.2 For periodic motion of small amplitude A, the time period t of this particle is proportional to

(A) $A\sqrt{\dfrac{m}{\alpha}}$ (B) $\dfrac{1}{A}\sqrt{\dfrac{m}{\alpha}}$

(C) $A\sqrt{\dfrac{\alpha}{m}}$ (D) $\dfrac{1}{A}\sqrt{\dfrac{\alpha}{m}}$

Q.3 The acceleration of this particle for $|x| > X_0$ is

(A) Proportional to V_0

(B) Proportional to $\dfrac{V_0}{mX_0}$

(C) Proportional to $\sqrt{\dfrac{V_0}{mX_0}}$

(D) Zero

Q.4 A small block is connected to one end of a massless spring of un-stretched length 4.9 m. The other end of the spring (see the figure) is fixed. The system lies on a horizontal frictionless surface. The block is stretched by 0.2 m and released from rest at t = 0. It then executes simple harmonic motion with angular frequency $\omega = \dfrac{\pi}{3}$ rad/s. Simultaneously at t = 0, a small pebble is projected with speed v form point P at an angle of 45° as shown in the figure. Point P is at a horizontal distance of 10 m from O. If the pebble hits the block at t = 1 s, the value of v is (take g = 10 m/s²) **(2012)**

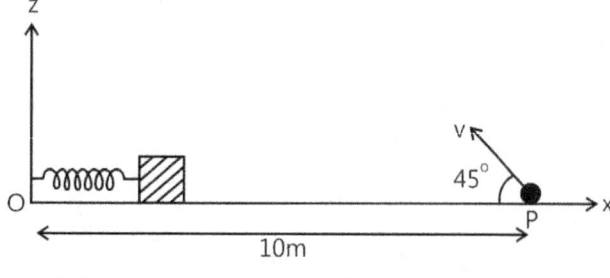

(A) $\sqrt{50}$ m/s (B) $\sqrt{51}$ m/s

(C) $\sqrt{52}$ m/s (D) $\sqrt{53}$ m/s

Q.5 A particle of mass m is attached to one end of a mass-less spring of force constant k, lying on a frictionless horizontal plane. The other end of the spring is fixed. The particle starts moving horizontally from its equilibrium position at time t = 0 with an initial velocity u_0. When the speed of the particle is 0.5 u_0. It collides elastically with a rigid wall. After this collision, **(2013)**

(A) The speed of the particle when it returns to its equilibrium position is u_0

(B) The time at which the particle passes through the equilibrium position for the first time is $t = \pi\sqrt{\dfrac{m}{k}}$.

(C) The time at which the maximum compression of the spring occurs is $t = \dfrac{4\pi}{3}\sqrt{\dfrac{m}{k}}$

(D) The time at which the particle passes through the equilibrium position for the second time is $t = \dfrac{5\pi}{3}\sqrt{\dfrac{m}{k}}$

Q.6 Two independent harmonic oscillators of equal mass are oscillating about the origin with angular frequencies ω_1 and ω_2 and have total energies E_1 and E_2, respectively. The variations of their momenta p with positions x are shown in the figures. If $\dfrac{a}{b} = n^2$ and $\dfrac{a}{R} = n$, then the correct equation(s) is(are) **(2015)**

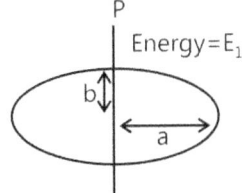
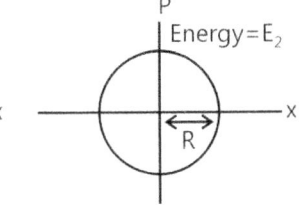

(A) $E_1\,\omega_1 = E_2\,\omega_2$

(B) $\dfrac{\omega_2}{\omega_1} = n^2$

(C) $\omega_1\,\omega_2 = n^2$

(D) $\dfrac{E_1}{\omega_1} = \dfrac{E_2}{\omega_2}$

Q.7 A block with mass M is connected by a massless spring with stiffness constant k to a rigid wall and moves without friction on a horizontal surface. The block oscillates with small amplitude A about an equilibrium position x_0. Consider two cases: (i) when the block is at x_0; and (ii) when the block is at $x = x_0 + A$. In both the cases, a particle with mass m (< M) is softly placed on the block after which they stick to each other. Which of the following statement(s) is (are) true about the motion after the mass m is placed on the mass M? **(2016)**

(A) The amplitude of oscillation in the first case changes by a factor of $\sqrt{\dfrac{M}{m+M}}$, whereas in the second case it remains unchanged

(B) The final time period of oscillation in both the cases is same

(C) The total energy decreases in both the cases

(D) The instantaneous speed at x_0 of the combined masses decreases in both the cases

Q.8 Column I describes some situations in which a small object moves. Column II describes some characteristics of these motions. Match the situations in column I with the characteristics in column II. **(2007)**

Column I	Column II
(A) The object moves on the x-axis under a conservative force in such a way that its speed and position satisfy $v = c_1\sqrt{c_2 - x^2}$, where c_1 and c_2 are positive constants.	(p) The object executes a simple harmonic motion.
(B) The object moves on the x-axis in such a way that its velocity and its displacement from the origin satisfy $v = -kx$, where k is a positive constant.	(q) The object does not change its direction.
(C) The object is attached to one end of a mass-less spring of a given spring constant. The other end of the spring is attached to the ceiling of an elevator. Initially everything is at rest. The elevator starts going upwards with a constant acceleration a. The motion of the object is observed from the elevator during the period it maintain this acceleration.	(r) The kinetic energy of the object keeps on decreasing.
(D) The object is projected from the earth's surface vertically upwards with a speed $2\sqrt{\dfrac{GM_e}{R_e}}$, where M_e is the mass of the earth and R_e is the radius of the earth. Neglect forces from objects other than the earth.	(s) The object can change its direction only once.

Q.9 A linear harmonic oscillator or force constant 2×10^6 N/m and amplitude 0.01m has a total mechanical energy of 160 J. Its *(1989)*

(A) Maximum potential energy is 100 J

(B) Maximum kinetic energy is 100J

(C) Maximum potential energy is 160J

(D) Maximum potential energy is zero

Q.10 Three simple harmonic motions in the same direction having the same amplitude and same period are superposed. If each differ in phase from the next by $45°$, then *(1999)*

(A) The resultant amplitude is $\left(1+\sqrt{2}\right)a$

(B) The phase of the resultant motion relative to the first is 90°

(C) The energy associated with the resulting motion is $\left(3+2\sqrt{2}\right)$ times the energy associated with any single motion

(D) The resulting motion is not simple harmonic

Q.11 Function $x = A\sin^2 \omega t + B\cos^2 \omega t + C\sin \omega t\cos \omega t$ represent SHM *(2006)*

(A) For any value of A, B and C (except C=0)

(B) If A=-B, C=2B, amplitude=$\left|B\sqrt{2}\right|$

(C) If A=B; C=0

(D) If A=B; C=2B, amplitude=$|B|$

Q.12 A metal rod of length L and mass m is pivoted at one end. A thin disk of mass M and radius R $\left(<L\right)$ is attached at its center to the free end of the rod. Consider two ways the disc is free to rotate about its center. The rod-disc system performs SHM in vertical plane after being released from the same displaced position. Which of the following statement(s) is/are true? *(2011)*

(A) Restoring torque in case A=Restoring torque in case B

(B) Restoring torque in case A<Restoring torque in case B

(C) Angular frequency for case A>Angular frequency for case B

(D) Angular frequency for case A<, angular frequency for case B

Important Questions

JEE Main/Boards

Exercise 1

Exercise 2

Previous Years' Questions

JEE Advanced/Boards

Exercise 1

Exercise 2

Answer Key

JEE Main/Boards

Exercise 1

Q.1 0.5 rad

Q.2 1.048 s

Q.3 0.61 rad

Q.4 0.5s

Q.5 0.71 a

Q.6 7.56J

Q.7 $0.35\,ms^{-1}$, 0.06 J

Q.8 1.5J

Q.9 86.4s

Q.10 141°.4′

Q.11 $0.612\,s^{-1}$

Q.18 (a) $y = 2\sin\left(\dfrac{2\pi t}{3} + \dfrac{\pi}{3}\right)$ (b) $y = 3\cos\left(\dfrac{\pi}{30}t\right)$

Q.19 $2\pi/3\,rad$

Q.20 (a) 6m (b) $-48.99\,ms^{-1}$

Q.21 19.67m

Q.22 4.86s; 5.06m

Q.23 $2\sqrt{3}\,cm; 2\sqrt{2}\,cm$

Q.24 0.1875 J; 0.0625 J, 0.25J

Q.25 3.2 Hz

Q.26 $\dfrac{1}{(2\pi)}\left[\dfrac{k_1k_2(k_3+k_4)}{\{(k_1+k_2)\times(k_3+k_4)+k_1k_2\}m}\right]^{1/2}$

Exercise 2

Single Correct Choice Type

Q.1 A	**Q.2** C	**Q.3** C	**Q.4** D	**Q.5** A	**Q.6** C
Q.7 C	**Q.8** B	**Q.9** C	**Q.10** B	**Q.11** B	**Q.12** C
Q.13 A	**Q.14** B	**Q.15** C			

Previous Years' Questions

Q.1 C	**Q.2** B	**Q.3** D	**Q.4** B	**Q.5** A	**Q.6** B
Q.7 D	**Q.8** A	**Q.9** A	**Q.10** A	**Q.11** A	**Q.12** A
Q.13 D	**Q.14** C	**Q.15** D	**Q.16** A	**Q.17** B	**Q.18** D
Q.19 B	**Q.20** C	**Q.21** C			

JEE Advanced/Boards

Exercise 1

Q.1 $1/\sqrt{3}$

Q.2 2 m/s

Q.3 1.8a

Q.4 $\dfrac{1}{2\pi}\sqrt{\dfrac{\beta}{\alpha}}$

Q.5 $x = 0.1\sin(4t + \pi/4)$

Q.6 $3cm$, $x = 10 - 3\sin 5t$; $\Delta E = 0.135J$

Q.7 (a) $k = \dfrac{2mg}{b-a}$: (c) $\dfrac{ab}{b-a}$, (b) $\dfrac{1}{2\pi}\sqrt{\dfrac{2mg}{(b-a)(M+m)}}$

Q.8 $f = \dfrac{1}{\pi}$; $E = 4\pi^2 \times 10^{-5}J$; $v = 2\pi \times 10^{-2}m/s$

Q.9 $0.8t + 0.12\sin 10t$

Q.10 (a) 0.4 m, (b) $\dfrac{\pi}{5}$ sec., (c) $x = 0.2 - 0.4\cos\omega t$

Q.11 (i) $x_0 = 2m$; (ii) $T = \sqrt{2}\pi$ sec.; (iii) $2\sqrt{3}$ m

Q.12 $25\sqrt{2}N$

Q.13 (a) $\dfrac{\pi}{10}$ sec, (b) $6cm$ (c) $2.40 kgm/s$.

Q.14 $\dfrac{3\sqrt{5}A}{8}\tan^{-1}\left(\dfrac{1}{2}\right)$

Q.15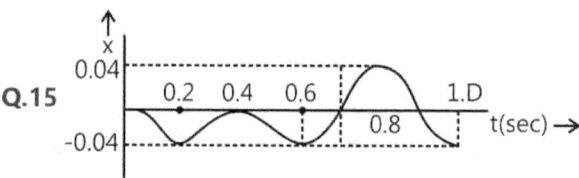

Q.16 $\pi\sqrt{m/k} + 2\sqrt{2E/mg^2}$

Q.17 $10\sqrt{6}cm$, $\dfrac{1}{\pi}\sin^{-1}\left(\sqrt{\dfrac{2}{3}}\right) - \dfrac{1}{6}$ sec

Q.18 (a) $25cm$, (b) $3\pi/56$ seconds

Q.19 $2\pi\sqrt{\dfrac{17L}{18g}}$

Q.20 (a) 2 sec, (b) $T = \dfrac{2}{5^{1/4}}$ sec

Q.21 $0.06m$

Q.22 $10\pi Hz$, $\dfrac{5\sqrt{37}}{6}cm$

Q.23 $60N$, $40N$ **Q.24** (a) $\dfrac{1}{T} = \dfrac{1}{2\pi}\sqrt{\dfrac{g}{L}}$, (b) $\dfrac{L}{3}$

Exercise 2

Single Correct Choice Type

Q.1 C	**Q.2** D	**Q.3** C	**Q.4** C	**Q.5** C	**Q.6** C
Q.7 B	**Q.8** C	**Q.9** A	**Q.10** C	**Q.11** A	**Q.12** D
Q.13 C					

Multiple Correct Choice Type

Q.14 B, C, D	**Q.15** A, B, C	**Q.16** B, C	**Q.17** A, C	**Q.18** B, C, D	**Q.19** A, B, C
Q.20 A, C	**Q.21** A, B	**Q.22** B, C, D	**Q.23** B, D	**Q.24** B, C	**Q.25** A, B, C, D
Q.26 B, C, D	**Q.27** B, D	**Q.28** A, B, C	**Q.29** B, C	**Q.30** A, B, C	

Assertion Reasoning Type

Q.31 D	**Q.32** C	**Q.33** A	**Q.34** D	**Q.35** A

Comprehension Type

Paragraph 1:   **Q.36** D   **Q.37** C

Paragraph 2:   **Q.38** A   **Q.39** D   **Q.40** D

Previous Years' Questions

Q.1 C	**Q.2** B	**Q.3** D	**Q.4** A
Q.5 A, D	**Q.6** B, D	**Q.7** A, B, D	**Q.8** A → p; B → q, r; C → p; D → r, q
Q.9 A	**Q.10** A, C	**Q.11** A, B, D	**Q.12** A, D

S.H.M Solutions

JEE Main/Boards

Exercise 1

Sol 1: $y(t) = 10 \sin(20t + 0.5)$

A = 10 m

ω = 20 rad./sec

ϕ = 0.5 radians

$f = \dfrac{w}{2\pi} = \dfrac{20}{2\pi} = \dfrac{10}{\pi}$ hz

$T = \dfrac{1}{f} = \dfrac{\pi}{10}$ sec

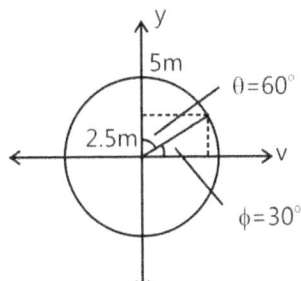

Sol 2: $V = \omega\sqrt{A^2 - y^2}$

$4 = \omega\sqrt{A^2 - 9}$

$3 = \omega\sqrt{A^2 - 16}$

$T = \dfrac{2\pi}{\omega} = 2\pi \sec \omega = 1$ sec

$t = \dfrac{\theta}{360°} \times T = \dfrac{60}{360°} \times 2\pi$ sec

$t = \dfrac{\pi}{3}$ sec

Sol 3: $y = 0.4 \sin(440\, t + 0.61)$

(i) Amplitude = 0.4 m

(ii) ω = 440 rad.sec

(iii) $f = \dfrac{\omega}{2\pi}, \dfrac{220}{\pi}$ hz

(iv) $T = \dfrac{1}{f} = \dfrac{2\pi}{\omega} = \dfrac{\pi}{220}$ sec

(v) Initial phase = 0.61 radians

Sol 4: A = 25 cm, T = 3s

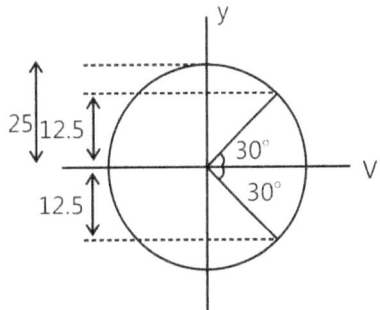

$\Rightarrow t = \dfrac{60°}{360°} \times = \dfrac{1}{2}$ sec.

Sol 5: Amplitude = 0

Total energy = $\dfrac{1}{2}$ Ka²

Potential energy = $\dfrac{1}{2}$ Kx²

$\dfrac{1}{2}$ Kx² = $\dfrac{1}{2} \times \dfrac{1}{2}$ Ka² $\Rightarrow$ x = $\dfrac{a}{\sqrt{2}}$

Sol 6: m = 8 kg

a = 30 cm

k × 0.3 = 60 $\Rightarrow$ k = $\dfrac{60}{0.3}$ = 200 n/m

T = $2\pi \sqrt{\dfrac{m}{k}}$ $\Rightarrow$ T = $2\pi \sqrt{\dfrac{8}{200}}$ = $\dfrac{2\pi}{5}$ = 0.4 π

(a) T = 0.4 π sec.

(b) a = $\dfrac{-k}{m}$ x

a = $\dfrac{-200}{84}$ × 0.12 = 3m/sec²

P.E. = $\dfrac{1}{2}$ kx² = $\dfrac{1}{2}$ × 200 × (0.12)² = 1.44 J

K.E. = $\dfrac{1}{2}$ k(A² – x²)

= 100 x (0.09 – 0.0144) = 7.56 J

Sol 7: k = 1200 Nm⁻¹

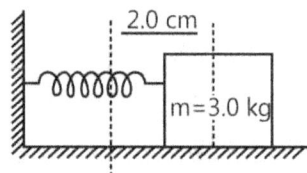

(a) w = $\sqrt{\dfrac{k}{m}}$ = $\sqrt{\dfrac{1200}{3}}$ = 20 rad/s.

v = $\omega \sqrt{A^2 - x^2}$ = 20 $\dfrac{\sqrt{4-1}}{100}$ = $\dfrac{\sqrt{3}}{5}$

v = 0.35 m/s

(b) P.E. = $\dfrac{1}{2}$ kx² = $\dfrac{1}{2}$ × 1200 × $\left(\dfrac{1}{100}\right)^2$

= 600 × $\dfrac{1}{100 \times 100}$

P.E. = 0.06 J

Sol 8:

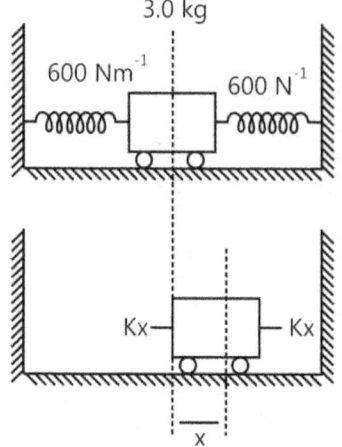

k_{eq} = 2k = 1200 Nm⁻¹

Total energy stored = $\dfrac{1}{2}$ kA²

= $\dfrac{1}{2}$ × 1200 × $\left(\dfrac{1}{2}\right)^2$ = 1.5 Joules.

Sol 9: T = $2\pi \sqrt{\dfrac{\ell}{g}}$; T ∝ $\ell^{1/2}$

$\ell \to$ 0.998 l

$\ell \to$ (0.998)$^{1/2}$ T

T → 0.999 T

Error in a day = 0.001 × (60 × 60 × 24) = 86.4 sec

Sol 10:

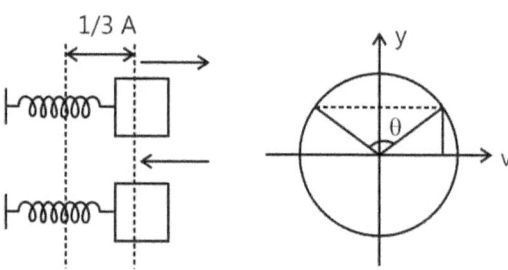

Phase Difference = θ = 2 cos⁻¹ (1/3) = 141.05°

Sol 11: Elevator moving up

Frequency of seconds pendulum = f_0 = 0.5 Hz

g_{eff} = g + $\dfrac{g}{2}$ = $\dfrac{3}{2}$ g

f = $\dfrac{1}{2\pi} \sqrt{\dfrac{g_{eff}}{\ell}}$ = $\sqrt{\dfrac{3}{2}} f_0$ = $\sqrt{\dfrac{3}{2}}$ × 0.5 Hz; f = 0.61 Hz

Sol 12: Periodic motion: A motion which repeats itself after equal intervals of time is called periodic motion

eq; motion of a pendulum

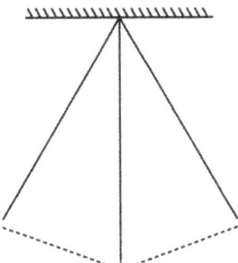

Oscillatory motion: A body is said to possess oscillatory or vibratory motion if it moves back and forth repeatedly about a mean position. For an oscillatory motion, a restoring force is required.

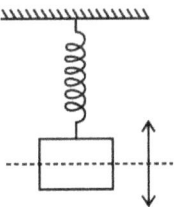

Sol 13: Simple Pendulum: A simple pendulum is a weight suspended from a pivot so that it can swing freely.

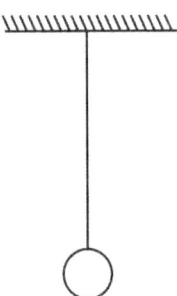

Time period = $2\pi \sqrt{\dfrac{\ell}{g}}$

$\ell \rightarrow$ length of pendulum

$g \rightarrow$ acceleration due to gravity

Frequency = $f = \dfrac{1}{2\pi} \sqrt{\dfrac{g}{\ell}}$

Sol 14: Refer spring mass system and ex.3

Sol 15: Spring factory: It is a mesure of the stiffness of a spring

Service:-parallel :

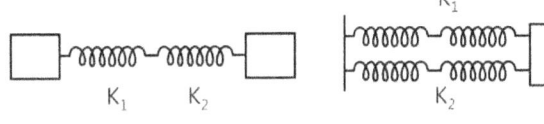

$k = \dfrac{k_1 k_2}{k_1 + k_2}$ $k = k_1 + k_2$

Sol 16: Phase: Phase of a vibrating particle at any instant is the state of the vibrating particle regarding it's displacement and direction of vibration at that particular instant. It is denoted by ϕ.

Phase difference is the difference in phases of two vibrating particles at a given time.

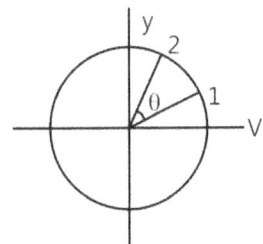

Particle 1 lags in phase by θ.

i.e. $\phi_2 - \phi_1 = \theta$

Angular frequency:- It is frequency f multiplied by a numerical quantity ω. It is denoted by ω.

$\omega = 2\pi f = \dfrac{2\pi}{T}$

$f \rightarrow$ frequency

$T \rightarrow$ Time period

Displacement in periodic motion: It is the displacement from the mean/equilibrium position.

Sol 17: $a = -\dfrac{d^2 x}{dt^2} = -\omega^2 x$

$a = -\omega^2 x\omega = \dfrac{2\pi}{T}$

$a = -\dfrac{4\pi^2 x}{T^2}$ $x \rightarrow$ displacement

$T \rightarrow$ time period

$v = \omega \sqrt{A^2 - x^2}$ $v \rightarrow$ velocity

$A \rightarrow$ Amplitude

$x \rightarrow$ Displacement

Sol 18: Figure (a) Initial phase = ϕ = 60° = $\dfrac{\pi}{3}$

$y = 2 \sin(\omega t + \phi)$

$y = 2 \sin\left(\dfrac{2\pi}{3}t + \dfrac{\pi}{3}\right)$

$\omega = \dfrac{2\pi}{T} = \dfrac{2\pi}{3}$

Figure (b) initial phase = ϕ = $\dfrac{\pi}{2}$

$A = 3$ cm

$\omega = \dfrac{2\pi}{60} = \dfrac{\pi}{30}$

$y = A \sin(\omega t + \phi)$

$y = 3 \sin\left(\dfrac{\pi t}{3} + \dfrac{\pi}{2}\right)$ cm

Sol 19:

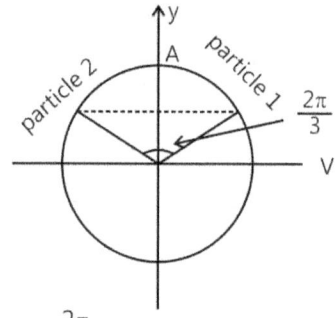

$\phi = \dfrac{2\pi}{3}$

Sol 20: $x = 6 \cos(3\pi t + \pi/3)$ metres

(a) $A = 6$m

(b) $V = \omega\sqrt{A^2 - x^2}$

$w = 3\pi$

$T = \dfrac{2\pi}{\omega} = 2/3$ sec.

At t = 2s particle will complete 3 orcillations

So the position will be same as at t = 0 s.

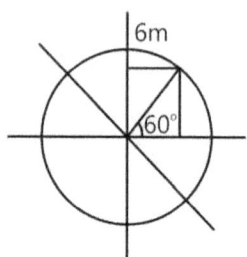

$x = 6 \sin(3xt + 5\pi/6)$

$x = 6 \cos 60° = 6 \sin 30°$

$x = 3$cm

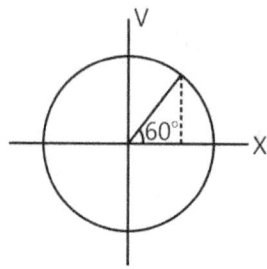

$v = \omega\sqrt{A^2 - x^2} = 3\pi\sqrt{36 - 9}$

$v = 6\sqrt{3}\,\pi$

$v = -48.97$ ms^{-1}

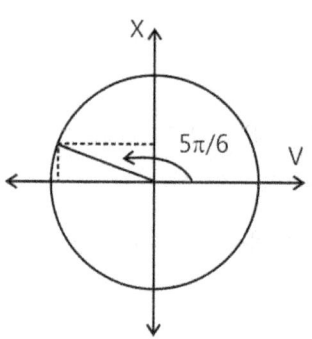

Sol 21: T = 20 s V = 5 ms^{-1}

$\Rightarrow \theta = \dfrac{2}{20} \times 360° \Rightarrow \theta = 36°$

$5 = A\omega \cos 36°$

$\omega = \dfrac{2\pi}{20} = \dfrac{\pi}{\omega}$

$5 = A\dfrac{\pi}{10} \cos 36°$

$A = \dfrac{50}{\pi \cos 36°} = 19.68$ m

Amplifide of SHM = 19.68 m

Sol 22: $x = 1$m $v = 3$ ms^{-1}

$x = 2$m$v = 2$ ms^{-1}

$v = 3 = \omega\sqrt{A^2 - 1}$

$3 = \omega\sqrt{A^2 - 1}$

$2 = \omega\sqrt{A^2 - 4}$

$$\Rightarrow \frac{9}{4} = \frac{A^2 - 1}{A^2 - 4} \Rightarrow 9A^2 - 36 = 4A^2 - 4$$

$$\Rightarrow 5A^2 = 32 \Rightarrow A = \sqrt{\frac{32}{5}} = \sqrt{6.4}$$

$$\Rightarrow A = 2.53 \text{ m}$$

$$3 = \omega \sqrt{6.4 - 1}$$

$$w = \frac{3}{\sqrt{5.4}} = 1.29 \text{ rad/s}$$

Period of motion: $T = \frac{2\pi}{\omega} = 4.86$ s

Length of path = 2A = 5.06 m

Sol 23: A = 4 cm

$$v_{max} = A\omega$$

$$v = \frac{v_{max}}{2} = \frac{A\omega}{2} = \omega\sqrt{A^2 - \omega^2}$$

$$A^2 - x^2 = \frac{A^2}{4}$$

$$x = \frac{\sqrt{3}}{2} A$$

$$x = \frac{\sqrt{3}}{2} \times 4 = 2\sqrt{3} = 3.464 \text{ cm}$$

P.E. = K.E.

$$\Rightarrow P.E. = \frac{1}{2} \times \frac{1}{2} kA^2$$

$$\frac{1}{2} kx^2 = \frac{1}{4} kA^2$$

$$x = 2\sqrt{2} = 2.828 \text{ m}$$

Sol 24: k = 50 Nm⁻¹

m = 1kg

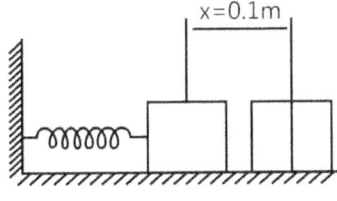

$$\omega^2 = \frac{k}{m} = \frac{50}{1} = 50$$

$$\omega = 5\sqrt{2} \text{ rad/s}$$

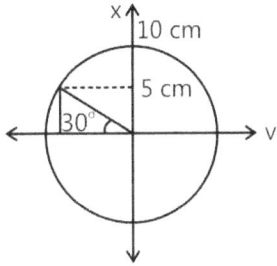

$$v = -A\omega \cos 30°$$

$$v = \frac{10}{100} \times 5\sqrt{2} \times \frac{\sqrt{3}}{2} \; ; \; v = \sqrt{\frac{3}{2}}$$

$$v = 0.61 \text{ ms}^{-1}$$

$$K.E. = \frac{1}{2} \times m \times v^2 = \frac{1}{2} \times 1 \times \frac{3}{8} = 0.1875 \text{ J}$$

$$P.E. = \frac{1}{2} \times 50 \times (0.05)^2 = 0.0625 \text{ J}$$

Total energy $= \frac{1}{2} \times 50 \times (0.1)^2 = 0.25$ J

Sol 25:

$$\boxed{1 \text{ kg}} \text{--} \text{0000000} \text{--} \boxed{3 \text{ kg}}$$
$$300 \text{ Nm}^{-1}$$

For two mass system.

We take effective mass instead of mass to calculate frequency.

$$\mu = \frac{m_1 m_2}{m_1 + m_2} = \frac{1 \times 3}{1 + 3} = \frac{3}{4} \text{ kg}$$

$$\omega^2 = \frac{k}{\mu} = \frac{300}{3/4} = 400$$

$$\omega = 20 \text{ rad/sec.}$$

$$f = \frac{10}{\pi} \text{ Hz} \cong 3.2 \text{ Hz}$$

Sol 26: $k_{34} = k_3 + k_4$

$$\frac{1}{k_{1234}} = \frac{1}{k_1} + \frac{1}{k_2} + \frac{1}{k_{34}}$$

$$= \frac{1}{k_1} + \frac{1}{k_2} + \frac{1}{(k_3 + k_4)}$$

$$= \frac{k_2(k_3 + k_4) + k_1(k_3 + k_4) + k_1 k_2}{k_1 k_2 (k_3 + k_4)}$$

$$\frac{1}{k_{1234}} = \frac{(k_1 + k_2)(k_3 + k_4) + k_1 k_2}{k_1 k_2 (k_3 + k_4)}$$

$$\omega = \left(\frac{k_{1234}}{m}\right)^{1/2}$$

$$f = \frac{1}{2\pi}\left(\frac{k_{1234}}{m}\right)^{1/2}$$

$$f = \frac{1}{2\pi}\left(\frac{k_1 k_2(k_3 + k_4)}{(k_1 + k_2)(k_3 + k_4) + (k_1 k_2)m}\right)^{1/2}$$

Exercise 2

Single Correct Choice Type

Sol 1: (A) $y = 5 \sin(\pi t + 4\pi)$

$A = 5$ $T = \frac{2\pi}{\omega} = \frac{2\pi}{\pi} = 2$ sec

$A = 5$; $T = 2$ sec

Sol 2: (C) $a_{max.} = A\omega^2$

$v_{max.} = A\omega$

Double ω; half the amplitude

Sol 3: (C) $A = \frac{0.8}{2} = 0.4$ m

$f = \frac{30}{60} = \frac{1}{2}$ hz $T = 2$ sec

Sol 4: (D)

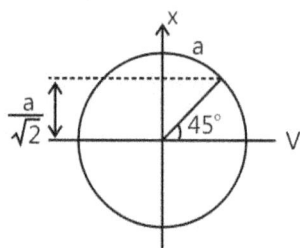

Sol 5: (A) $2\pi t + \frac{\pi}{3} = \pi$

$2\pi t = \frac{2\pi'}{3}$

$t = 1/3$ sec

Sol 6: (C) $T = 1.2$ sec

$A = 8$ cm

$\theta = \cos^{-1}\frac{5}{8}$; $\theta = 51.31°$

$t = \frac{\theta}{360} \times 1.2$; $t = 0.17$ sec

Sol 7: (C) $x = A + a \sin \omega t$

$t = \frac{5}{4}T$

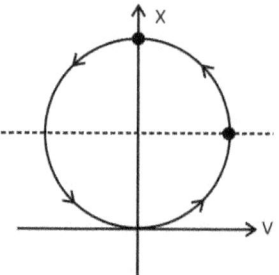

Distance in one rev. = 4A

Total distance covered = 4A + A = 5A

Sol 8: (B) $k_1 = \frac{k \times k}{k + k} = \frac{k}{2}$

$k_2 = k + k = 2k$

$t \propto k^{-1/2}$ $\frac{T_1}{T_2} = \sqrt{\frac{k_2}{k_1}} = 2$

Sol 9: (C) $y_1 = \sin\left(\omega t + \frac{\pi}{3}\right)$ $y_2 = \sin \omega t$

$y_1 + y_2$ $2 \sin\left(\omega t + \frac{\pi}{6}\right) \cos\frac{\pi}{6}$

$= \sqrt{3} \sin\left(\omega t + \frac{\pi}{6}\right)$

Sol 10: (B) $y = A \sin \omega t + A \cos \omega t$

$= 2A\left(\sin \omega t + \sin\left(\frac{\pi}{2} + \omega t\right)\right)$

$= 2A \sin\left(\omega t + \frac{\pi}{4}\right) \sin\left(\frac{\pi}{4}\right) = \sqrt{2}\,A \sin\left(\omega t + \frac{\pi}{4}\right)$

T.E. $= \frac{1}{2} \times m\omega^2 \times \left(\sqrt{2}A\right)^2$

T.E. $= m\omega^2 A^2$

Sol 11: (B) $\frac{1}{2}kA^2 = \frac{1}{2}mv^2 \Rightarrow A = \sqrt{\frac{mv^2}{k}}$

Sol 12: (C) Amplitude dose not depend on frame of reference.

Sol 13: (A) $s = ut + \dfrac{1}{2} at^2 \Rightarrow -a = 0 + \dfrac{1}{2} \times (-5) \times t^2$

$t = \sqrt{\dfrac{2a}{5}}$

$T = 4t = 4\sqrt{\dfrac{2a}{5}}$

Sol 14: (B) $\dfrac{1}{2} kA^2 \left(1 - \dfrac{3}{4}\right) = \dfrac{1}{2} mv^2$

$\Rightarrow v = \left(\dfrac{k}{m}\right)^{1/2} \dfrac{A}{2}$

$\Rightarrow v' = \left(\dfrac{k}{m}\right)^{1/2} \dfrac{A}{4}, \; \omega_2 = \dfrac{\omega_1}{\sqrt{2}}$

T.E. $= \dfrac{3}{8} kA^2 + \dfrac{2m}{2}\left(\dfrac{k}{m}\right)\dfrac{A^2}{16} = \dfrac{7}{16} kA^2$

$\Rightarrow \dfrac{7}{16} kA^2 = \dfrac{1}{2} kA'^2$

$A' = \sqrt{\dfrac{7}{8}} A$

Sol 15: (C)

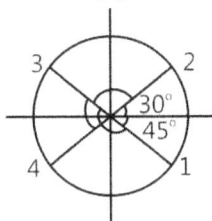

$\theta_{24} = 195°$

$\theta_{12} = 75°$

$q_{31} = 165°$

Previous Years' Questions

Sol 1: (C) In SHM frequency with which kinetic energy oscillation is two times the frequency of oscillation of displacement.

Sol 2: (B) $(v_M)_{max} = (v_N)_{max}$

$\therefore \omega_M A_M = \omega_N A_N$

or $\dfrac{A_M}{A_N} = \dfrac{\omega_N}{\omega_M} = \sqrt{\dfrac{k_2}{k_1}} \quad \left(\because \omega = \sqrt{\dfrac{k}{m}}\right)$

Sol 3: (D)

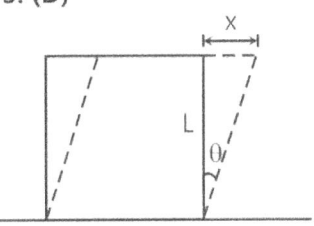

Modulus of rigidity, $\eta = F/A\theta$

Here, $A = L^2$ and $\theta = \dfrac{x}{L}$

Therefore, restoring force force is

$F = -\eta A\theta = -\eta L x$

Or acceleration, $a = \dfrac{F}{M} = -\dfrac{\eta L}{M} x$

Since, $a \propto -x$, oscillations are simple harmonic in nature, time period of which is given by

$T = 2\pi \sqrt{\left|\dfrac{displacement}{acceleration}\right|} = 2\pi \sqrt{\left|\dfrac{x}{a}\right|}$

$= 2\pi \sqrt{\dfrac{M}{\eta L}}$

Sol 4: (B) $Ke\theta = \dfrac{k_1 k_2}{k_1 + k_2} = \dfrac{\dfrac{YA}{L}}{\dfrac{YA}{L} + k} = \dfrac{YAk}{YA + Lk}$

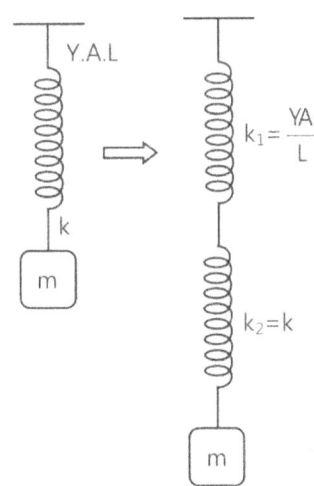

$\therefore T = 2\pi \sqrt{\dfrac{m}{k_{eq}}} = 2\pi \sqrt{\dfrac{m(YA + Lk)}{YAk}}$

Note Equivalent fore constant for a wire is given by k = $\frac{AY}{L}$. Because in case of a wire. F = $\left(\frac{AY}{L}\right)\Delta L$ and in case of spring F = k.Δx. Comparing these two, we find

k of wire = $\frac{AY}{L}$

Sol 5: (A) U(x) = k|x|³

$\therefore$ [k] = $\frac{[U]}{[x^3]}$ = $\frac{[ML^2T^{-2}]}{[L^3]}$ = [ML^{-1}T^{-2}]

Now, time period may depend on

T $\propto$ (mass)^x(amplitude)^y(k)^z

[M^{0}L^{0}T] = [M]^x[L]^y[ML^{-1}T^{-2}]^z = [M^{x+z}L^{y-z}T^{-2z}]

Equating the powers, we get

$-2z = 1$ or $z = -1/2$

$y - z = 0$ or $y = z = -\frac{1}{2}$

Hence, T $\propto$ (amplitude)$^{-1/2}$ $\propto$ (a)$^{-1/2}$

or T $\propto \frac{1}{\sqrt{a}}$

Sol 6: (B)

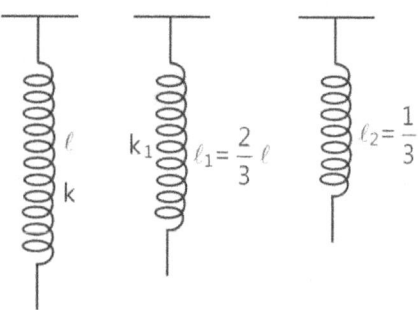

$\lambda_1 = 2\lambda_2$

$\therefore \lambda_1 = \frac{2}{3}$l

Force constant k $\propto \frac{1}{\text{length of spring}}$

$\therefore k_1 = \frac{3}{2}k$

Sol 7: (D) U(x) = k(1 $-$ e$^{-x^2}$)

It is an exponentially increasing graph of potential energy (U) with x². Therefore, U versus x graph will be as shown. At origin.

Potential energy U is minimum (therefore, kinetic energy will be maximum) and force acting on the particle is zero because.

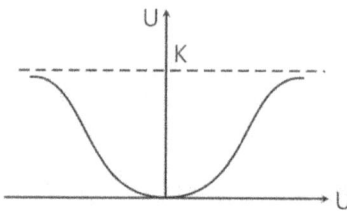

F = $\frac{-dU}{dx}$ = $-$ (slope of U-x graph) = 0.

Therefore, origin is the stable equilibrium position. Hence, particle will oscillate simple harmonically about x = 0 for small displacement. Therefore, correct option is (d).

(a), (b) and (c) options are wrong due to following reasons.

(a) At equilibrium position F = $\frac{-dU}{dx}$ = 0 i.e., slope of U-x graph should be zero and from the graph we can see that slope is zero at x = 0 and x = $\pm\infty$

Now among these equilibriums stable equilibrium position is that where U is minimum (Here x=0). Unstable equilibrium position is that where U is maximum (Here none).

Neutral equilibrium position is that where U is constant (Here x = $\pm\infty$)

Therefore, option (a) is wrong.

(b) For any finite non-zero value of x, force is directed towards the origin because origin is in stable equilibrium position. Therefore, option (b) is incorrect.

(c) At origin, potential energy is minimum, hence kinetic energy will be maximum. Therefore, option (c) is also wrong.

Sol 8: (A) Free body diagram of bob of the pendulum with respect to the accelerating frame of reference is as follows

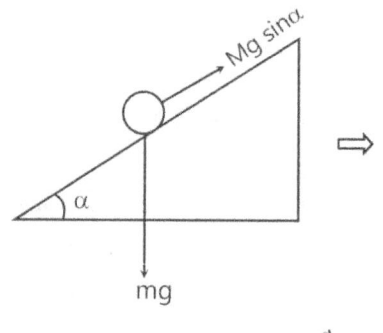

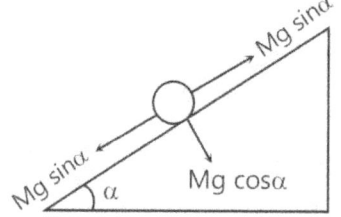

∴ Net force on the bob is F_{net} = mg cosα

or Net acceleration of the bob is g_{eff} = g cosα

$$T = 2\pi \sqrt{\frac{L}{g_{eff}}}$$

or $T = 2\pi \sqrt{\frac{L}{g\cos\alpha}}$

Note: Whenever point of suspension is accelerating

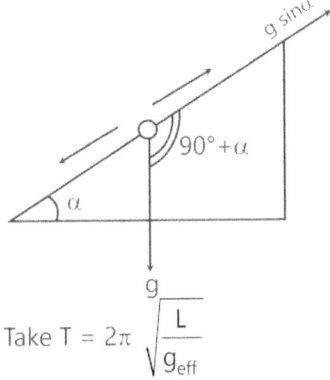

Take $T = 2\pi \sqrt{\frac{L}{g_{eff}}}$

Where $\vec{g}_{eff} = \vec{g} - \vec{a}$

$\vec{a}$ = acceleration of point of suspension.

In this question $\vec{a}$ = g sin α (down the plane)

$\therefore |\vec{g} - \vec{a}| = g_{eff}$

$= \sqrt{g^2 + (g\sin\alpha)^2 + 2(g)(g\sin\alpha)\cos(90° + \alpha)}$

= g cosα

Sol 9: (A) In SHM, velocity of particle also oscillates simple harmonically. Speed is more near the mean position and less near the extreme positions. Therefore, the time taken for the particle to go from O to A/2 will be less than the time taken to go it from A/2 to A, or $T_1 < T_2$

Note From the equation of SHM we can show that

$t_1 = T_{0-A/2} = T/12$

and $t_2 = T_{A/2-A} = T/6$

So, that $t_1 = t_2 = T_{0-A} = T/4$

Sol 10: (A) Potential energy is minimum (in this case zero) at mean position (x = 0) and maximum at extreme positions (x = ± A).

At time t = 0, x = A. Hence, PE should be maximum. Further in graph III, PE is minimum at x = 0 Hence, this is also correct.

Sol 11: (A) Angular frequency of the system,

$$\omega = \sqrt{\frac{k}{m+m}} = \sqrt{\frac{k}{2m}}$$

Maximum acceleration of the system will be, $\omega^2 A$ or $\dfrac{kA}{2m}$

This acceleration to the lower block is provided by friction.

Hence, $f_{max} = ma_{max}$

$= m\omega^2 A = m\left(\dfrac{kA}{2m}\right) = \dfrac{kA}{2}$

Sol 12: (A) $y = kt^2$

$\dfrac{d^2y}{dt^2} = 2k$ or a_y = 2m/s²(as k = 1 m/s²)

$T_1 = 2\pi \sqrt{\dfrac{\ell}{g}}$ and $T_2 = 2\pi \sqrt{\dfrac{\ell}{g + a_y}}$

$\therefore \dfrac{T_1^2}{T_2^2} = \dfrac{g + a_y}{g} = \dfrac{10 + 2}{10} = \dfrac{6}{5}$

Sol 13: (D) T = 8s, $\omega = \dfrac{2\pi}{T} = \left(\dfrac{\pi}{4}\right)$ rads⁻¹

x = A sin ωt

$\therefore a = -\omega^2 x = -\left(\dfrac{\pi^2}{16}\right)\sin\left(\dfrac{\pi}{4}t\right)$

Substituting $t = \dfrac{4}{3}$ s, we get

$a = -\left(\dfrac{\sqrt{3}}{32}\pi^2\right)$ cms⁻²

Sol 14: (C)

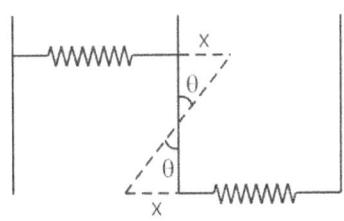

$x = \dfrac{L}{2}\theta$

Restoring torque $= -(2kx) \cdot \dfrac{L}{2}$

$$\alpha = -\dfrac{kL(L/2\theta)}{I} = -\left[\dfrac{kL^2/2}{ML^2/12}\right]\theta = -\left(\dfrac{6k}{M}\right)\theta$$

$$\therefore f = \dfrac{1}{2\pi}\sqrt{\left|\dfrac{\alpha}{\theta}\right|} = \dfrac{1}{2\pi}\sqrt{\dfrac{6k}{M}}$$

Sol 15: (D) $x_1 + x_2 = A$ and $k_1 x_1 = k_2 x_2$

or $\dfrac{x_1}{x_2} = \dfrac{k_2}{k_1}$

Solving these equations, we get

$$x_1 = \left(\dfrac{k_2}{k_1 + k_2}\right) A$$

Sol 16: (A) Frequency or time period of SHM depends on variable forces. It does not depend on constant external force. Constant external force can only change the mean position. For example, in the given question mean position is at natural length of spring in the absence of electric field. Whereas in the presence of electric field mean position will be obtained after a compression of x_0. Where x_0 is given by

$Kx_0 = QE$

or $x_0 = \dfrac{QE}{K}$

Sol 17: (B)

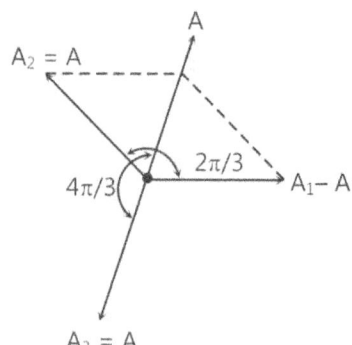

Resultant amplitude of x_1 and x_2 is A at angle $\left(\dfrac{\pi}{3}\right)$ from A_1. To make resultant of x_1, x_2 and x_3 to be zero. A_3 should be equal to A at angle $\phi = \dfrac{4\pi}{3}$ as shown in figure.

Alternate solution: It we substitute, $x_1 + x_2 + x_3 = 0$

or $A \sin \omega t + A \sin\left(\omega t + \dfrac{2\pi}{3}\right) + B \sin(\omega t + \phi)$

Then by applying simple mathematics we can prove that

$B = A$ and $\phi = \dfrac{4\pi}{3}$.

Sol 18: (D) As retardation $= bv$

$\therefore$ Retarding force $= mbv$

$\therefore$ Net restoring torque when angular displacement is θ is given by

$= -mg\,\ell \sin\theta + mbv\,\ell$

$\therefore I\alpha = -mg\,\ell \sin\theta + mbv\,\ell$

Where, $I = m\,\ell^2$

$$\therefore \dfrac{d^2\theta}{dt^2} = \alpha = -\dfrac{g}{\ell}\sin\theta + \dfrac{bv}{\ell}$$

for small damping, the solution of the above differential equation will be

$$\therefore \theta = \theta_0\, e^{-\frac{bt}{2}} \sin(wt + \phi)$$

$\therefore$ Angular amplitude will be $= \theta \cdot e^{\frac{-bt}{2}}$

According to question, in τ time (average life–time),

Angular amplitude drops to $\dfrac{1}{e}$ value of its original value (θ)

$$\therefore \dfrac{\theta_0}{e} = \theta_0\, e^{-\frac{6\tau}{2}} \Rightarrow \dfrac{6\tau}{2} = 1$$

$$\therefore \tau = \dfrac{2}{b}$$

Sol 19: (B) $A = A_0\, e^{-kt}$

$\Rightarrow 0.9\, A_0 = A_0 e^{-5k}$

and $\alpha\, A_0 = A_0\, e^{-15k}$

Solving $\Rightarrow \alpha = 0.729$

At mean position, K.E. is maximum where as P.E. is minimum.

Sol 20: (C) $3\omega\sqrt{A^2 - \left(\dfrac{2A}{3}\right)^2} = \omega\sqrt{A_1^2 - \left(\dfrac{2A}{3}\right)^2}$

$$\therefore A_1 = \dfrac{7A}{3}$$

Sol 21: (C) $v = \omega\sqrt{A^2 - \left(\dfrac{2A}{3}\right)^2}$

$v = \sqrt{5}\,\dfrac{A\omega}{3}$

$v_{new} = 3v = \sqrt{5}\,A\omega$

So the new amplitude is given by

$V_{new} = \omega\sqrt{A_{new}^2 - x^2} \Rightarrow \sqrt{5}\,A\omega = \omega\sqrt{A_{new}^2 - \left(\dfrac{2A}{3}\right)^2}$

$A_{new} = \dfrac{7A}{3}$

JEE Advanced/Boards

Exercise 1

Sol 1: $T \propto \dfrac{1}{k^{1/2}}$; $T = 2\pi\sqrt{\dfrac{m}{k}}$

$k_1 = 4k$; $k_2 = \dfrac{4k}{3}$

By $k_1\lambda_1 = k_2\lambda_2 = kl$

$T_1 = \dfrac{T}{2}$; $T_2 = \dfrac{T\sqrt{3}}{2}$

$\dfrac{T_1}{T_2} = \dfrac{1}{\sqrt{3}}$

Sol 2: $x = 0.2\cos 5\pi t$

velocity $= \dfrac{dx}{dt} = -\pi\sin 5\pi t$

speed $= \pi\,|\sin 5\pi t|$

$v_{avg} = \dfrac{\pi\int_0^{0.7}|\sin 5\pi t|\,dt}{0.7}$

$= \dfrac{\pi}{0.7} \times 7 \times \int_0^{0.1}\sin 5\pi t\,dt = \dfrac{10\pi}{5\pi}\left[-\cos 5\pi t\right]_0^{0.1}$

$v_{avg} = 2$ m/s

Sol 3: $\phi = 2\sin^{-1}(0.9)$

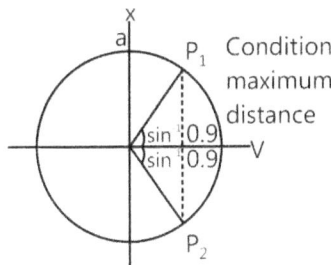

$P_1P_2 \,||\,$ y-axis

Max. Distance $= 1.8\,a$

Sol 4:

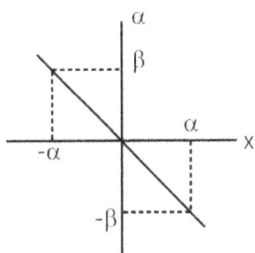

$a = -\omega^2 x$

$-\omega^2 = \dfrac{\beta}{\alpha}$ = slope of a-x graph

$\omega = \sqrt{\dfrac{\beta}{\alpha}}$

Frequency $= \dfrac{\omega}{2\pi} = \dfrac{1}{2\pi}\sqrt{\dfrac{\beta}{\alpha}}$

Sol 5: $m = 0.1$ kg

$A = 0.1$ m

$\dfrac{1}{2} \times mv_{max}^2 = 8 \times 10^{-3}$ J

$0.1\times v_{max}^2 = 16 \times 10^{-3} \Rightarrow v_{max} = 0.4$ m/s

$A\omega = 0.4$

$0.1\times \omega = 0.4 \Rightarrow \omega = 4$

$x = A\sin(\omega t + \phi)$

$x = 0.1\sin(4t + \pi/4)$

Sol 6: (i)

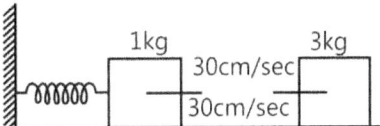

x = 10 + 3 sin 10 t

At t = 0 s block 1 is at equilibrium position.

$v_1 = A\omega = 3 \times 10 = 30$ cm/s

$v_2 = 30$ cm/s

Conservation of momentum

$m_1 v_1 + m_2 v_2 = (m_1 + m_2) v$

$- 1 \times 30 + 3 \times 30 = 4 \times v$

v = 15 cm/s

Final velocity is in opposite direction of initial velocity of block 1. This causes a phase change of π.

$\omega \propto m^{-1/2}$

$\omega' = 5$ rad/s

$A'\omega' = 15$; $A' = 3$ cm

New amplitude = 3 cm

(ii) New equation

$x = 10 + 3 \sin (5t + \pi)$

(iii) Loss of energy

$= \left(\dfrac{1}{2} \times 1 \times 30^2 + \dfrac{1}{2} \times 3 \times 30^2 - \dfrac{1}{2} \times 4 \times 15^2 \right) \times 10^{-4}$ J

$= \dfrac{1}{2} (900 + 2700 - 900) \times 10^{-4}$ J $= 1350 \times 10^{-4}$ J

$\Delta E_{loss} = 0.1350$ J

Sol 7:

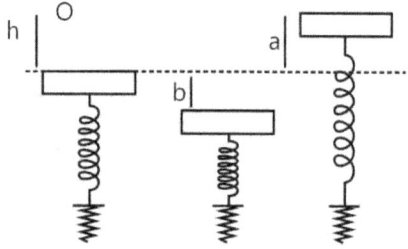

(a) $mgh + \dfrac{1}{2} k \left(\dfrac{Mg}{k} \right)^2$

$= \dfrac{1}{2} k \left(\dfrac{Mg}{k} + b \right)^2 - (M + m) gb$

$= \dfrac{1}{2} k \left(\dfrac{Mg}{k} - a \right)^2 + (M + m) ga$

Equalising energies in 3 states

$\dfrac{1}{2} k \left(\dfrac{Mg}{k} + b \right)^2 - (M + m) gb$

$= \dfrac{1}{2} k \left(\dfrac{Mg}{k} - a \right)^2 + (M + m) ga$

$k \left(\dfrac{2mg}{k} (b + a) + b^2 - a^2 \right) = 2 (M + m) g (a + b)$

$2Mg (b + a) + k (b^2 - a^2) = 2 (M + m) g (a + b)$ $k = \dfrac{2mg}{b - a}$

Constant of force of spring $= \dfrac{2mg}{b - a}$

(b) $\omega = \sqrt{\dfrac{k}{(M + m)}} = \sqrt{\dfrac{2mg}{(M + m)(b - a)}}$

$f = \dfrac{1}{2\pi} \sqrt{\dfrac{2mg}{(M + m)(b - a)}}$

(c) $mgh = \dfrac{1}{2} k \left[\left(\dfrac{Mg}{k} - a \right)^2 - \left(\dfrac{Mg}{k} \right)^2 \right]$

$+ (M + m) ga$

$mgh = -\dfrac{1}{2} k \left[a \times \left(\dfrac{2Mg}{k} - a \right) \right]$

$+ (M + m) ga$

$mgh = \dfrac{-ka}{2} \left(\dfrac{2mg}{k} - a \right) + (M + m) ga$

$mgh = -Mga + \dfrac{ka^2}{2} + (M + m) ga$

$mgh = mga + \dfrac{2mga^2}{(b - a)2}$

$h = a + \dfrac{a^2}{(b - a)} = \dfrac{ab}{b - a}$

Sol 8:

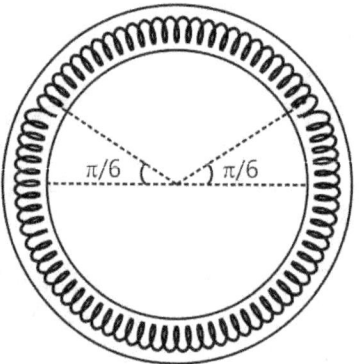

(a) Frequency

Displace by $d\theta$

$\Delta x = 2Rd\theta$

$d\alpha = -2\dfrac{k}{m} \times \dfrac{2Rd\theta}{R}$

$d\alpha = -\omega^2\, d\theta$

$\omega^2 = \dfrac{4k}{m}$ $\omega = 2\sqrt{\dfrac{k}{m}} = 2$

$f = \dfrac{2}{2\pi} = \dfrac{1}{\pi}$

(b) Total energy $= 2 \times \dfrac{1}{2} k \left(R\dfrac{\pi}{3} \right)^2$

$= 2 \times \dfrac{1}{2} \times 0.1 \times \left(\dfrac{0.06 \times \pi}{3} \right)^2$

$= 3.94 \times 10^{-4}\ J/4\pi^2 \times 10^{-5}\ J$

(c) $2 \times \dfrac{1}{2} mv^2 = 4\pi^2 \times 10^{-5}$

$v^2 = \dfrac{4\pi^2 \times 10^{-5}}{0.1}$

$v^2 = 4\pi^2 \times 10^{-4}$

$v = 2\pi \times 10^{-2} = 0.02\ \pi\ m/sec$

Sol 9:

$V_{com} = \dfrac{2 \times 2 + 3 \times 0}{5} = 0.8\ m/s$

$x_A = v_{com}\, t + A \sin \omega t$

At maximum expansion

$\dfrac{1}{2} \times 5 \times (0.8)^2 + \dfrac{1}{2} kx^2 = \dfrac{1}{2} \times 2 \times 2^2$

$kx^2 = 8 - 3.2 = 4.8$

$x = 0.2$

$A = \dfrac{3}{5} x = \dfrac{3}{5} \times 0.2 = 0.12$

$\mu = \dfrac{3 \times 2}{3 + 2} = 1.2$

$\omega = \sqrt{\dfrac{k}{\mu}} = \sqrt{\dfrac{120}{1.2}} = 10$

$x_A = 0.8\, t + 0.12 \sin 10\, t$

Sol 10: (a) m = 0.1 kg

$F = 10\, x + 2$

Only variable force causes SHM

(a) $F(x) = 10\, x + 2$

$a(x) = 100\, x + 20$

$v(x) = 50\, x^2 + 20\, x + c$

$v(0.2) = 0$

$50 \times 0.04 + 20 \times 0.2 + c = 0$

$c = -6$

$v = 50\, x^2 + 20\, x - 6$ $\begin{cases} x = 0.2 \\ x = -0.6 \end{cases}$

$A = \dfrac{0.2 - (-0.6)}{2} = 0.4\ m$

Amplitude = 0.4 m

(b) $\omega = \sqrt{\dfrac{10}{0.1}} = 10\ rad/sec$

$T = \dfrac{2\pi}{\omega} = \dfrac{\pi}{5}\ sec.$

(c) $x = 0.2 - A \cos \omega t$

$x = 0.2 - 0.4 \cos \dfrac{5t}{\pi}$

Sol 11: $u = (x^2 - 4x + 3)$

(i) $F = -\dfrac{dU}{dx}$

$F = -2x + 4$

At equilibrium F = 0

$-2x + 4 = 0 \Rightarrow x = 2\ m$

(ii) $dF = -2dx$ similar to $dF = -\omega^2 dx$ as in SHM

$2 = \dfrac{\omega^2}{m} = \omega^2$

$\omega = \sqrt{2}$

$T = \dfrac{2\pi}{\omega} = \sqrt{2}\ \pi\ sec$

(iii) $A\omega = 2\sqrt{6}$

$A = \dfrac{2\sqrt{6}}{\sqrt{2}} \Rightarrow A = 2\sqrt{3}\ m$

Sol 12: $F_{max} = m\omega^2 A$

P.E. $= \dfrac{1}{2}$ K.E.

$\Rightarrow \dfrac{1}{2} kx^2 = \dfrac{1}{2} \times \dfrac{1}{2} kA^2$

$\Rightarrow x = \dfrac{A}{\sqrt{2}}$

$F = m\omega^2 \dfrac{A}{\sqrt{2}} = \dfrac{F_{max}}{\sqrt{2}}$

$F = 25\sqrt{2}$ N

Sol 13: (a) $T = 2\pi \sqrt{\dfrac{\mu}{k}}$

$\mu = \dfrac{3 \times 6}{3 + 6} = \dfrac{18}{9} = 2$ kg

$T = 2\pi \sqrt{\dfrac{1}{400}} = \dfrac{\pi}{10}$ sec

(b) A = 6 cm

(c) $v_{cm} = 0;$ $\quad v_B = \dfrac{-1}{2} v_A$

$\dfrac{1}{2} \times k \times A^2 = \dfrac{1}{2} \times 3 \times (-2v_B)^2 + \dfrac{1}{2} \times 6 \times v_B^2$

$800 \times (0.06)^2 = 12 v_B^2 + 6v_B^2$

$v_B^2 = \dfrac{8 \times 0.36}{18}$

$v_B = \dfrac{2 \times 0.6}{3}$

$V_{Bmax.} = 0.4$ m/s

$P_{Bmax} = 0.4 \times 6$

$P_{Bmax} = 2.4$ kg ms^{-1}

Sol 14: $s = \left(A - \dfrac{A}{4}\right) \cos \omega t - \left(\dfrac{A}{2} - \dfrac{A}{8}\right) \sin \omega t$

$s = \dfrac{3A}{4} \cos \omega t - \dfrac{3A}{8} \sin \omega t$

$s = \dfrac{3A}{8} (2 \cos \omega t - \sin \omega t)$

$s = \dfrac{3\sqrt{5}}{8} A \left(\dfrac{2}{\sqrt{5}}\cos\omega t - \dfrac{1}{\sqrt{5}}\sin\omega t\right)$

$s = \dfrac{3\sqrt{5}}{8} A \cos\left(\omega t + \sin^{-1}\left(\dfrac{1}{\sqrt{5}}\right)\right)$

$A' = \dfrac{3\sqrt{5}}{8} A;$ $\quad \delta = \sin^{-1}\left(\dfrac{1}{\sqrt{5}}\right)$

Sol 15: $T = 2\pi \sqrt{\dfrac{m}{k}} = 0.4$ sec

$\omega = 5\pi$

For $0 < t < 0.6$ sec

$x = -\dfrac{mg}{2k} + \dfrac{mg}{2k} \sin\left(5\pi t + \dfrac{\pi}{2}\right)$

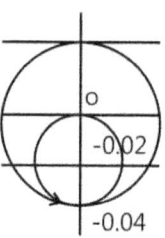

$\dfrac{mg}{2k} = \dfrac{4 \times 10}{2 \times 1000} = 0.02$ m

for $0 < t < 0.6$ sec

$x = -0.02 + 0.02 \sin (5\pi t + \pi/2)$

for $0.6 < t\ 1$ sec

$x = -0.04 + 0.04 \sin (5\pi t)$

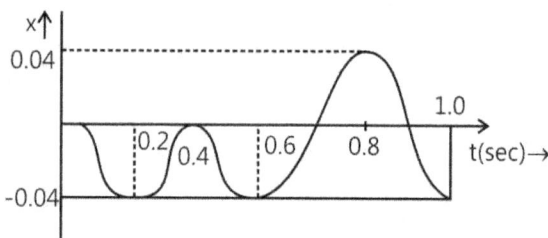

Sol 16:

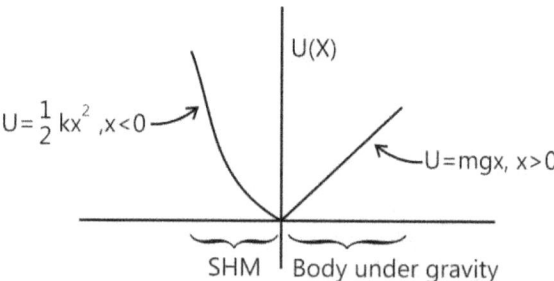

$T = \dfrac{1}{2} \times 2\pi \sqrt{\dfrac{m}{k}} + \dfrac{2v}{g}$

$\because E = \dfrac{1}{2} mv^2;\ T = \pi\sqrt{\dfrac{m}{k}} + \dfrac{2}{g}\sqrt{\dfrac{2E}{m}}$

$\because v = \sqrt{\dfrac{2E}{m}};\ T = \pi\sqrt{\dfrac{m}{k}} + \dfrac{2\sqrt{2}}{g}\sqrt{\dfrac{E}{m}}$

Sol 17: $x = 30 \sin\left(\pi t + \dfrac{\pi}{6}\right)$

$T = \dfrac{2\pi}{\pi} = 2$

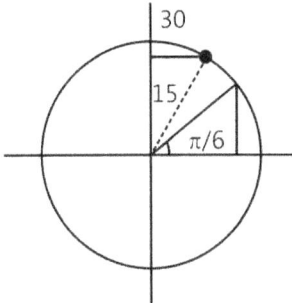

P.E. = 2 K.E.

P.E. = $\dfrac{2}{3}$ T.E.

$x = \sqrt{\dfrac{2}{3}}\ A$

Position: $x = \sqrt{\dfrac{2}{3}} \times 30$

$x = 10\ \sqrt{6}\ $ cm

$t = \dfrac{\left(\sin^{-1}\sqrt{\dfrac{2}{3}} - \dfrac{\pi}{6}\right)}{2\pi} \times 2$ sec

$t = \dfrac{1}{\pi}\left(\sin^{-1}\sqrt{\dfrac{2}{3}} - \dfrac{\pi}{6}\right)$ sec

Sol 18: (a)

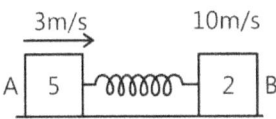

$v_{cm} = \dfrac{5 \times 3 + 10 \times 2}{7} = 5\,ms^{-1}$

$\dfrac{1}{2}\,5 \times 3^3 + \dfrac{1}{2} \times 2 \times 10^2$

$= \dfrac{1}{2}\,7 \times 5^2 + \dfrac{1}{2}\,kx^2$

$45 + 200 = 175 + kx^2$

$kx^2 = 70$

$x^2 = \dfrac{70}{1120}\quad x = \dfrac{1}{4}\ $ m

Maximum extension = 0.25 m

(b) $t = \dfrac{3}{4}\ T$

$T = 2\pi\ \sqrt{\dfrac{\mu}{k}}$

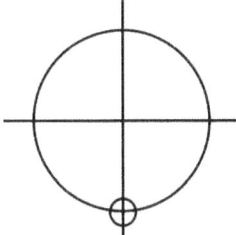

$\mu = \dfrac{5 \times 2}{7} = \dfrac{10}{7}$

$T = 2\pi\ \sqrt{\dfrac{10}{7 \times 1120}} = \dfrac{2\pi}{28} = \dfrac{\pi}{14}$

Time for first maximum compression

$= \dfrac{3}{4}\ \times\ \dfrac{\pi}{14}\ =\ \dfrac{3\pi}{56}$ sec

Sol 19: $T = 2\pi\ \sqrt{\dfrac{I}{mgx}}$

$I = \dfrac{m\ell^2}{3} + \left(\dfrac{m\ell^2}{12} + m\ell^2\right)$

$x = \dfrac{mx\dfrac{1}{2} + mx\ell}{2m} = \dfrac{3\ell}{4} = \dfrac{4m\ell^2}{12} + \dfrac{13}{12}m\ell^2$

$I = \dfrac{17}{12}\ m\ell^2$

$T = 2\pi\ \sqrt{\dfrac{\dfrac{17}{12}m\ell^2}{2mg\dfrac{3\ell}{4}}}\ \ ;T = 2\pi\ \sqrt{\dfrac{17\ell}{18g}}$

Sol 20: (a) $I\alpha = -\,k\theta$

$\alpha = -\,\dfrac{k}{I}\ \theta$

$\omega^2 = \dfrac{k}{2}\quad \omega = \sqrt{\dfrac{k}{I}}\ = \pi$

$T = \dfrac{2\pi}{\omega}\ = \dfrac{2\pi}{\pi}\ ;T = 2$ sec

(b)

$g_{eff.} = \sqrt{10^2 + 5^2} = \sqrt{125}$

$g_{eff.} = 5\sqrt{5}$

$T = 2\pi\sqrt{\dfrac{\ell}{g}} = 2\pi\sqrt{\dfrac{0.5}{5\sqrt{5}\times 10}}$

$T = \dfrac{2}{5^{1/4}}$ sec

Sol 21: $m = 0.2$ kg $f = \dfrac{25}{\pi}$ Hz

$P.E. = \dfrac{4}{9}$ T.E.

$\dfrac{1}{2}kx^2 = \dfrac{4}{9}\times\dfrac{1}{2}kA^2; x = \dfrac{2}{3}A$

$A = \dfrac{3}{2}x = \dfrac{3}{2}\times 0.04$

$A = 0.06$ m

Sol 22: $0.5 \times 3 = 1.5 \times v$

$v = 1$ m/s

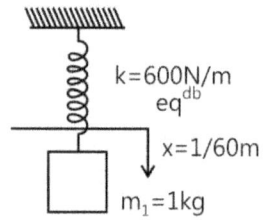

$\omega = \sqrt{\dfrac{600}{1.5}} = \sqrt{400} = 20$ rad./sec

$f = \dfrac{20}{2\pi} = \dfrac{10}{\pi}$ Hz

$\dfrac{1}{2}kx^2 + \dfrac{1.5\times 1^2}{2}$

$= \dfrac{1}{2}kh^2 - 1.5\times 10\times\left(h - \dfrac{1}{60}\right)$

$\dfrac{1}{2}\times 600\times\dfrac{1}{60^2} + \dfrac{1.5}{2}$

$= \dfrac{1}{2}\times 600\times h^2 - 15\times\left(h - \dfrac{1}{60}\right)$

$60h^2 - 3h - 7/60 = 0$

$h = \dfrac{1.5}{60} + \dfrac{\sqrt{37}}{120}$ m

$A = h - \dfrac{1.5}{60} = \dfrac{\sqrt{37}}{120}$ m

$A = \dfrac{\sqrt{37}}{120}\times 100$ cm $= \dfrac{5\sqrt{37}}{60}$ cm

Sol 23: $m_1 = 1$kg; $m_2 = 4$kg

$a = 1.6$ cm

$kx = m_1 g$

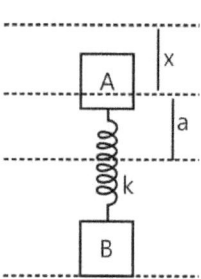

$k = \omega^2 m_1 = 25^2\times 1 = 625$ N/m

$N_{max} = m_2 g + k(x + a)$

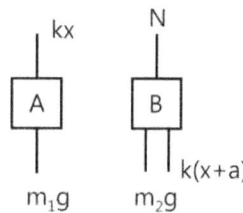

$= (m_1 + m_2)g + ka = 50 + \dfrac{625\times 1.6}{100}$

$N_{max} = 60$ N

$N_{min} = (M_1 + M_2)g - ka$

$N_{min} = 40$N

2.65

Sol 24: $k\ell = mg$; $\omega = \sqrt{\dfrac{k}{m}} = \sqrt{\dfrac{9}{\ell}}$

$f = \dfrac{1}{2\pi}\left(\dfrac{g}{\ell}\right)^{1/2}$

$A = \ell/3$; $m \times \dfrac{2g}{3} = k \times x$

$x = \dfrac{2}{3}\dfrac{mg}{k} = \dfrac{2}{3}\ell$

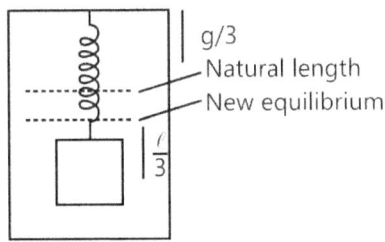

g/3
— Natural length
— New equilibrium
$\dfrac{\ell}{3}$

Exercise 2

Single Correct Choice Type

Sol 1: (C)

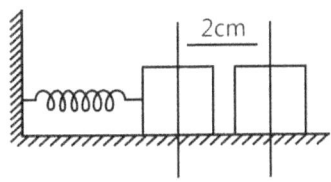

2cm

$\omega\sqrt{4-1} = \omega^2 \times 1$

$\omega = \sqrt{3}$

$F = \dfrac{\sqrt{3}}{2\pi}$ Hz

Sol 2: (D)

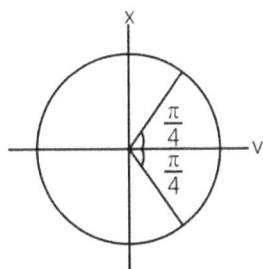

x

$\dfrac{\pi}{4}$

$\dfrac{\pi}{4}$

v

$V_{avg} = \dfrac{\text{displacement}}{\text{time}} = \dfrac{\sqrt{2}\times A}{T/4}$

$V_{avg} = \dfrac{4\sqrt{2}A}{T}$

Sol 3: (C) $V_{mean} = \dfrac{a/2}{T/6} = \dfrac{3a}{T}$

Sol 4: (C)

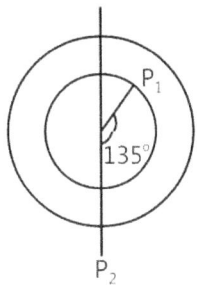

P_1

135°

P_2

Sol 5: (C)

A B C D E

$V_B^2 = \dfrac{1}{4} V_A^2$

$\dfrac{1}{4} R^2\omega^2 = \omega^2(R^2 - x^2) \Rightarrow R^2 = (R^2 - x^2)4$

$x = \dfrac{\sqrt{3}}{2} R$

$d_{BD} = 2x = \sqrt{3} R$

Sol 6: (C) $T_S = 2\pi\sqrt{\dfrac{m}{k}}$ T_S doesn't depend on g.

$T_p = 2\pi\sqrt{\dfrac{\ell}{g}}$; $T_p \propto g^{1/2}$

$\therefore T_p$ decreases

Sol 7: (B) $V_{max} = A\omega = \dfrac{A\sqrt{k}}{\sqrt{m}}$

$\dfrac{A_1\sqrt{k_1}}{\sqrt{m}} = \dfrac{A_2\sqrt{k_2}}{\sqrt{m}}$

$\dfrac{A_1}{A_2} = \sqrt{\dfrac{k_2}{k_1}}$

Sol 8: (C) $k_A = k/3$; $k_B = 3k$

$T_A \propto k^{-1/2}$; $\dfrac{T_A}{T_B} = 3$

Sol 9: (A) $T = 2\pi\sqrt{\dfrac{m}{k}} \times \dfrac{2\pi/3}{2\pi}$

$T = \dfrac{2\pi}{3}\sqrt{\dfrac{m}{k}}$

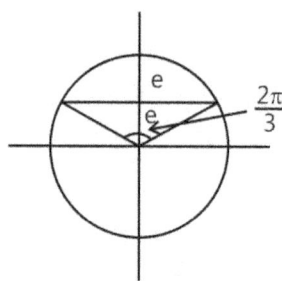

Sol 10: (C) $t = \dfrac{T}{4}$

$t = \dfrac{\pi}{2} \sqrt{\dfrac{\mu}{k}}$

$t = \dfrac{\pi}{2} \sqrt{\dfrac{1}{\pi^2}}$; $\qquad t = \dfrac{1}{2}$ sec

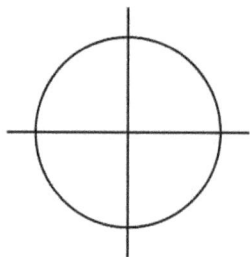

Sol 11: (A) Both block have speed same as

$v_{cm} = 5 m/s$

Sol 12: (D)

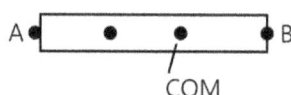

$T = 2\pi \sqrt{\dfrac{I}{mg\ell}}$

$\dfrac{T_A}{T_B} = \dfrac{3}{4} = \sqrt{\dfrac{9}{4} \times \dfrac{\ell_B}{\ell_A}}$

$\dfrac{3}{4} = \dfrac{3}{2} \sqrt{\dfrac{\ell_B}{\ell_A}}$

$\dfrac{\ell_B}{\ell_A} = \dfrac{1}{4}$

$\ell_A = \dfrac{4}{5} \times 25 = 20$ cm

Sol 13: (C)

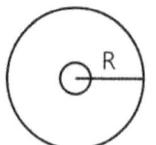 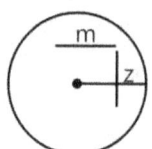

$T = 2\pi \sqrt{\dfrac{I}{mgz}} \qquad \Rightarrow I = \dfrac{mR^2 + 2mz^2}{2}$

$T = \sqrt{\dfrac{2\pi}{g}} \sqrt{m\dfrac{R^2}{2} + 2mz}$

$\dfrac{mR^2}{2} = 2mz$ for minimum T

$z = \dfrac{R}{\sqrt{2}}$

Multiple Correct Choice Type

Sol 14: (B, C, D) $v = 0$ at $t = T/2$

a is maximum at extremes

$F = 0$ at $t = \dfrac{3T}{4}$

K.E. $= 0$ at $t = T/2$

Sol 15: (A, B, C) K.E. $= 0.64 \, KE_{max.}$

$v = 0.8 \, v_{max.}$

$\therefore x = 0.6 \, A = 6$ cm

$x = \dfrac{A}{2} \quad P.E. = \dfrac{PE_{max}}{4} \quad KE = \dfrac{3}{4} PE_{max}$

KE_{max} = TE at mean position

$x = \dfrac{A}{2} \quad v = \dfrac{\sqrt{3} v_{max}}{2}$

Sol 16: (B, C) (A) KE_{avg} is never zero in SHM

(B) $PE_{avg} = \dfrac{1}{2}$ TE $= m\pi^2 t^2 A^2$

(C) Frequency of occurrence of mean position $=2f$

(D) Acceleration leads

Sol 17: (A, C) $v_{rms} = \sqrt{\dfrac{\int_0^T v^2 dt}{T}} = \dfrac{v}{\sqrt{2}}$

$$v_{mean} = \frac{\int_0^T v\, dt}{T} = \frac{\sqrt{8}}{\pi}\, V$$

Sol 18: (B, C, D) A = 3cm

$$\frac{1}{2}\, m v_m^2 = \frac{1}{2} \times 500 \times 9$$

$v_m = 3 \times 10\sqrt{5}$ cm/s ; $\omega = \sqrt{500} = 10\sqrt{5}$

$a_{max.} = \omega\, v_m$

$= 10\sqrt{5} \times 30\sqrt{5}$ cm/s^2

$= 15$ m/s^2

$PE_{min} = 0$ at mean position

Sol 19: (A, B, C) A = 2.5

$\omega = \dfrac{v_{max}}{A} = 4;\ T = \dfrac{2\pi}{4} = \dfrac{\pi}{2} = 1.57$ s

$v_{max} = 16 \times 2.5 = 40$ cm/s^2

$v = \omega\sqrt{A^2 - x^2} = 4\sqrt{2.5^2 - 1^2}$

$= 4\sqrt{5.25}$ cm/s $= 2\sqrt{21}$ cm/s

Sol 20: (A, C) Energy conservation:

$$\frac{1}{2} \times 900 \times 2^2 = \frac{1}{2} \times 900 \times 1^2 + \frac{1}{2} \times 3\, v_1^2$$

$2700 = 3 \times v_1^2$

$v_1^2 = 900$

$v_1 = 30$ m/s

Conservation of momentum:-

$3 \times 30 + 6 \times 0 = 9 \times v$

$v = 10$ ms^{-1}

Energy conservation:-

$$\frac{1}{2} \times 900 \times 1^2 + \frac{1}{2} \times 9 \times 10^2 = \frac{1}{2} \times 900 \times A^2$$

$A^2 = 2;\quad A = \sqrt{2}$ m

Sol 21: (A, B) $t = \dfrac{\pi/6}{2\pi} \times T \Rightarrow t = \dfrac{T}{12}$

$v = \dfrac{\sqrt{3}}{2}\, V_0 \Rightarrow a \propto x \Rightarrow a = a_0/2$

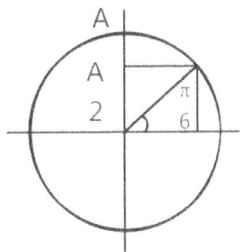

Sol 22: (B, C, D) $v^2 = \omega^2 (A^2 - x^2)$

$a = -\omega^2 x$

$$v^2 = \omega^2 \left(A^2 - \frac{a^2}{\omega^4} \right)$$

Sol 23: (B, D) $x = A\sin\left(\dfrac{2\pi t}{T} + \dfrac{5\pi}{6} \right) = A\cos\left(\dfrac{2\pi t}{T} + \dfrac{\pi}{3} \right)$

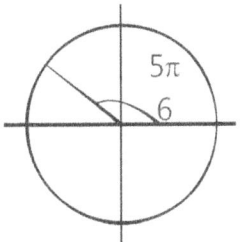

Sol 24: (B, C)

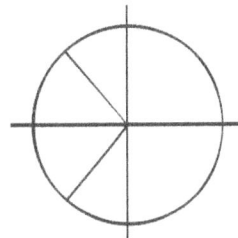

Initial phase difference $= 0,\ \dfrac{2\pi}{3},\ \dfrac{4\pi}{3}$

Sol 25: (A, B, C, D) $x = \dfrac{mg}{k} = \dfrac{0.2 \times 10}{200} = 1$ cm

Amplitude = 1 cm

$\omega = \sqrt{\dfrac{k}{m}} = \sqrt{\dfrac{200}{0.2}} = 10\sqrt{10}$

$f = \dfrac{10\sqrt{10}}{2\pi} = \dfrac{5\sqrt{10}}{\pi} \cong 5$Hz

Amplitude changes, frequency remains the same.

Sol 26: (B, C, D) m = 0.1 kg

$U = 5x(x - 4)$

$F = -\dfrac{dU}{dx} = 20 - 10x$

P.E. minimum at x = 2 m

Force is linear function of x with negative slope.

$$\omega^2 = \frac{10}{m}$$

$$\omega = \sqrt{\frac{10}{0.1}} = 10 \text{ rad/s}$$

$$T = \frac{2\pi}{10} = \frac{\pi}{5} \text{ sec}$$

Sol 27: (B, D) $x = 3 \sin 100 t + 8 \cos^2 50 t$

$= 3 \sin 100 t + 4 \cos 100 t + 4$

$x = 5 \sin (100 t + \sin^{-1} 4/5) + 4$

Sol 28: (A, B, C) $x = 5 \sin (4\pi t + \sin^{-1} 4/5)$ mm

$$T = \frac{2\pi}{4\pi} = 0.5 \text{ s}$$

$A = 5$ mm

$\phi = \sin^{-1}(4/5)$

Sol 29: (B, C) $k = 2 \times 10^6$ Nm^{-1}

$A = 0.01$ m

T.E. = 160 J

PE_{max} = 160 J

when KE = 0 J

i.e. at equilibrium

$$KE_{max} = \frac{1}{2} \times 2 \times 10^6 \times 10^{-4} = 100 \text{ J}$$

PE_{min} = 60 J

Sol 30: (A, B, C) $t = n\dfrac{T}{2}$

$$T = 2\pi\sqrt{\frac{m}{k}} = 2\pi\sqrt{\frac{2}{3 \times 24}} = \frac{\pi}{3}$$

Assertion Reasoning Type

Sol 31: (D) The motion is SHM with $\omega = \sqrt{\dfrac{a}{m}}$

If the force is linear w.r.t. x and slope is negative. The motion is always SHM.

Sol 32: (C) When particle moves from extreme to mean position velocity and acceleration have same direction.

Sol 33: (A) Statement-II is the correct explanation.

Sol 34: (D) Phase remains same and SHMs are perpendicular.

Sol 35: (A) Statement-II is the correct explanation.

Comprehension Type

Paragraph 1:

Sol 36: (D) $\omega = 25$ rad/s

$k = m\omega^2 = 1 \times 625 = 625$ Nm^{-1}

$$F_{max} = 1 \times 9.8 + 625 \times \frac{16}{100 \times 10} + 4.1 \times 9.8$$

$= 59.98 \text{ N} \cong 60 \text{ N}$

$F_{min} = 5.1 \times 9.8 - 10 \cong 40$ N

Sol 37: (C) Minimum force on the surface = (50 − 10) N = 40 N

Sol 38: (A) TE of system is constant

Sol 39: (D) $d = A \sin (\omega t + \phi)$

Sol 40: (D) $F = -kx + c$

$k > 0$

Match the Columns

Sol 41: (B) (a) $y = A \sin (t)$

$v = A \cos (t)$

$KE = c \times \cos^2 (t)$

(a) → (ii)

(b) → (i) PE + KE = const.

$PE = c \times \sin^2 t$

(c) → (iii) TE constant always

(d) → (iv) $v = A \cos t$

Sol 42: (A) $PE \propto x^2$ (A) → p, s

(B) $s = ut + \dfrac{1}{2}at^2$

q, r when a = 0 ;   S when a ≠ 0

(C) Range = $\dfrac{v^2 \sin 2\theta}{g}$

(D) $T^2 = \dfrac{4\pi^2 \ell}{g}$

Previous Year's Questions

Sol 1: (C) If $E > V_B$, particle will escape. But simultaneously for oscillations, $E > 0$

Hence, the correct answer is $V_0 > E > 0$

Or the correct option is (c)

Sol 2: (B) $[\alpha] = \left[\dfrac{PE}{x^4}\right] = \left[\dfrac{ML^2T^{-2}}{L^4}\right] = [ML^{-2}T^{-2}]$

$\therefore \left[\dfrac{m}{\alpha}\right] = [L^2T^2]$; $\therefore \left[\dfrac{1}{A}\sqrt{\dfrac{m}{\alpha}}\right] = [T]$

As dimensions of amplitude A is [L]

Sol 3: (D) For $|x| > x_0$, potential energy is constant. Hence, kinetic energy, speed or velocity will also remain constant.

$\therefore$ Acceleration will be zero

Sol 4: (A) $\dfrac{2v\sin 45°}{g} = 1$

$\therefore v = \sqrt{50}$ m/s

Sol 5: (A, D)

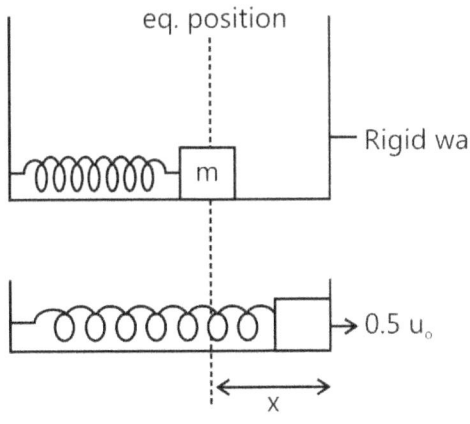

$$\frac{1}{2}mu_0^2 = \frac{1}{2}kx^2 + \frac{1}{2} \times m0.25\, u_0^2 \qquad ....(i)$$

After elastic collision

Block speed is $0.5\, u_0$

So when it will come back to equilibrium point its speed will be u_0 as (A)

Amplitude $\dfrac{1}{2}mu_0^2 = \dfrac{1}{2}kA^2$

$A = \dfrac{u_0}{\sqrt{k}}$

Value of x from eq. (i)

$$\frac{3}{4} \times \frac{1}{2}mu_0^2 = \frac{1}{2}kx^2$$

$$x = \frac{\sqrt{3}u_0}{2}\sqrt{\frac{m}{k}}$$

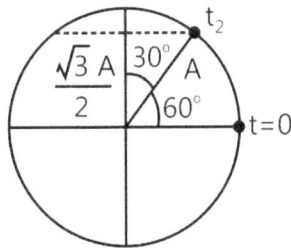

Time to reach eq. position first time $\Rightarrow \dfrac{2\pi}{3}\sqrt{\dfrac{m}{k}}$

Second time it will reach at time $\Rightarrow$

$$\frac{2\pi}{3}\sqrt{\frac{m}{k}} + \frac{T}{2} \Rightarrow \frac{2\pi}{3}\sqrt{\frac{m}{k}} + \frac{2\pi\sqrt{m}}{\sqrt{k}\times 2} \Rightarrow \frac{5\pi}{3}\sqrt{\frac{m}{k}}$$ as (D)

For max. compression time is t_2

$$t_2 = \frac{2\pi}{3}\sqrt{\frac{m}{k}} + \frac{T}{4}$$

$$= \frac{2\pi}{3}\sqrt{\frac{m}{k}} + \frac{2\pi}{\sqrt{k}}\frac{\sqrt{m}}{k} = \frac{7\pi}{6}\sqrt{\frac{m}{k}}$$

Sol 6: (B, D)

$$E_1 = \frac{1}{2}m\omega_1^2 a^2 = \frac{b^2}{2m} \qquad \frac{a}{b} = \frac{1}{m\omega_1} = n^2 \qquad ...(i)$$

$$E_2 = \frac{1}{2}m\omega_2^2 R^2 = \frac{R^2}{2m} \qquad m\omega_2 = 1 \qquad ...(ii)$$

From (i) and (ii) $\dfrac{\omega_2}{\omega_1} = n^2$

$$\frac{E_1}{E_2} = \left(\frac{\omega_2}{\omega_1}\right)^2 \left(\frac{a}{R}\right)^2 = \frac{1}{n^2}\cdot\frac{\omega_1}{\omega_2}\cdot n^2 \Rightarrow \frac{E_1}{\omega_1} = \frac{E_2}{\omega_2}$$

Sol 7: (A, B, D)

(A) $\omega_i = \left(\dfrac{K}{M}\right)$ and $\omega_f = \left[\dfrac{K}{(M+m)}\right]$

Case I: $v_f = \dfrac{Mv_i}{M+m} \quad \dfrac{1}{2}Mv^2 = \dfrac{1}{2}KA_i^2$

2.70

$\Rightarrow A_i^2 = \dfrac{M}{K} v_1^2$ and $\dfrac{1}{2}(M+m)v_f^2 = \dfrac{1}{2}KA_i^2$

$\Rightarrow A_f^2 = \dfrac{Mv^2}{K} \cdot \dfrac{M}{M+m} \Rightarrow \dfrac{A_f}{A_i} = \sqrt{\dfrac{M}{M+m}}$

(B) $T_f = 2\pi\sqrt{\dfrac{M}{M+m}}$ for both

(C) $TE_{case\,I} = \dfrac{1}{2}(M+m)v_f^2 = \dfrac{1}{2}Mv^2\left(\dfrac{M}{M\ m}\right)$

$TE_{case\,II} = \dfrac{1}{2}KA_f^2 = \dfrac{1}{2}KA_i^2$

(D) $VEP = A_f\omega_f$: Decreases in both cases.

Sol 8: A → p; B → q, r; C → p; D → r, q

Sol 9: (A) The total mechanical energy = 160 J

The maximum PE will be 160 J at the instant when KE = 0

Sol 10: (A, C) By principle of superposition $y = y_1 + y_2 + y_3$

$= a\sin(\omega t + 45°) + a\sin\omega t + a\sin(\omega t - 45°)$

$= a\sin(\omega t + 45°) + a\sin(\omega t - 45°) + a\sin\omega t$

$= 2a\sin\omega t\cos 45° + a\sin\omega t$

$= \sqrt{2}a\sin\omega t + a\sin\omega t = (1+\sqrt{2})a\sin\omega t$

∴ Amplitude of resultant motion $= (1+\sqrt{2})\,a$ ...(i)

(b) The option is incorrect as the phase of the resultant motion relative to the first is 45°.

(c) Energy is SHM is proportional to (amplitude)^2

$\therefore \dfrac{E_R}{E_S} = \dfrac{(1+\sqrt{2})^2 a^2}{a^2} \quad \therefore \dfrac{E_R}{E_S} = \dfrac{(1+2+2\sqrt{2})}{1}$

or $E_R = (3+2\sqrt{2})E_S$

(d) Resultant motion is $y = (1+\sqrt{2})a\sin\omega t$

li is SHM.

Sol 11: (A, B, D)

$x = \dfrac{A}{2}(1 - \cos 2\omega t) + \dfrac{B}{2}(1 + \cos 2\omega t) + \dfrac{C}{2}\sin 2\omega t$

For A = 0, B = 0

$x = \dfrac{C}{2}\sin 2\omega t$

A = -B and C = 2B

$X = B\cos 2\omega t + B\sin 2\omega t$

Amplitude $= \left|B\sqrt{2}\right|$

For A = B; C = 0

$X = A$,

Hence this is not correct option.

For A = B, C = 2B

$X = B + B\sin 2\omega t$

It is also represent SHM.

Sol 12: (A, D) Restoring torque is same in both cases

$\alpha = \dfrac{T}{I} = -\omega^2\theta$

In case A the moment of inertia is more as compared to B, so $\omega_B > \omega_A$

ELASTICITY

1. INTRODUCTION

We have learnt that the shape and size of a rigid body does not change but this is an ideal concept. Actually a rigid solid does experience some kind of deformation under the action of external forces and if the magnitude of forces cross a certain limit, the deformation is so severe that the material of the solid loses its rigidity. We say that the material has broken-down or failure has happened. In this chapter we learn about the properties of solid bodies by virtue of which they resist the deformation in their shape and size. These properties constitute the strength of a material and the knowledge of these is very essential in constructing small and large structures like houses, tall buildings, bridges, railway tracks etc.

2. MOLECULAR STRUCTURE OF A MATERIAL

Matter is made up of atoms and molecules. An atom is made up of a nucleus and electrons. Nucleus contains protons and neutrons (collectively known as "nucleons"). Nuclear forces are responsible for the structure of nucleus. Likewise, forces between different atoms and molecules are responsible for the structure of a material.

2.1 Interatomic and Intermolecular Forces

The forces that are responsible for holding the atoms/molecules in place in a solid or liquid are called interatomic and intermolecular forces. The interaction between any isolated pair of atoms and molecules may be represented by a curve that shows how the potential energy varies with the separation between them as shown in the Fig. 8.32

We see that as the distance R decreases, the attractive force first increases and then decreases to zero at a separation R_0 where the potential energy is the minimum. For smaller distance, force is repulsive.

The above picture of interatomic or intermolecular force is an over simplification on the actual situation. However, it provides a reasonable visualisation.

The force between the atoms can be found from the potential energy using the relation,

$$F(R) = -\frac{dU}{dR}$$

The resulting force curve is shown in Fig. 8.33.

Force is along the line joining the atoms or molecules, and is shown negative for attraction & positive for repulsion.

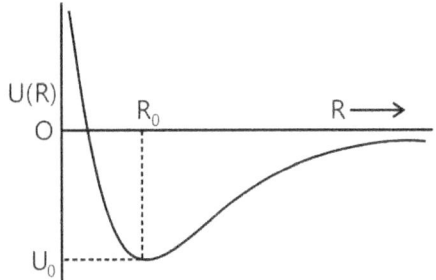

Figure 8.32: Potential energy versus separation

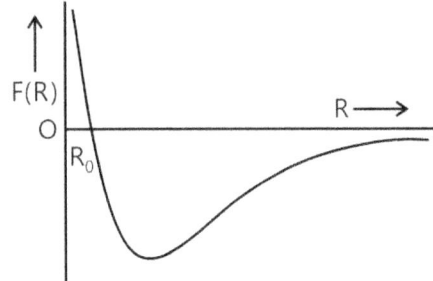

Figure 8.33: Graph of force versus separation

2.2 Classification of Matter

Matter can be classified into three states:- solids, liquids and gases.

Solids: A solid is that state of matter whose atoms and molecules are strongly bound so as to preserve their original shape and volume. Solids are of two types-crystalline & amorphous.

(a) **Crystalline solid:** A crystalline solid is one which has regular & periodic arrangement of atoms or molecules in three dimensions. Examples of crystalline solids are diamond, rock salt, mica, sugar etc.

(b) **Amorphous solids or glassy solids:** The word 'amorphous' literally means 'without any form'. There is no 'order' in arrangement of atoms in such a solid. Example - glass.

In solids, the intermolecular forces are so strong that there is no change in shape and size easily.

Liquids: The intermolecular forces are comparably less than that in solids, so the shape can easily be changed.

But volume of a given mass of a liquid is not easy to change. It needs quite an effort to change the density of liquids.

Liquids are not able to produce reaction forces to applied forces in arbitrary directions.

Gases: This is the third state of matter which cannot support compressive, tensile, or sharing forces. Densities of gases change very rapidly with increase in temperature.

Liquids and gases are together classified as fluids: The word "fluid" comes from a Latin word meaning "to flow".

On an average, the atoms or molecules in a gas are far apart, typically about ten atomic diameters at room temperature and pressure. They collide much less frequently than those in a liquid. Gases in general are compressible.

3. INTRODUCTION TO ELASTICITY

When external forces are applied to a body which is fixed to a rigid support, there is a change in its length, volume or shape. When the external forces are removed, the body tends to regain its original shape and size. Such a property of a body by virtue of which a body tends to regain its original shape or size, when the external forces are removed, is called elasticity.

If a body completely regains its shape and size, it is called perfectly elastic. If it does not regain its shape and size completely, it is called inelastic material. Those materials which hardly regain their shape are called plastic material.

An **elastic** body is one that returns to its original shape after a deformation. Eg- golf ball, rubber band, soccer ball.

An **inelastic** body is one that does not return to its original shape after a deformation. Eg – dough or bread, clay, inelastic ball.

NOMORECLASS CONCEPTS

Microscopic reason of elasticity

Each molecule in a solid body is acted upon by forces due to neighboring molecules. When all molecules are in a state of stable equilibrium, the solid takes a particular shape. When the body is deformed, molecules are displaced from their stable equilibrium positions. The intermolecular distances change and restoring forces start acting which drives the molecules to come back to its original shape.

One can compare this situation to a spring-mass system. Consider a particle connected to several particles through spring. If this particle is displaced a little, the spring exerts a resultant force which tries to bring the particle towards its natural position. In fact, the particle will oscillate about this position. In due course, the oscillations will be damped out and the particle will regain its original position.

3.1 Stress and Strain

Stress: Elastic bodies regain their original shape due to internal restoring forces. This internal restoring force, acting per unit area of a deformed body is called a stress.

i.e. Stress $= \dfrac{\text{Restoring force}}{\text{Area}}$

SI unit of stress is N/m^2 and Dimensional formula of stress is [ML^{-1}T^{-2}]

An object can be deformed in different ways.

Misconception: People often get confused between pressure and stress.

Difference between pressure v/s stress:

S. No.	Pressure	Stress
1	Pressure is always normal to the area	Stress can be normal or tangential
2	Always compressive in nature	May be compressive or tensile in nature

3.1.1 Types of Stress

There are 2 types of stresses – NORMAL stress and SHEAR stress

Normal stress – When the force applied is perpendicular to the area of application of force, it is called normal stress. Normal stress usually leads to a change in length (**longitudinal stress**) or a change in volume.

Normal stress can be of two types – tensile stress and compressive stress.

(a) Tensile Stress: Pulling force per unit area. It is applied parallel to the length.

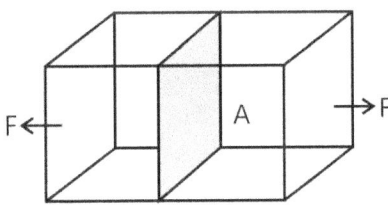

Figure 8.34: Tensile stress

It causes increase in length or volume.

(b) Compressive Stress: Pushing force per unit area. It is applied parallel to the length.

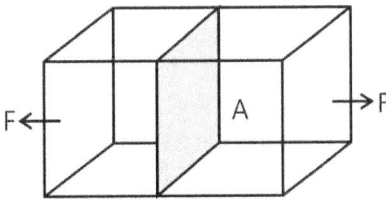

Figure 8.35: Compressive stress

It causes decreases in length or volume.

If the force is applied tangentially to one face of a rectangular body keeping the opposite face fixed, the stress is called tangential or shearing stress.

Stress is measured in units of 1N/m^2. 1N/m^2 = 10 dynes/cm^2.

Strain: The fractional or relative change in shape, size or dimensions of body is called the strain.

$$\text{Strain} = \frac{\text{change in dimension}}{\text{original dimension}}$$

There are three types of strains:

(i) **Longitudinal strain**: It is the ratio of the change in length, $\Delta\ell$, to the original length, ℓ i.e. $\frac{\Delta\ell}{\ell}$.

(ii) **Volume strain**: It is the ratio of change in volume, ΔV, to the original volume V i.e. $\frac{\Delta V}{V}$

(iii) **Shearing strain**: The angular deformation, θ, in radians of a face of a rectangular body is called shearing strain.

If a tangential force F is used to displace upper face of rectangular body

through a small angle θ such that the upper face is displaced through distance

Δx where ℓ is height of the body, then shearing strain $= \theta \approx \tan\theta = \frac{\Delta x}{\ell}$

Strain is a ratio of two similar quantities and does not have any units.

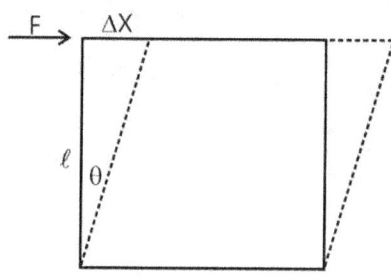

Figure 8.36: Shearing strain

Illustration 1: A 4.0 m long copper wire of cross sectional area 1.2 cm^2 is stretched by a force of 4.8×10^3 N. Stress will be **(JEE MAIN)**

(A) 4.0×10^7 N/mm^2 (B) 4.0×10^7 kN/m^2 (C) 4.0×10^7 N/m^2 (D) None

Sol: (C) Stress is restoring force per unit area of cross-section.

$$\text{Stress} = \frac{F}{A} = \frac{4.8 \times 10^3 \text{N}}{1.2 \times 10^{-4} \text{m}^2} = 4.0 \times 10^7 \text{ N/m}^2$$

Illustration 2: A copper rod 2m long is stretched by 1mm. Strain will be **(JEE MAIN)**

(A) 10^{-4}, volumetric   (B) 5×10^{-4}, volumetric   (C) 5×10^{-4}, longitudinal   (D) 5×10^{-3}, volumetric

Sol: Longitudinal strain is equal to change in length per unit length.

(C) $\text{Strain} = \frac{\Delta\ell}{\ell} = \frac{1 \times 10^{-3}}{2} = 5 \times 10^{-4}$, Longitudinal

Illustration 3: A lead of 4.0 kg is suspended from a ceiling through a steel wire of radius 2.0 mm. Find the tensile stress developed in the wire when equilibrium is achieved. Take $g = 3.1\pi$ ms^{-2}. **(JEE MAIN)**

Sol: Stress is restoring force per unit area of cross-section.

Tension in the wire is $F = 4.0 \times 3.1 \pi$ N.

The area of cross section is $A = \pi r^2 = \pi \times (2.0 \times 10^{-3} \text{ m})^2 = 4.0 \pi \times 10^{-6}$ m^2.

Thus, the tensile stress developed $\frac{F}{A} = \frac{4.0 \times 3.1\pi}{4.0\pi \times 10^{-6}}$ N/m^2 = 3.1×10^6 N/m^2.

Illustration 4: Find the stress on a bone (1 cm in radius and 50 cm long) that supports a mass of 100kg. Find the strain on the bone if it is compressed 0.15 mm by this weight. Find the proportionality constant C for this bone.
(JEE MAIN)

Sol: Stress is restoring force per unit area of cross-section. Strain is equal to change in length per unit length. Strain $\propto$ Stress

Stress $= F/A = (100\text{kg}) (9.8 \text{ m/s}^2) / \pi \times (0.01 \text{ m})^2 = 3.1 \times 10^6 \text{ N/m}^2$

Strain $= \Delta L/L_0 = (0.15 \times 10^{-3}\text{m}) / (0.5\text{m}) = 3.0 \times 10^{-4}$

Since strain $= C \times$ stress, $C =$ strain / stress $= 0.96 \times 10^{-10} \text{ m}^2/\text{N}$.

4. HOOKE'S LAW AND MODULI OF ELASTICITY

Hooke's Law: It states that for small deformations, stress is directly proportional to strain within elastic limits and the ratio is a constant called modulus of elasticity.

$$\frac{\text{Stress}}{\text{Strain}} = \text{modulus of Elasticity} = E$$

4.1 Young's Modulus

Young's modulus is a measure of the resistance of a solid to a change in its length when a force is applied perpendicular to its surface. Consider a rod with an unstressed length L_0 and cross-sectional area A, as shown in the Fig. 8.37. When it is subjected to equal and opposite forces F_n along its axis and perpendicular to the end faces, its length changes by ΔL. These forces tend to stretch the rod. The tensile stress on the rod is defined as $\sigma = \dfrac{F_n}{A}$

Forces acting in the opposite direction, as shown in Fig. 8.37, would produce a compressive stress. The resulting strain is defined as the dimensionless ratio, $\varepsilon = \dfrac{\Delta L}{L_0}$ Young's modulus Y for the material of the rod is defined as the ratio of tensile stress to tensile strain.

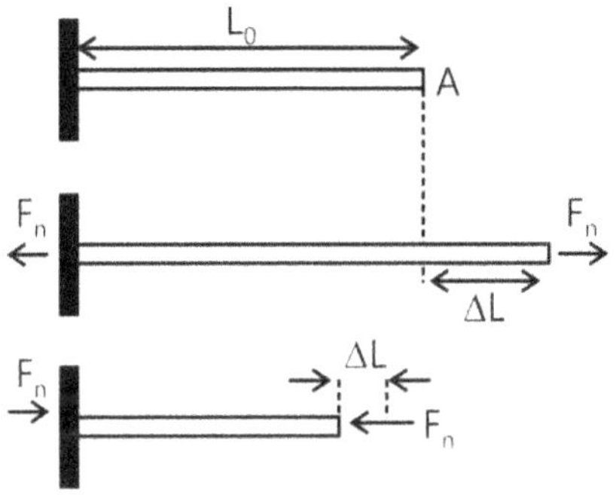

Figure 8.37: Variation in length of rod

So Young's Modulus $= \dfrac{\text{Tensile stress}}{\text{Tensile strain}}$; $Y = \dfrac{\sigma}{\varepsilon} = \dfrac{F_n / A}{\Delta L / L_0} = \dfrac{F_n L_0}{A \Delta L}$

A force applied normal to the end face of a rod cause a change in length.

NOMORECLASS CONCEPTS

(a) For loaded wire: $\Delta L = \dfrac{FL}{\pi r^2 \gamma}$

For rigid body $\Delta L = 0$ so $Y = \infty$ i.e. Elasticity of rigid body is infinite.

(b) If same stretching force is applied to different wires of same material, $\Delta L \propto \dfrac{L}{r^2}$ [As F and Y are const.]

Greater the value ΔL, greater will be elongation.

Following conclusions can be drawn from $\gamma =$ stress/strain:

(i) $E \propto$ stress (for same strain), i.e. if we want the equal amount of strain in two different materials, the one which needs more stress is having more E.

(ii) $E \propto \dfrac{1}{\text{strain}}$ (for same stress), i.e., if the same amount of stress is applied on two different materials, the one having less strain is having more Elasticity. Rather we can say that, the one which offers more resistance to the external forces is having greater value of E. So, we can see that modulus of elasticity of steel is more than that of rubber or $E_{steel} > E_{rubber}$

(iii) E = stress for unit strain $\left(\dfrac{\Delta x}{x} = 1 \text{ or } \Delta x = x \right)$, i.e. suppose the length of a wire is 2m, then the Young's modulus of elasticity (Y) is the stress applied on the wire to stretch the wire by the same amount of 2m.

Illustration 5: Two wires of equal cross section but one made of steel and the other of copper, are joined end to end. When the combination is kept under tension, the elongations in the two wires are found to be equal. Find the ratio of the lengths of the two wires. Young's modulus of steel = 2.0×10^{11} Nm^{-2}.   **(JEE ADVANCED)**

Sol: The wires joined together have same stress and same elongation. Ratio of stress and young's modulus is strain. As young's modulus for steel and copper is different, strains of the wires will be different.

As the cross sections of the wires are equal and same tension exists in both, the stresses developed are equal. Let the original lengths of the steel wire and the copper wire be L_s and L_c respectively and the elongation in each wire be ℓ.

$$\frac{\ell}{L_S} = \frac{\text{stress}}{2.0 \times 10^{11} \text{Nm}^{-2}} \qquad \text{... (i)}$$

And $\dfrac{\ell}{L_C} = \dfrac{\text{stress}}{1.1 \times 10^{11} \text{Nm}^{-2}}$ ... (ii)

Dividing (ii) by (i), $L_s/L_c = 2 \cdot 0 / 1 \cdot 1 = 20{:}11$.

Illustration 6: A solid cylindrical steel column is 4.0 m long and 9.0 cm in diameter. What will be decrease in length when carrying a load of 80000 kg? Y = 1.9×10^{11} Nm^{-2}.   **(JEE MAIN)**

Sol: The stress will be equal to load per unit cross section. Strain is the ratio of stress and young's modulus.

Let us first calculate the cross-sectional area of column = $\pi r^2 = \pi(0.045\text{m})^2 = 6.36 \times 10^{-3}$ m^2

Then, from $Y = \dfrac{F/A}{\Delta L/L}$ we have $\Delta L = \dfrac{FL}{AY} = \dfrac{[(8 \times 10^4)(9.8\text{N})](4.0\text{m})}{(6.36 \times 10^{-3}\text{m}^2)(1.9 \times 10^{11}\text{Nm}^{-2})} = 2.6 \times 10^{-3}$ m.

Illustration 7: A load of 4.0 kg is suspended from a ceiling through a steel wire of length 20 m and radius 2.0 mm. It is found that the length of the wire increases by 0.031 mm as equilibrium is achieved. Find Young's modulus of steel. Take g = 3.1 π m/s^2.   **(JEE MAIN)**

Sol: The stress will be equal to load per unit cross section. Strain is the change in length per unit length. Young's modulus is the ratio of stress and strain.

The longitudinal stress = $\dfrac{(4.0\text{kg})(3.1\pi\text{ms}^{-2})}{\pi(2.0 \times 10^{-3}\text{m})^2} = 3.1 \times 10^6$ N/m^2

The longitudinal strain $= \dfrac{0.031 \times 10^{-3}\,m}{2.0\,m} = 0.0155 \times 10^{-3}$

Thus $Y = \dfrac{3.1 \times 10^{6}\,Nm^{-2}}{0.0155 \times 10^{-3}} = 2.0 \times 10^{11}\,N/m^{2}.$

Illustration 8: A bar of mass m and length ℓ is hanging from point A as shown in Fig. 8.38. Find the increase in its length due to its own weight. The Young's modulus of elasticity of the wire is Y and area of cross-section of the wire is A. **(JEE ADVANCED)**

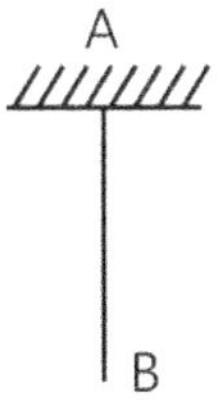

Sol: Find the elongation for an elementary length dx of the wire due to tension in the wire at the location of the element.

Consider a small section dx of the bar at a distance x from B. The weight of the bar for a length x is,

$W = \left(\dfrac{mg}{\ell}\right)x$

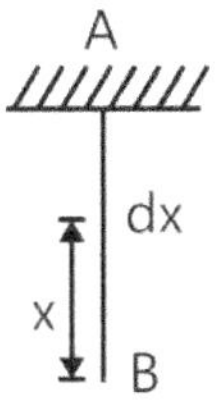

Elongation in section dx will be $d\ell = \left(\dfrac{W}{AY}\right)dx = \left(\dfrac{mg}{\ell AY}\right)x\,dx$

Total elongation in the bar can be obtained by integrating this expression for $x = 0$ to $x = \ell$.

$\therefore \quad \Delta\ell = \int\limits_{x=0}^{x=\ell} d\ell = \left(\dfrac{mg}{\ell AY}\right)\int\limits_{0}^{\ell} x\,dx$ or $\quad \Delta\ell = \dfrac{mg\ell}{2AY}$

A
/////////
B

Figure 8.38

A
/////////
dx
x
B

Figure 8.39

Illustration 9: One end of a metal wire is fixed to a ceiling and a load of 2 kg hangs from the other end. A similar wire is attached to the bottom of the load and another load of 1 kg hangs from this lower wire. Find the longitudinal strain in both the wires. Area of cross section of each wire is 0.005 cm² and Young modulus of the metal is $2.0 \times 10^{11}\,N\,m^{-2}$. Take $g = 10\,ms^{-2}$. **(JEE ADVANCED)**

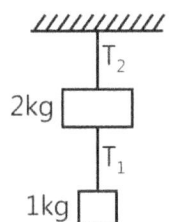

Figure 8.40

Sol: Find the tension in each wire. Stress is tension per unit area of cross section. Strain is the ratio of stress and Young's modulus.

The situation is described in Fig. 8.40. As the 1kg mass is in equilibrium, the tension in the lower wire equals the weight of the load.

Thus $T_1 = 10N$; Stress $= 10N/0.005\,cm^2 = 2 \times 10^7\,N/m^2$

Longitudinal strain $= \dfrac{stress}{Y} = \dfrac{2 \times 10^7\,N/m^2}{2 \times 10^{11}\,N/m^2} = 10^{-4}$

Considering the equilibrium of the upper block, we can write, $T_2 = 20N + T_1$ or $T_2 = 30N$

Stress $= 30\,N/0.005\,cm^2 = 6 \times 10^7\,N/m^2$

Longitudinal strain $= \dfrac{6 \times 10^7\,N/m^2}{2 \times 10^{11}\,N/m^2} = 3 \times 10^{-4}.$

Illustration 10: Each of the three blocks P, Q and R shown in Figure has a mass of 3 kg. Each of the wires A and B has cross-sectional area 0.005 cm² and Young modulus $2 \times 10^{11}\,N/m^2$. Neglect friction. Find the longitudinal strain developed in each of the wires. Take $g = 10\,m/s^2$. **(JEE ADVANCED)**

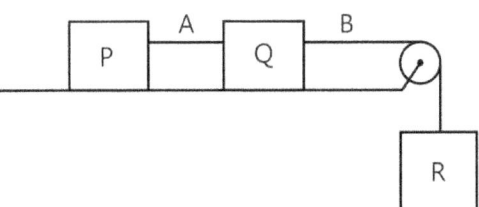

Figure 8.41

Sol: Find the tension in each wire. Stress is tension per unit area of cross section. Strain is the ratio of stress and Young's modulus.

The block R will descend vertically and the blocks P and Q will move on the frictionless horizontal table. Let the common magnitude of the acceleration be a. Let the tensions in the wires A and B be T_A and T_B respectively.

Writing the equations of motion of the blocks P, Q and R, we get,

$$T_A = (3kg) a \qquad \qquad \text{.... (i)}$$

$$T_B - T_A = (3kg) a \qquad \qquad \text{.... (ii)}$$

And $(3kg)g - T_B = (3kg)a$ .... (iii)

By (i) and (ii), $T_B = 2T_A$; By (i) and (iii), $T_A + T_B = (3kg) g = 30$ N

or $3T_A = 30$N or, $T_A = 10$N and $T_B = 20$ N.

$$\text{Longitudinal strain} = \frac{\text{Longitudinal stress}}{\text{Young modulus}}$$

Strain in wire A $= \dfrac{10N / 0.005 \text{ cm}^2}{2 \times 10^{11} \text{ N} / \text{m}^2} = 10^{-4}$;   And strain in wire B $= \dfrac{20N / 0.005 \text{ cm}^2}{2 \times 10^{11} \text{ N} / \text{m}^2} = 2 \times 10^{-4}$.

NOMORECLASS CONCEPTS

In practical life, we often hear something like elastic band is usually referred to a rubber band because it is easily stretchable and a steel rod is not.

However, here elasticity has some different meaning. Being more elastic means, the material will resist more to any external force which tries to change its configuration.

That is why $E_{steel} > E_{rubber}$.

4.2 Shear Modulus

The shear modulus of a solid measures its resistance to a shearing force, which is a force applied tangentially to a surface, as shown in the Fig. 8.42. (Since the bottom of the solid is assumed to be at rest, there is an equal and opposite force on the lower surface). The top surface is displaced by x relative to the bottom surface.

The shear stress is defined as, Shear stress $= \dfrac{\text{Tangential force}}{\text{Area}} = \tau = \dfrac{F_t}{A}$ where A is the area of the surface.

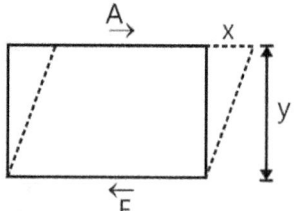

Figure 8.42: Shearing stress

The shear strain is defined as   Shear strain $= \dfrac{x}{y}$

where y is the separation between the top and the bottom surfaces.

The shear modulus G is defined as

Shear modulus $= \dfrac{\text{Shear Stress}}{\text{Shear Strain}}$; $G = \dfrac{F_t / A}{x / y} = \dfrac{Fy}{Ax}$

Illustration 11: A box shaped piece of gelatin dessert has a top area of 15 cm^2 and a height of 3cm. When a shearing force of 0.50 N is applied to the upper surface, the upper surface displaces 4 mm relative to the bottom surface. What are the shearing stress, the shearing strain and the shear modulus for the gelatin? **(JEE MAIN)**

Sol: Shearing stress is tangential force per unit area of surface. Shearing Strain is the ratio of displacement of the surface to the distance of the surface from the fixed surface. Shear modulus is the ratio of shearing stress to shearing strain.

$$\text{Shear stress} = \frac{\text{tangential force}}{\text{area of face}} = \frac{0.50 \text{N}}{15 \times 10^{-4} \text{m}^2} = 333 \text{ N/m}^2$$

$$\text{Shear stress} = \frac{\text{Displacement}}{\text{height}} = \frac{0.4 \text{cm}}{3 \text{cm}} = 0.133$$

$$\text{Shear modulus } G = \frac{\text{stress}}{\text{strain}} = \frac{333}{0.133}$$

$$= 2.5 \times 10^3 \text{ N/m}^2 \quad (1 \text{ Pa} = 1 \text{ N/m}^2)$$

4.3 Bulk Modulus

The bulk modulus of a solid or a fluid indicates its resistance to a change in volume. Consider a cube of some material, solid or fluid, as shown in the Fig. 8.43. We assume that all faces experience the same force F_n normal to each face. (One way to accomplish this is to immerse the body in a fluid-as long as the change in pressure over the vertical height of the cube is negligible). The pressure on the cube is defined as the normal force per unit area $p = \dfrac{F_n}{A}$

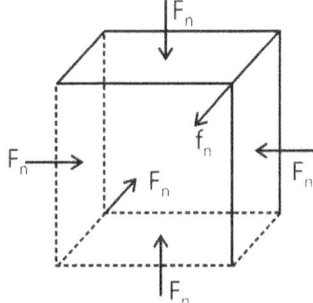

Figure 8.43: Determination of bulk modulus of an object

The SI unit of pressure is N/m^2 and is given the name Pascal (Pa).

The change in pressure ΔP is called the volume stress and the fractional change in volume $\Delta V / V$ called the volume strain. The bulk modulus B of the material is defined as

$$\text{Bulk modulus} = \frac{\text{Volume stress}}{\text{Volume strain}} \quad \text{or} \quad B = \frac{-\Delta P}{\Delta V / V}$$

The negative sign is included to make B a positive number since an increase in pressure $(\Delta p > 0)$ leads to a decrease in volume $(\Delta V < 0)$.

The inverse of B is called the compressibility factor $k = \dfrac{1}{B}$

Elastic properties of matter

Sate	Shear Modulus	Bulk Modulus
Solid	Large	Large
Liquid	Zero	Large
Gas	Zero	Small

Bulk Modulus has very important applications in case of fluids. Actually, it has various applications in adiabatic expansion of gases. Also, while calculating speed of sound through air, one would find that it would come out to be directly proportional to square root of bulk modulus of air. (In general, speed of sound depends of elastic properties of matter. A more general statement is that mechanical waves' speed depends on elastic properties of matter)

Illustration 12: Find the decrease in the volume of a sample of water from the following data. Initial volume = 1000 cm^3, initial pressure = 10^5 Nm^{-2}, final pressure = 10^6 Nm^{-2}, compressibility of water = 50×10^{-11} m^2N^{-1}. **(JEE MAIN)**

Sol: Using the formula for bulk modulus deduce the value for decrease in volume.

The change in pressure = $\Delta P = 10^6$ $Nm^{-2} - 10^5$ $Nm^{-2} = 9 \times 10^5$ Nm^{-2}.

$$\text{Compressibility} = \frac{1}{\text{Bulk modulus}} = -\frac{\Delta V / V}{\Delta P} \text{ or,}$$

$$50 \times 10^{-11} \text{ m}^2\text{N}^{-1} = -\frac{\Delta V}{(10^{-3}\text{ m}^3) \times (9 \times 10^5\text{ Nm}^{-2})}$$

or, $\Delta V = -50 \times 10^{-11} \times 10^{-3} \times 9 \times 10^5$ $m^3 = -4.5 \times 10^{-7}$ $m^3 = -0.45$ cm^3.

Thus the decrease in volume is 0.45 cm^3.

A solid will have all the three moduli of elasticity Y, B and η. But in case of a liquid or a gas, only B can be defined because a liquid or a gas cannot be framed into a wire or no shear force can be applied on

them. For a liquid or a gas, $\qquad B = \left(\dfrac{-dP}{dV / V} \right)$

So, instead of P, we are more interested in change in pressure dP.

In case of a gas, $\qquad B = XP$

5. THE STRESS-STRAIN CURVE

The stress-strain graph of a ductile metal is shown in Fig. 8.44 Initially, the strain graph is linear and it obeys the Hooke's Law up to the point P called the proportional limit. After the proportional limit, the $\sigma - \varepsilon$ graph is non-linear but it still remains elastic up to the yield point Y where the slope of the curve is zero. At the yield point, the material starts deforming under constant stress and it behaves like a viscous liquid. The yield point is the beginning of the plastic zone. After the yield point, the material starts gaining strength due to excessive deformation and this phenomenon is called strain hardening. The point U shows the

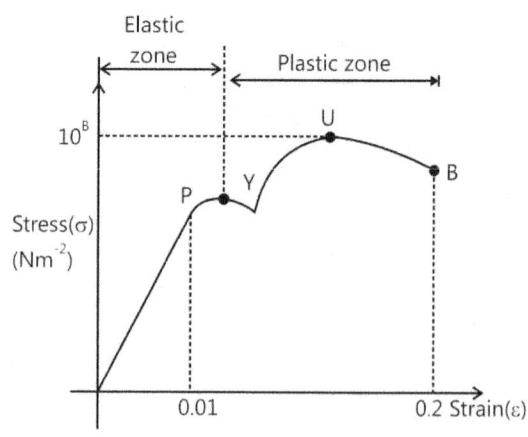

Figure 8.44

ultimate strength of the material. It is the maximum stress that the material can sustain without failure. After the point U the curve goes down towards the breaking point B because the calculation of the stress is based on the original cross-sectional area whereas the cross-sectional area of the sample actually decreases.

NOMORECLASS CONCEPTS

It is generally thought that strain results from stress, or many say that Hooke's law states wrong statement that stress is directly proportional to strain.

However, we must not worry because Hooke's law is correct. Going deeper to a microscopic level will help us understand better. It appears that external force cause strain in the body on which it is applied. However, stress is defined as restoring force (at equilibrium) per unit area. There can be no restoring force if there is no strain. Hence, strain is the cause and not stress. The only glitch here is that restoring force is equal to the force applied because (again not to forget) body is in equilibrium. So, it creates confusion but we must not take it for granted and understand the minute concepts.

6. RELATION BETWEEN LONGITUDINAL STRESS AND STRAIN

For small deformations, longitudinal stress is directly proportional to the longitudinal strain. What if the deformation is large? The stress-strain relation gets more complicated in that case and depends on the material under study. Let's take a metal wire and a rubber piece as example and study the same.

Metal Wire: The Fig. 8.45 shows the relation between stress and strain as the deformation gradually decreases in a stretched wire.

Up to a strain < 0.01, Hooke's law is valid and Young's modulus is defined. Point a represents proportional limit up to which stress is proportional to strain.

Point b is called the yield point or elastic limit up to which stress is not proportional to strain (a to b) but elasticity still holds true.

The wire shows plastic behavior after point b where there is a permanent deformation in the wire and it does not return back to its original dimensions.

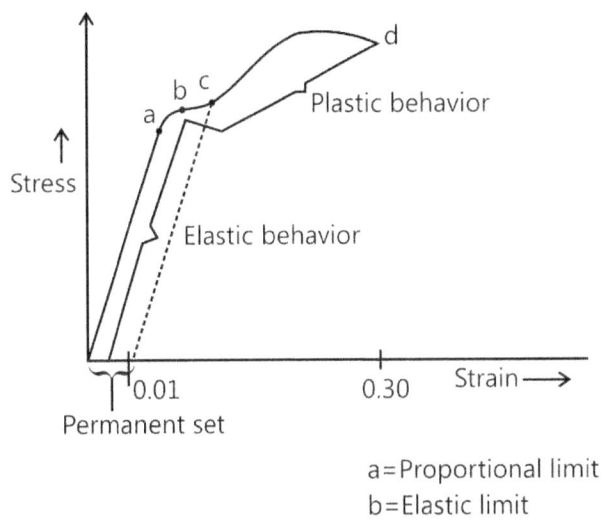

a = Proportional limit
b = Elastic limit
d = Fracture point

Figure 8.45: Graph of Stress versus Strain

The wire breaks at d which is the fracture point if stretched beyond point c. The corresponding stress is called breaking stress.

NOMORECLASS CONCEPTS

If large deformation takes place between the elastic limit and the fractured point, the material is called ductile. If it breaks soon after the elastic limit is crossed, it is called brittle.

Rubber: Vulcanized rubber shows a very different stress-strain behavior. It remains elastic even if it is stretched to 8 times its original length. There are 2 important phenomena to note from the above Fig 8.46. Firstly, stress is nowhere proportional to strain during deformation. Secondly, when external forces are removed, body comes back to original dimensions but it follows a different retracing path.

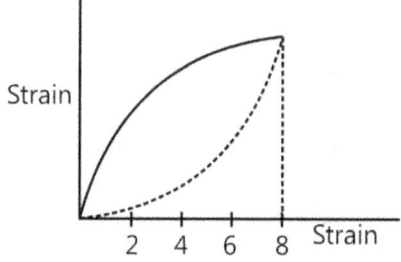

The work done by the material in returning to its original shape is less than the work done by the deforming force when it was deformed. A particular amount of energy is, thus, absorbed by the material in the cycle which appears as heat. This phenomenon is called elastic hysteresis.

Figure 8.46: Stress versus Strain curve for rubber

Elastic hysteresis has an important application in shock absorbers.

> **NOMORECLASS CONCEPTS**
>
> The material which has smaller value of Y is more ductile, i.e., it offers less resistance in framing it into a wire. Similarly, the material having the smaller value of B is more malleable. Thus, for making wire, we choose a material having less value of Y.

7. POISSON'S RATIO

When a longitudinal force is applied on a wire, its length increases but its radius decreases. Thus two strains are produced by a single force.

(a) Longitudinal strain = $\dfrac{\Delta l}{l}$ and (b) Lateral strain = $\dfrac{\Delta R}{R}$

The ratio of these two strains is called the Poisson's ratio.

Thus, the Poisson's ratio $\sigma = \dfrac{\text{Lateral strain}}{\text{Longitudinal strain}} = -\dfrac{\Delta R / R}{\Delta l / l}$

Negative sign in σ indicates that radius of the wire decreases as the length increases.

> **NOMORECLASS CONCEPTS**
>
> Relation between Y, B, η and σ : Following are some relations between the four
>
> (a) $B = \dfrac{Y}{3(1 - 3\sigma)}$ (b) $\eta = \dfrac{Y}{2(1 + \sigma)}$ (c) $\sigma = \dfrac{3B - 2\eta}{2\eta + 6B}$ (d) $\dfrac{9}{Y} = \dfrac{1}{B} + \dfrac{3}{\eta}$

8. ELASTIC POTENTIAL ENERGY OF A STRAINED BODY

When a body is in its natural shape, potential energy due to molecular forces is minimum and assumed to be zero. When deformed, internal forces come into existence and work is done against these forces. Thus potential energy of the body increases. This is called elastic potential energy.

8.1 Work Done in Stretching a Wire

If a force F is applied along the length of a wire of length l, area of cross-section A and Young's modulus Y, such that the wire is extended through a small length x, then $Y = \dfrac{Fl}{Ax}$ or $F = \dfrac{YAx}{l}$

The work done, W, in extending the wire through length Dl is given by

$$W = \int_0^{\Delta l} F dx = \frac{YA}{l}\int_0^{\Delta l} x dx = \frac{YA(\Delta l)^2}{2l} = \frac{1}{2}\left(\frac{Y\Delta l}{l}\right)\left(\frac{\Delta l}{l}\right)(Al) = \frac{1}{2} \times \text{stress} \times \text{strain} \times \text{volume}$$

Also $W = \dfrac{1}{2}\left(\dfrac{YA\Delta l}{l}\right)\Delta l = \dfrac{1}{2}\times \text{force}\times \text{extension}$

This work is stored in the wire as elastic potential energy.

Work done per unit volume $= \dfrac{1}{2}\left[\dfrac{Y\Delta l}{l}\right]\times \dfrac{\Delta l}{l} = \dfrac{1}{2}\times Y \times (\text{strain})^2 = \dfrac{1}{2}\times \text{stress}\times \text{strain}$.

Illustration 13: Spring is stretched by 3 cm when a load of 5.4×10^6 dyne is suspended from it. Work done will be

(A) 8.1×10^6 J   (B) 8×10^6 J   (C) 8.0×10^6 ergs   (D) 8.1×10^6 ergs   **(JEE MAIN)**

Sol: Work done in stretching the spring is equal to the elastic potential energy stored in the spring.

(D) $W = \dfrac{1}{2} \times$ load $\times$ elongation $W = 8.1 \times 10^6$ ergs $= 0.81$ J

Illustration 14: A steel wire of length 2.0 m is stretched through 2.0 mm. The cross-sectional area of the wire is 4.0 mm². Calculate the elastic potential energy stored in the wire in the stretched condition. Young modulus of steel $= 2.0 \times 10^{11}$ N/m². **(JEE MAIN)**

Sol: We know the formula to find the elastic potential energy stored per unit volume of the wire. Calculate the volume of the wire and find the energy stored in the entire wire.

The strain in the wire $\dfrac{\Delta l}{l} = \dfrac{2.0\,\text{mm}}{2.0\,\text{m}} = 10^{-3}$.

The stress in the wire = Y × strain = 2.0×10^{11} N m⁻² × 10^{-3} = 2.0×10^8 N/m².

The volume of the wire = $(4 \times 10^{-6}$ m²$) \times (2.0$ m$) = 8.0 \times 10^{-6}$ m³.

The elastic potential energy stored $= \dfrac{1}{2} \times$ stress × strain × volume

$$= \frac{1}{2} \times 2.0 \times 10^8 \text{ Nm}^{-2} \times 10^{-3} \times 8.0 \times 10^{-6} \text{ m}^3 = 0.8 \text{ J}$$

NOMORECLASS CONCEPTS

This energy can also be thought of as elastic potential energy of a spring. You just need to calculate spring constant.

A simple way would be considering $\Delta l = x$ and rearranging terms of Hooke's law in the form of F=-kx.

Remember F here is restoring force. Now energy is simply $\dfrac{1}{2}kx^2$

9. THERMAL STRESS AND STRAIN

A body expands or contracts whenever there is an increase or decrease in temperature. No stress is induced when the body is allowed to expand and contract freely. But when deformation is obstructed, stresses are induced. Such stresses are called thermal/ temperature stresses. The corresponding strains are called thermal/temperature strains.

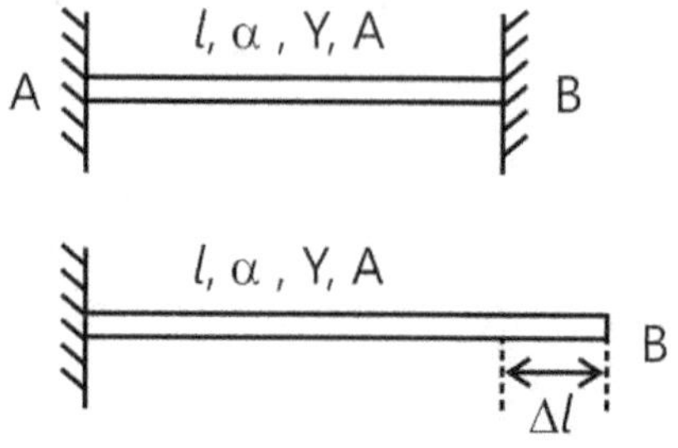

Consider a rod AB fixed at two supports as shown in Fig. 8.47.

Let l = Length of rod

A = Area of cross-section of the rod

Y = Young's modulus of elasticity of the rod

And α = Thermal coefficient of linear expansion of the rod

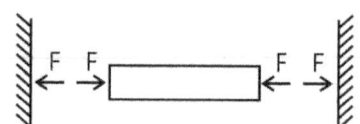

Figure 8.47: Thermal expansion of a rod

Let the temperature of the rod is increased by an amount t. The length of the rod would had increased by an amount Δl, if it were not fixed at two supports. Hence $\Delta l = l\alpha t$

But since the rod is fixed at the supports, a compressive strain will be produced in the rod. Because at the increased temperature, the natural length of the rod is $l + \Delta l$, while being fixed at two supports its actual length is l.

Hence, thermal strain $\varepsilon = \dfrac{\Delta l}{l} = \dfrac{l\alpha t}{l} = \alpha t$ or $\varepsilon = \alpha t$

Therefore, thermal stress $\sigma = Y\varepsilon$ (stress = Y × strain)

or $\sigma = Y\alpha t$ or force on the supports, $F = \sigma A = YA\alpha t$

This force F is in the direction shown:

Figure 8.48: Thermal stress on a rod

Illustration 16: A wire of cross sectional area 3 mm² is just stretched between two fixed points at a temperature of 20°C. Determine the tension when the temperature falls to 20°C. Coefficient of linear expansion $\alpha = 10^{-5}$ / °C and $Y = 2 \times 10^{11}$ N/m². **(JEE MAIN)**

(A) 120 kN   (B) 20 N   (C) 120 N   (D) 12 N

Sol: Thermal stress is equal to product of young's modulus and thermal strain. Tension is product of area of cross-section and stress.

(C) $F = Y A \alpha \Delta t = 2 \times 10^{11} \times 3 \times 10^{-6} \times 10^{-5} \times 20$;   F = 120 N

10. DETERMINATION OF YOUNG'S MODULUS IN LABORATORY

The given Fig. 8.49 shows an experimental set up of a simple method to determine Young's modulus in laboratory. A 2-3 metres long wire is suspended from a fixed support. It carries a graduated scale and below it a heavy fixed load. This load keeps the wire straight. Wire A is the reference wire whereas wire B serves as the experimental wire. A Vernier scale is placed at the end of the experimental wire.

Now the stress due to the weight Mg at the end is

Stress $= \dfrac{Mg}{\pi r^2}$ and strain $= \dfrac{l}{L}$; Thus, $Y = \dfrac{MgL}{\pi r^2 l}$

All the quantities on the right-hand side are known and hence Young's modulus Y may be calculated.

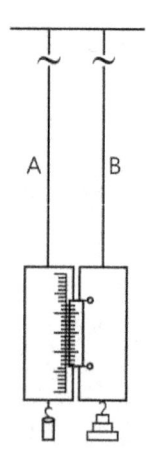

Figure 8.49: Searle's method for determination of young's modulus

PROBLEM-SOLVING TACTICS

- Be careful while using the Hooke's law of elasticity. Always remember that this law is not valid for an elastic material when it is stretched beyond its elastic limit. Stress is proportional to strain only when the material is stretched up to a certain limit.

- Always keep the stress-strain graph in mind while solving elasticity problems.

- The extent of ductility of a material can be calculated using the strain formulae. Greater the elongation, greater the ductility of the material. This concept can be used in questions where one is asked to arrange the elastic material in the order of increasing brittleness or ductility.

- Conservation of energy principle can be used to solved many problems where elastic potential energy gets converted to other forms of energy in the given problem system.

- Elongation and compression can be thought as analogous to a spring (refer to Plancess concept to how to do it) in appropriate limits.

- Direct questions may be asked on relation between Poisson's ratio and modulus of elasticity, so it would be nice if you learn them.

FORMULAE SHEET

Elasticity:

Stress: Stress $(\sigma) = \dfrac{\text{Restoring force}}{\text{Area}}$

SI units = N/m²

Normal/ longitudinal stress $S_n = \dfrac{F_n}{A}$

F_n is the normal force

A is the cross-sectional area

Tangential / shearing stress $S_t = \dfrac{F_t}{A}$

F_t is the tangential force

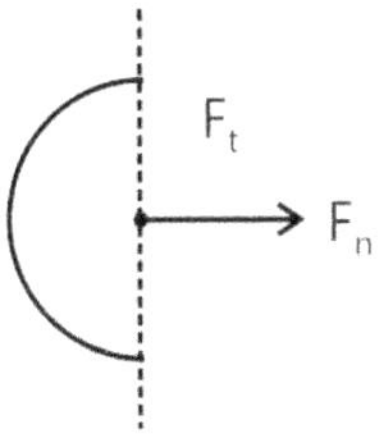

Volume stress $S_v = \dfrac{F}{A}$

Figure 8.50

Note: This is the stress developed when body is immersed in a liquid.

Strain: Longitudinal strain $\varepsilon = \dfrac{\Delta l}{l}$

Volumetric strain $\varepsilon = \dfrac{\Delta V}{V}$

Δl and ΔV are change in length and volume respectively.

Shearing strain $\varepsilon = \dfrac{\Delta X}{X}$

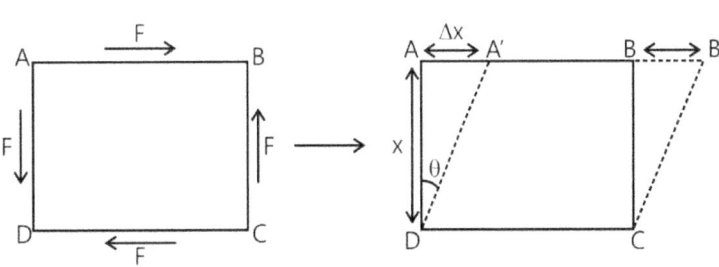

Figure 8.51

Hooke's Law: stress $\propto$ strain

stress = (E) (strain)   (E is modulus of elasticity)

E is constant for a particular type of strain for a particular material. SI unit of E is N/m^2.

Young's modulus of elasticity (Y) $Y = \dfrac{\text{longitudinal stress}}{\text{Longitudinal strain}} = \dfrac{(F_n / A)}{(\Delta l / l)}$

Bulk modulus of elasticity (B) $B = \dfrac{-\text{Volume stress}}{\text{Volume strain}} = \dfrac{-F / A}{\Delta V / V} = \dfrac{-P}{\Delta V / V}$

For a liquid or gas $B = -\dfrac{dp}{(dV / V)}$

Compressibility $= \left(\dfrac{1}{\beta}\right)$

Modulus of rigidity (η) $\eta = \dfrac{\text{shearing stress}}{\text{shearing strain}} = \dfrac{F_t / A}{(\Delta x / x)} = \dfrac{F_t / A}{\theta}$ (See Fig. 8.52)

Elastic potential energy stored per unit volume in a stretched wire

$u = \dfrac{1}{2}$ (stress x strain)

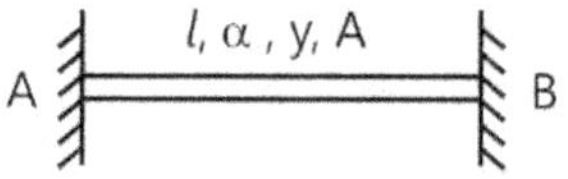

l, α, y, A

Figure 8.52

Thermal stress and strain $\in = \dfrac{\Delta l}{l} = \alpha \Delta T$

$Y = \dfrac{\sigma}{\in} = \dfrac{F}{A \in} = \dfrac{F}{A \alpha \Delta T}$

α is thermal coefficient of linear expansion of rod. ΔT is change in temperature of the rod.

Variation of density with pressure: As pressure on a body increases, its density also increases. When pressure increases by dp, the new density ρ' in terms of the previous density ρ is $\rho' = \dfrac{\rho}{1 - \dfrac{dp}{B}}$ where B is the Bulk modulus.

Poisson's ratio: As the length of a wire of circular cross-section increases, its radius decreases.

Poisson's ratio is defined as $\sigma = \dfrac{\text{lateral strain}}{\text{longitudinal strain}} = -\dfrac{\Delta R / R}{\Delta l / l}$

Relation between Y, B, η and σ

$B = \dfrac{Y}{3(1 - 2\sigma)}$; $\sigma = \dfrac{3B - 2\eta}{2\eta + 6B}$; $\eta = \dfrac{Y}{2(1 + \sigma)}$; $\dfrac{9}{Y} = \dfrac{1}{B} + \dfrac{3}{\eta}$

Solved Examples

JEE Main/Boards

Example 1: A steel wire of length 4 m and diameter 5 mm is stretched by 5 kg-wt. Find the increase in its length, if the Young's modulus of steel is 2.4×10^{12} dyne/cm².

Sol: From the formula for Young's modulus deduce the change in length.

Here, l = 4 m = 400 cm, 2r = 5 mm

or r = 2.5 mm = 5mm

F = 5 kg-wt = 5000 g-wt = 5000 × 980 dyne

Δl = ?, Y = 2.4×10^{12} dyne/cm²

As $Y = \dfrac{F}{\pi r^2} \times \dfrac{l}{\Delta l}$

$\Delta l = \dfrac{(5000 \times 980) \times 400}{(22/7) \times (0.25)^2 \times 2.4 \times 10^{12}} = 0.0041$ cm

$\Delta l = 4.1 \times 10^{-5}$ m

Example 2: One end of a wire 2 m long and 0.2 cm² in cross section is fixed in a ceiling and a load of 4.8 kg is attached to the free end. Find the extension of the wire. Young's modulus of steel = 2.0×10^{11} N/m².

Take g = 10 m/s².

Sol: From the formula for Young's modulus deduce the extension in wire.

We have $Y = \dfrac{\text{stress}}{\text{strain}} = \dfrac{T/A}{l/L}$

With symbols having their usual meanings. The extension is $l = \dfrac{TL}{AY}$

As the load is in equilibrium after the extension, the tension in the wire is equal to the weight of the load = 4.8 kg × 10 ms⁻² = 48 N

Thus, $l = \dfrac{(48N)(2m)}{(0.2 \times 10^{-4} m^2) \times (2.0 \times 10^{11} Nm^{-2})}$

$= 2.4 \times 10^{-5}$ m.

Example 3: A steel wire 4.0 m in length is stretched through 2.0 mm. The cross-sectional area of the wire is 2.0 mm². If Young's modulus of steel is 2.0×10^{11} N/m².

Find

(a) The energy density of wire,

(b) The elastic potential energy stored in the wire.

Sol: Find the stress and strain and use the formula for energy density. Product of energy density and volume is energy stored in entire wire.

Here, l = 4.0 m, $\Delta l = 2 \times 10^{-3}$ m,

A = 2.0×10^{-6} m², Y = 2.0×10^{11} N/m²

(a) The energy density of stretched wire

$U = \dfrac{1}{2} \times \text{stress} \times \text{strain} = \dfrac{1}{2} \times Y \times (\text{strain})^2$

$= \dfrac{1}{2} \times 2.0 \times 10^{11} \times \left(\dfrac{(2 \times 10^{-3})}{4} \right)^2$

$= 0.25 \times 10^5 = 2.5 \times 10^4$ J/m³.

(b) Elastic potential energy = energy density × volume = $2.5 \times 10^4 \times (2.0 \times 10^{-6}) \times 4.0$ J

$= 20 \times 10^{-2} = 0.20$ J

Example 4: The bulk modulus of water is

2.3×10^9 N/m².

(a) Find its compressibility.

(b) How much pressure in atmosphere is needed to compress a sample of water by 0.1%?

Sol: Compressibility is inverse of bulk modulus. From the formula for bulk modulus deduce the change in pressure required to produce the given change in volume.

Here, B = 2.3×10^9 N/m²

$= \dfrac{2.3 \times 10^9}{1.01 \times 10^5}$ atm = 2.27×10^4 atm

(a) Compressibility = $\dfrac{1}{B}$

$= \dfrac{1}{2.27 \times 10^4} = 4.4 \times 10^{-5}$ atm⁻¹

(b) Here, $\dfrac{\Delta V}{V} = -0.1\% = -0.001$

Required increase in pressure,

$$\Delta P = B \times \left(-\frac{\Delta V}{V}\right)$$

$$= 2.27 \times 10^4 \times 0.001 = 22.7 \text{ atm}$$

Example 5: One end of a nylon rope of length 4.5 m and diameter 6 mm is fixed to a tree-limb. A monkey weighing 100 N jumps to catch the free end and stays there. Find the elongation of the rope and the corresponding change in the diameter. Young's modulus of nylon = 4.8×10^{11} Nm^{-2} and Poisson ratio of nylon = 0.2.

Sol: From the formula for Young's modulus deduce the change in length of the rope. From the formula for Poisson ratio deduce the change in diameter.

As the monkey stays in equilibrium, the tension in the rope equals the weight of the monkey. Hence,

$$Y = \frac{\text{stress}}{\text{strain}} = \frac{T/A}{l/L} \text{ or } l = \frac{TL}{AY}$$

or, elongation

$$l = \frac{(100 \text{ N}) \times (4.5 \text{ m})}{(\pi \times 9 \times 10^{-6} \text{ m}^2) \times (4.8 \times 10^{11} \text{ Nm}^{-2})}$$

$$= 3.32 \times 10^{-5} \text{ m}$$

Again, Poisson ratio $= \dfrac{\Delta d/d}{l/L} = \dfrac{(\Delta d)L}{ld}$

or, $0.2 = \dfrac{\Delta d \times 4.5 \text{ m}}{(3.32 \times 10^{-5} \text{ m}) \times (6 \times 10^{-3} \text{ m})}$

or, $\Delta d = \dfrac{0.2 \times 6 \times 3.32 \times 10^{-8} \text{ m}}{4.5} = 8.8 \times 10^{-9} \text{ m}$

Example 6: A solid lead sphere of volume 0.5 m^3 is taken in the ocean to a depth where the water pressure is 2×10^7 N/m^2. If the bulk modulus of lead is 7.7×10^9 N/m^2. Find the fractional change in the radius of the sphere.

Sol: From the formula for bulk modulus deduce the change in volume for the given increase in pressure.

$$V = \frac{4}{3}\pi r^3 \Rightarrow \frac{\Delta r}{r} = \frac{1}{3}\frac{\Delta V}{V}$$

Bulk modulus $K = -\dfrac{\Delta P}{(\Delta V / V)}$

or $\dfrac{\Delta V}{V} = -\dfrac{\Delta P}{K}$

or $\dfrac{\Delta r}{r} = -\dfrac{1}{3}\dfrac{\Delta P}{K} = -\dfrac{1}{3} \times \dfrac{2 \times 10^7}{7.7 \times 10^9}$

$$= -0.87 \times 10^{-3}.$$

The negative sign indicates that the radius decreases.

Example 7: Find the greatest length of steel wire that can hang vertically without breaking. Breaking stress of steel $= 8.0 \times 10^8$ N/m^2.

Density of steel $= 8.0 \times 10^3$ kg/m^3.

Take g = 10 m/s^2.

Sol: Breaking stress gives the maximum weight per unit area of cross-section that the wire can withstand.

Let l be the length of the wire that can hang vertically without breaking. Then the stretching force on it is equal to its own weight. If therefore, A is the area of cross-section and ρ is the density, then

$$\text{Maximum stress } (s_m) = \frac{\text{weight}}{A}$$

$$\left(\text{stress} = \frac{\text{force}}{\text{area}}\right) \text{ or } \sigma_m = \frac{(A/\rho)g}{A}$$

$\therefore \quad l = \dfrac{\sigma_m}{\rho g}$ Substituting the values

$$l = \frac{8.0 \times 10^8}{(8.0 \times 10^3)(10)} = 10^4 \text{ m}$$

Example 8: A copper wire of negligible mass, length 1 m and cross-sectional area 10^{-6} m^2 is kept on a smooth horizontal table with one end fixed. A ball of mass 1 kg is attached to the other end. The wire and the ball are rotating with an angular velocity of 20 rad/s. If the elongation in the wire is 10^{-3} m, obtain the Young's modulus of copper. If on increasing the angular velocity to 100 rad/s, the wire breaks down, obtain the breaking stress.

Sol: The stress developed in the wire will be due to the centrifugal force. Ratio of stress and strain is the Young's modulus. The breaking stress will be due to the centrifugal force at increased angular velocity.

The stretching force developed in the wire due to rotation of the ball is

$$F = mr w^2 = 1 \times 1 \times (20)^2 = 400 \text{ N}$$

Stress in the wire $= \dfrac{F}{A} = \dfrac{400}{10^{-6}}$ N/m^2 Strain in the wire

$$= \frac{10^{-3}}{1} = 10^{-3}$$

$$Y = \frac{\text{Stress}}{\text{Strain}} = \frac{400}{10^{-6} \times 10^{-3}} = 4 \times 10^{11} \text{ N/m}^2$$

Breaking stress = $\dfrac{1 \times 1 \times (100)^2}{10^{-6}} = 10^{10}$ N/m².

Example 9: (a) A wire 4 m long and 0.3 mm in diameter is stretched by a force of 100 N. If extension in the wire is 0.3 mm, calculate the potential energy stored in the wire.

(b) Find the work done in stretching a wire of cross-section 1 mm² and length 2 m through 0.1 mm, Young's modulus for the material of wire is 2.0×10^{11} N/m².

Sol: Work done in stretching the wire is equal to the elastic potential energy stored in the wire. (a) Energy stored

$U = \dfrac{1}{2}$(stress)(strain)(volume)

or $\quad U = \dfrac{1}{2}\left(\dfrac{F}{A}\right)\left(\dfrac{\Delta l}{l}\right)(Al) = \dfrac{1}{2}F \cdot \Delta l$

$= \dfrac{1}{2}(100)\left(0.3 \times 10^{-3}\right) = 0.015$ J

(b) Work done = potential energy stored

$= \dfrac{1}{2}k(\Delta l)^2 = \dfrac{1}{2}\left(\dfrac{YA}{l}\right)(\Delta l)^2 \left(\text{as } k = \dfrac{YA}{l}\right)$

Substituting the values, we have

$W = \dfrac{1}{2}\dfrac{(2.0 \times 10^{11})(10^{-6})}{(2)}(0.1 \times 10^{-3})^3$

$= 5.0 \times 10^{-4}$ J

Example 10: A steel wire of diameter 0.8 mm and length 1 m is clamped firmly at two points A and B which are 1 m apart and in the same horizontal plane. A body is hung from the middle point of the wire such that the middle point sags 1 cm lower from the original position. Calculate the mass of the body. Given Young's modulus of the material of wire = 2×10^{12} dynes/cm².

Sol: Tension in the wire is the product of stress and area of cross-section. Stress is the product of Young's modulus and strain. The vertical components of tensions in the two parts of the wire will balance the weight of the body hung from the wire.

Let the body be hung from the middle point C so that it sags through 1cm to the point D as shown in the figure.

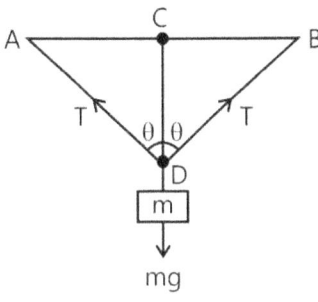

$\therefore$ AD² = AC² + CD² = (50)² + (1)²

or AD = 50.01 cm

Increase in length = 0.01 cm

Strain $= \dfrac{0.01}{50} = 2 \times 10^{-4}$

Stress = 2 × 10¹² × 2 × 10⁻⁴

$\therefore$ Stress = 4 × 10⁸ dynes/cm²

Tension T = Stress × Area of cross-section

$\qquad = 4 \times 10^8 \times \pi \times (0.08)^2$

Since the mass m is in equilibrium

mg = 2T cos θ or m $= \dfrac{2T\cos\theta}{g}$

$= \dfrac{2 \times 4 \times 10^8 \times \pi (0.08)^2 \times (1/50.01)}{980} = 82$ gm.

JEE Advanced/Boards

Example 1: A light of rod of length 200 cm is suspended from the ceiling horizontally by means of two vertical wires of equal length tied to its ends. One of the wires is made of steel and is of cross-section 0.1 sq cm and the other is of brass of cross-section 0.2 sq. cm. Find the position along the rod at which a weight may be hung to produce (a) equal stresses in both wires and (b) equal strains in both wires.

(Y_brass = 10 × 10¹¹ dynes/cm².

Y_steel = 20 × 10¹¹ dynes/cm²).

Sol: Net torque of the tensions in the wires about the point of suspension of the weight on the rod must be zero.

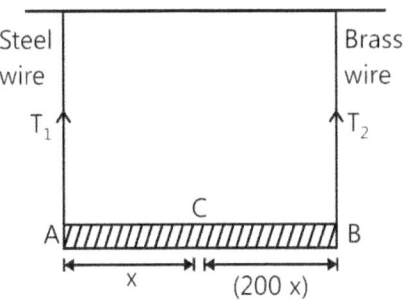

Let AB be the rod and let C be the point at which the weight is hung.

(a) Stress in steel wire $= \dfrac{T_1}{0.1}$

Stress in brass wire $= \dfrac{T_2}{0.2}$

As the two stresses are equal,

$$\dfrac{T_1}{0.1} = \dfrac{T_2}{0.2} \text{ or } \dfrac{T_1}{T_2} = 0.5 \qquad \ldots \text{(i)}$$

Taking moments about C,

$$T_1 = T_2 (200 - x) \text{ or } \dfrac{T_1}{T_2} = \dfrac{200 - x}{x} \qquad \ldots \text{(ii)}$$

Equations (i) and (ii) give

$$\dfrac{200 - x}{x} = 0.5$$

Or $x = 133.3$ cm $= 1.33$ m

(b) Strain $= \dfrac{\text{Stress}}{Y} = \dfrac{T}{AY}$

As the strain in both wires are equal,

$$\dfrac{T_1}{A_1 Y_1} = \dfrac{T_2}{A_2 Y_2} \text{ or } \dfrac{T_1}{T_2} = \dfrac{A_1 Y_1}{A_2 Y_2} = \dfrac{0.1 \times 20 \times 10^{11}}{0.2 \times 10 \times 10^{11}}$$

$\therefore T_1 = T_2$

Now, $T_1 x = T_2 (200 - x) \Rightarrow x = 200 - x$

or $x = 100$ cm $= 1$ m.

Example 2: A rod AD, consisting of three segments AB, BC and CD joined together, is hanging vertically from a fixed support at A. The lengths of the segments are respectively 0.1 m, 0.2 m and 0.15 m. The cross-section of the rod is uniformly equal to 10^{-4} m². A weight of 10 kg is hung from D. Calculate the displacements of the points B, C and D using the data on Young's moduli given below (neglect the weight of the rod).

$Y_{AB} = 2.5 \times 10^2$ N/m,

$Y_{BC} = 4.0 \times 10^2$ N/m and

$Y_{CD} = 1.0 \times 10^2$ N/m

Sol: From the formula for Young's modulus deduce the elongation in each segment of the wire.

We know that

$$\Delta l = \dfrac{mgl}{AY} = \dfrac{10 \times 9.8 \times 0.1}{10^{-4} \times 2.5 \times 10^{10}} = 3.92 \times 10^{-6} \text{ m}$$

This is the displacement of B.

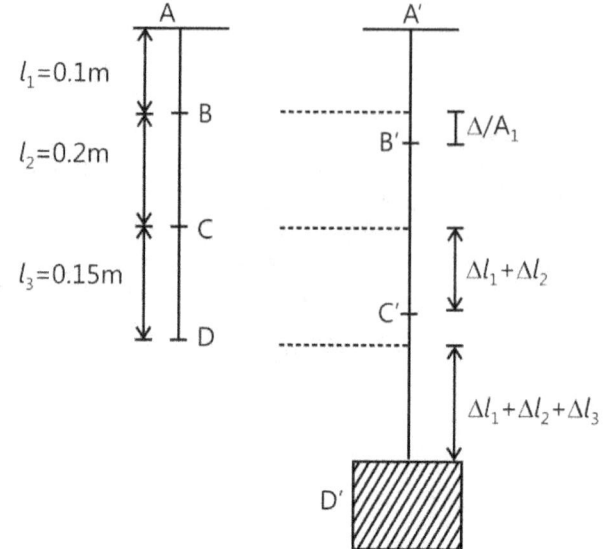

For segment BC:

$$\Delta l_2 = \dfrac{10 \times 9.8 \times 0.2}{10^{-4} \times 4.0 \times 10^{10}} = 4.9 \times 10^{-6} \text{ m}$$

Displacement of C: $= \Delta l_1 + \Delta l_2$

$= 4.9 \times 10^{-6}$ m

For segment CD:

$$\Delta l_3 = \dfrac{10 \times 9.8 \times 0.15}{10^{-4} \times 1.0 \times 10^{10}} = 14.7 \times 10^{-6} \text{ m}$$

Displacement of D $= \Delta l_1 + \Delta l_2 + \Delta l_3$

$= 23.52 \times 10^{-6}$ m.

Example 3: A steel rod of length 6.0 m and diameter 20 mm is fixed between two rigid supports. Determine the stress in the rod, when the temperature increases by 80° C if

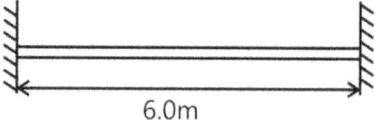

(a) The ends do not yield

(b) The ends yield by 1 mm.

Take Y $= 2.0 \times 10^6$ kg/cm²

And $\alpha = 12 \times 10^{-6}$ per °C.

Sol: Rise in temperature causes thermal strain and thermal stress. Use the formula for coefficient of thermal expansion to obtain thermal strain. Thermal stress is the product of Young's modulus and thermal strain.

Given, length of the rod $l = 6$ m

Diameter of the rod $d = 20$ mm $= 2$ cm

Increase in temperature $t = 80°C$

Young's modulus $Y = 2.0 \times 10^6$ kg/cm^2

And thermal coefficient of linear expansion

$\alpha = 12 \times 10^{-6}$ per °C

(a) When the ends do not yield

Let, s_1 = stress in the rod

Using the relation $\sigma = \alpha t Y$

$\therefore s_1 = (12 \times 10^{-6})(80)(2 \times 10^6)$

$= 1920$ kg/cm^2 $= 19.2 \times 10^6$ N

(b) When the ends yield by 1 mm

Increase in length due to increase in temperature $\Delta l = l\alpha t$

Of this 1mm or 0.1 cm is allowed to expand. Therefore, net compression in the rod

$\Delta l_{net} = (l t - 0.1)$

or compressive strain in the rod,

$$\varepsilon = \frac{\Delta l_{net}}{l} = \left(\alpha t - \frac{0.1}{l} \right)$$

$\therefore$ Stress $s_2 = Y\varepsilon = Y \left(\alpha t - \frac{0.1}{l} \right)$

Substituting the values,

$$s_2 = 2 \times 10^6 \left(12 \times 10^{-6} \times 80 - \frac{0.1}{600} \right)$$

$= 1587$ kg/cm^2 $= 15.8 \times 10^6$ N

Example 4: Two blocks of masses 1 kg and 2 kg are connected by a metal wire going over a smooth pulley as shown in figure The breaking stress of the metal is 2×10^9 Nm^{-2}. What should be the minimum radius of the wire used if it is not to break?

Take $g = 10$ ms^{-2}.

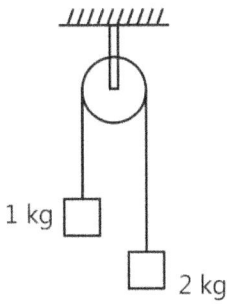

Sol: Find the tension in the metal wire due to the masses connected to it. The stress due to tension should not exceed the breaking stress.

The stress in the wire

$$= \frac{\text{Tension}}{\text{Area of cross section}}$$

To avoid breaking, this stress should not exceed the breaking stress.

Let the tension in the wire be T. The equations of motion of the two blocks are,

$T - 10$ N $= (1kg)$ a and 20 N $- T = (2kg)$ a

Eliminating a from these equations,

$T = (40/3)$ N

The stress $= \dfrac{(40/3)\,\text{N}}{\pi r^2}$

If the minimum radius needed to avoid breaking is r,

$$2 \times 10^9 \frac{N}{m^2} = \frac{(40/3)\,\text{N}}{\pi r^2}$$

Solving this, $r = 4.6 \times 10^{-5}$ m.

Example 5: A steel rod of cross-sectional area 16 cm^2 and two brass rods each of cross-sectional area 10 cm^2 together support a load of 5000 kg as shown in figure Find the stress in the rods. Take Y for steel $= 2.0 \times 10^6$ kg/cm^2 and for brass $= 1.0 \times 10^6$ kg/cm

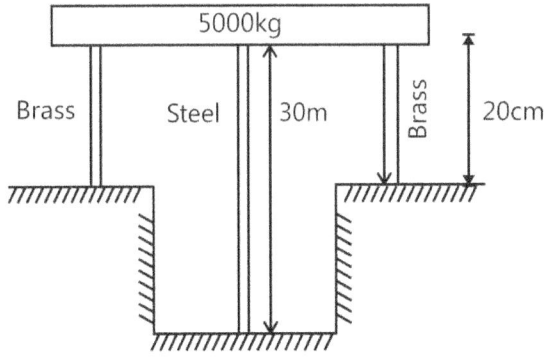

Sol: Compression in the length of steel and brass rods is equal. From the formula for Young's modulus deduce the compression in length of each rod and equate them to get the relation between respective stresses.

Given area of steel rod $A_s = 16$ cm^2

Area of two brass rods

$A_B = 2 \times 10 = 20$ cm^2

Load, $F = 5000$ kg

Y for steel $Y_s = 2.0 \times 10^6$ kg/cm^2 Y

for brass $Y_B = 1.0 \times 10^6$ kg/cm²

Length of steel rod $l_S = 30$ cm

Length of steel rod $l_B = 20$ cm

Let s_S = stress in steel

and σ_B = stress in brass

Decrease in length of steel rod = decrease in length of brass rod

or $\quad \dfrac{\sigma_S}{Y_S} \times l_S = \dfrac{\sigma_B}{Y_B} \times l_B$

or $\quad s_S = \dfrac{Y_S}{Y_B} \times \dfrac{l_B}{l_S} \times \sigma_B$

$\quad = \dfrac{2.0 \times 10^6}{1.0 \times 10^6} \times \dfrac{20}{30} \times \sigma_B$

$\therefore \quad s_S = \dfrac{4}{3} \sigma_B$ .... (i)

Now, using the relation,

$F = s_S A_S + \sigma_B A_B$ or

$5000 = s_S \times 16 + \sigma_B \times 20$ .... (ii)

Solving eq. (i) and (ii), we get

$\sigma_B = 120.9$ kg/cm² and $\quad s_S = 161.2$ kg/cm²

Example 6: A sphere of radius 0.1 m and mass 8π kg is attached to the lower end of a steel wire length 5.0 m and diameter 10^{-3} m. The wire is suspended from 5.22 m high ceiling of a room. When sphere is made to swing as a simple pendulum, it just grazes the floor at its lowest point. Calculate velocity of the sphere at the lowest position. Young's modulus of steel is 1.994 $\times 10^{11}$ N/m².

Sol: The elongation in the wire is known, thus the corresponding stress can be calculated. The stress in turn gives the tension in the wire. At the lowest point the net acceleration of the sphere is centripetal, i.e. directed vertically upwards. Apply Newton's second law at the lowest point to find the speed of the sphere.

Let Dl be the extension of wire when the sphere is at mean position. Then, we have

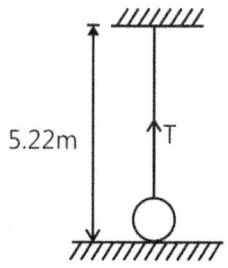

5.22m ↑T

$l + \Delta l + 2r = 5.22$

or $\quad \Delta l = 5.22 - l - 2r$

$\quad 5.22 - 5 - 2 \times 0.1 = 0.02$ m

Let T be the tension in the wire at mean position during oscillations, $Y = \dfrac{T/A}{\Delta l / l}$

$\therefore \quad T = \dfrac{YA\Delta l}{l} = \dfrac{Y\pi r^2 \Delta l}{l}$

Substituting the values, we have

$T = \dfrac{(1.994 \times 10^{11}) \times \pi \times (0.5 \times 10^{-3})^2 \times 0.02}{5}$

$= 626.43$ N

The equation of motion at mean position is,

$T - mg = \dfrac{mv^2}{R}$

Hence, $\quad R = 5.22 - r = 5.22 - 0.1 = 5.12$ m

and $\quad m = 8\pi$ kg = 25.13 kg

Substituting the proper values in Eq. (i), we have

$(626.43) - (25.13 \times 9.8) = \dfrac{(25.13)v^2}{5.12}$

Solving this equation, we get $\quad V = 8.8$ m/s

Example 7: A thin ring of radius R is made of a material of density ρ and Young's modulus Y. If the ring is rotated about its center in its own plane with angular velocity ω, find the small increase in its radius.

Sol: As the ring rotates each element of the ring of infinitesimal length experiences a centrifugal force, due to which the ring slightly expands, thus increasing its radius. The longitudinal strain in the ring produces a tensile stress or tension in the ring.

Consider an element PQ of length dl. Let T be the tension and A the area of cross-section of the wire.

Mass of element dm = volume × density = A (dl)ρ

The component of T, towards the center provides the necessary centripetal force

$\therefore \quad 2T \sin\left(\dfrac{\theta}{2}\right) = (dm)R\omega^2$ ... (i)

For small angles $\sin\dfrac{\theta}{2} \approx \dfrac{\theta}{2} = \dfrac{(dl/R)}{2}$

Substituting in eq. (i), we have

$T.\dfrac{dl}{R} = A(dl)\rho R\omega^2 \quad$ or $\quad T = Arw^2R^2$

Let DR be the increase in radius,

Longitudinal strain

$$\frac{\Delta l}{l} = \frac{\Delta(2\pi R)}{2\pi R} = \frac{\Delta R}{R}$$

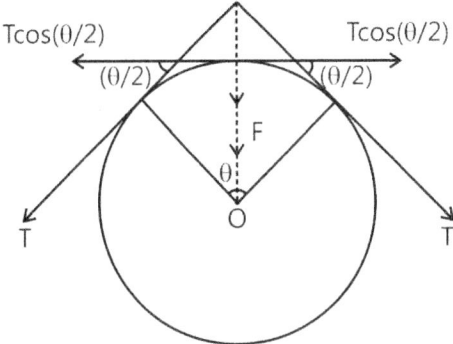

Now, $Y = \dfrac{T/A}{\Delta R/R}$

$\therefore \Delta R = \dfrac{TR}{AY} = \dfrac{(A\rho\omega^2 R^2)R}{AY}$ or $DR = \dfrac{\rho\omega^2 R^3}{Y}$

Example 8: A member ABCD is subjected to point loads F_1, F_2, F_3 and F_4 as shown in figure Calculate the force F_2 for equilibrium if $F_1 = 4500$ kg,

$F_3 = 45000$ kg and $F_4 = 13000$ kg.

Determine the total elongation of the member, assuming modulus of elasticity to be 2.1×10^6 kg/cm².

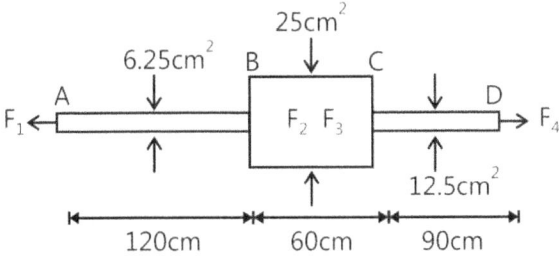

Sol: Find the tension in each segment of the member ABCD. From the formula of Young's modulus, find the elongation in each segment.

Given

Area of part AB, $A_1 = 6.25$ cm²

Area of part AB, $A_2 = 25$ cm²

Area of part CD, $A_3 = 12.5$ cm²

Length of part AB, $l_1 = 120$ cm

Length of part BC, $l_2 = 60$ cm

Length of part CD, $l_3 = 90$ cm

Young's modulus of elasticity

$Y = 2.1 \times 10^6$ kg/cm²

Magnitude of the force F_2 for equilibrium

The magnitude of force F_2 may be found by equating the forces acting towards right to those acting towards left,

$$F_2 + F_4 = F_1 + F_3$$

$$F_2 + 13000 = 4500 + 45000$$

$$\therefore \quad F_2 = 36500 \text{ kg}$$

Total Elongation of the member

For the sake of simplicity, the force of 36500 kg (acting at B) may be split up into two forces of 4500 kg and 32000 kg. The force of 45000 kg acting at C may be split into two forces of 32000 kg and 13000 kg. Now, it will be seen that the part AB of the member is subjected to a tension of 4500 kg, part BC is subjected to a compression of 32000 kg and part CD is subjected to a tension of 13,000 kg. Using the relation.

$$\Delta l = \frac{1}{Y}\left(\frac{F_1 l_1}{A_1} - \frac{F_2 l_2}{A_2} + \frac{F_3 l_3}{A_3} \right)$$

With usual notation

$$\Delta l = \frac{1}{2.1 \times 10^6} \times$$

$$\left(\frac{4500 \times 120}{6.25} - \frac{32000 \times 60}{25} + \frac{13000 \times 90}{12.5} \right) \text{cm}$$

$$= 0.049 \text{ cm or } Dl = 0.49 \text{ mm}$$

Exercise 1

Q.1 A wire is replaced by another wire of same length and material but of twice diameter.

(i) What will be the effect on the increase in its length under a given load?

(ii) What will be the effect on the maximum load which it can bear?

Q.2 Two wires are made of same metal. The length of the first wire is half that of the second wire and its diameter is double that of the second wire. If equal loads are applied on both wires, find the ratio of increase in their lengths.

Q.3 The breaking force for a wire is F. What will be the breaking forces for

(i) Two parallel wires of this size and

(ii) For a single wire of double thickness?

Q.4 What force is required to stretch a steel wire 1 sq. cm in cross section to double its length? $Y_{steel} = 2 \times 10^{11}$ Nm^{-2}.

Q.5 A structural steel rod has a radius of 10 mm and length of 1.0 m. A 100 kN force stretches it along its length. Calculate (a) stress, (b) elongation, and (c)% strain in the rod. Young's modulus of structural steel is 2.0×10^{11} Nm^{-2}.

Q.6 Find the maximum length of a steel wire that can hang without breaking.

Breaking stress = 7.9×10^{12} dyne/cm^2.

Density of steel = 7.9 g/cc.

Q.7 A spherical ball contracts in volume by 0.01%, when subjected to a normal uniform pressure of 100 atmosphere. Calculate the bulk modulus of the material.

Q.8 A sphere contracts in volume by 0.02% when taken to the bottom of sea 1 km deep. Find bulk modulus of the material of sphere. Density of sea water is 1000 kg/m^3.

Q.9 A metal cube of side 10 cm is subjected to a shearing stress of 10^4 Nm^{-2}. Calculate the modulus of rigidity if the top of the cube is displaced by 0.05 cm with respect to its bottom.

Q.10 Calculate the increase in energy of a brass bar of length 0.2 m and cross sectional area 1cm^2 when combined with a load of 5kg weight along its length. Young's modulus of brass = 1.0×10^{11} Nm^{-2} and g = 9.8 ms^{-2}.

Q.11 A wire 30m long and of 2 mm^2 cross-section is stretched due to a 5kg-wt by 0.49 cm. Find

(i) The longitudinal strain

(ii) The longitudinal stress and

(iii) Young's modulus of the material of the wire.

Exercise 2

Single Correct Choice Type

Q.1 A wire of length 1m is stretched by a force of 10N. The area of cross-section of the wire is 2×10^{-6} m^2 & γ is 2×10^{11} N/m^2. Increase in length of the wire will be-

(A) 2.5×10^{-5} cm   (B) 2.5×10^{-5} mm

(C) 2.5×10^{-5} m   (D) None

Q.2 A uniform steel wire of density 7800 kg/m^3 is 2.5 m long and weighs 15.6×10^{-3} kg. It extends by 1.25 mm when loaded by 8kg. Calculate the value of young's modulus for steel.

(A) 1.96×10^{11} N/m^2 (B) 19.6×10^{11} N/m^2

(C) 196×10^{11} N/m^2 (D) None

Q.3 The work done in increasing the length of a one meter long wire of cross-sectional area 1mm^2 through 1mm will be (Y = 2×10^{11} N/m^2)

(A) 250 J   (B) 10 J   (C) 5 J   (D) 0.1 J

Q.4 The lengths and radii of two wires of same material are respectively L, 2L, and 2R, R. Equal weights are applied on them. If the elongations produced in them are l_1 and l_2 respectively, then their ratio will be

(A) 2 : 1   (B) 4 : 1   (C) 8 : 1   (D) 1 : 8

Q.5 What is the density of lead under a pressure of 2.0×10^8 N/m², if the bulk modulus of lead is 8.0×10^9 N/m² and initially the density of lead is 11.4g/cm³ ?

(A) 11.69g/cm³   (B) 11.92g/cm³

(C) 11.55g/cm³   (D) 11.862g/cm³

Q.6 A rubber rod of density 1.3×10^3 kg/m³ and Young's modulus 6×10^6 N/m² hangs from the ceiling of a room. Calculate the deviation in the value of its length from the original value 10m.

(A) 10.9 cm   (B) 5.8 cm   (C) 9.3 cm   (D) 10.6 cm

Q.7 A metal rod is trapped horizontally between two vertical walls. The coefficient of linear expansion of the rod is equal to 1.2×10^{-5} /°C and its Young's modulus 2×10^{11} N/m². If the temperature of the rod is increased by 5°C, calculate the stress developed in it.

(A) 2.2×10^7 N/m²   (B) 3.1×10^7 N/m²

(C) 1.2×10^7 N/m²   (D) 1.2×10^4 N/m²

Previous Years' Questions

Q.1 The following four wires are made of the same material. Which of these will have the largest extension when the same tension is applied?   *(1981)*

(A) Length=50 cm, diameter=0.5 mm

(B) Length=100 cm, diameter = 1 mm

(C) Length=200cm, diameter= 2 mm

(D) Length=300 cm, diameter=3 mm

Q.2 A given quantity of an ideal gas is at pressure p and absolute temperature T. The isothermal bulk modulus of the gas   *(1998)*

(A) $\dfrac{2}{3}p$ (B) p   (C) $\dfrac{3}{2}p$ (D) 2p

Q.3 The pressure of a medium is changed from 1.01×10^5 Pa to 1.165×10^5 Pa and change in volume is 10% keeping temperature constant. The bulk modulus of the medium is   *(2005)*

(A) 204.8×10^5 Pa   (B) 102.4×10^5 Pa

(C) 51.2×10^5 Pa   (D) 1.55×10^5 Pa

Q.4 A pendulum made of a uniform wire of cross sectional area A has time period T. When an additional mass M is added to its bob, the time period changes to T_M. If the Young's modulus of the material of the wire is Y then $\dfrac{1}{Y}$ is equal to: (g = gravitational acceleration   *(2015)*

(A) $\left[\left(\dfrac{T_M}{T} \right)^2 - 1 \right] \dfrac{Mg}{A}$ (B) $\left[1 - \left(\dfrac{T_M}{T} \right)^2 \right] \dfrac{A}{Mg}$

(C) $\left[1 - \left(\dfrac{T}{T_M} \right)^2 \right] \dfrac{A}{Mg}$ (D) $\left[\left(\dfrac{T_M}{T} \right)^2 - 1 \right] \dfrac{A}{Mg}$

JEE Advanced/Boards

Exercise 1

Q.1 A rubber cord has a cross-sectional area 1mm² and total unstretched length 10.0 cm. It is stretched to 12.0 cm and then released to project a missile of mass 5.0g. Taking Young's modulus Y for rubber as 5.0×10^8 N/m². Calculate the velocity of projection.

Q.2 Calculate the pressure required to stop the increase in volume of a copper block when it is heated from 50°C to 70°C. Coefficient of linear expansion of copper = 8.0×10^{-6}/°C and the bulk modulus of elasticity = 10^{11} N/m².

Q.3 Calculate the increase in energy of a brass bar of length 0.2m and cross-sectional area 1.0 cm², when compressed with a load of 5kg-weight along its length. Young's modulus of brass = 1.0×10^{11} N/m² and g = 9.8 m/s².

Exercise 2

Q.1 A steel wire of uniform cross-section of 2mm² is heated upto 50° and clamped rigidly at two ends. If the temperature of wire falls to 30° then change in tension in the wire will be, if coefficient of linear expansion of steel is 1.1×10^{-5} /°C and young's modulus of elasticity of steel is 2×10^{11} N/m².

(A) 44 N   (B) 88 N   (C) 132 N   (D) 22 N

Q.2 A metallic wire is suspended by suspending weight to it. If S is longitudinal strain and Y its young's modulus of elasticity. Potential energy per unit volume will be

(A) $\frac{1}{2}Y^2S^2$ (B) $\frac{1}{2}Y^2S$ (C) $\frac{1}{2}YS^2$ (D) $2YS^2$

Q.3 The compressibility of water is 5×10^{-10} m² / N. Find the decrease in volume of 100 ml of water when subjected to a pressure of 15 mPa.

(A) 0.75 ml   (B) 0.75 mm

(C) 0.75 mm   (D) 7.5 mm

Q.4 The upper end of a wire 1 meter long and 2mm radius is clamped. The lower end is twisted through an angle of 45°. The angle of shear is

(A) 0.09°   (B) 0.9°   (C) 9°   (D) 90°

Previous Years' Questions

Q.1 Two rods of different materials having coefficient of thermal expansion a_1, a_2 and Young's moduli Y_1, Y_2 respectively are fixed between two rigid massive walls. The rods are heated such that they undergo the same increase in temperature. There is no bending of the rods. If $a_1 : a_2$ = 2:3, the thermal stresses developed in the two roads are equal provided $Y_1 : Y_2$ is equal to

(1989)

(A) 2 : 3   (B) 1 : 1   (C) 3 : 1   (D) 4 : 9

Q.2 The adjacent graph shows extension (Dl) of a wire of length 1m suspended from the top of a roof at one end and with a load W connected to the other end. If the cross-sectional area of the wire is 10^{-6}m², calculate from the graph the Young's modulus of the material of the wire. *(2003)*

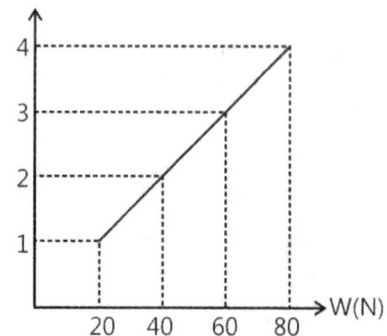

(A) 2×10^{11} N/m²   (B) 2×10^{-11} N/m²

(C) 2×10^{12} N/m²   (D) 2×10^{13} N/m²

Q.3 In Searle's experiment, which is used to find Young's modulus of elasticity, the diameter of experimental wire is D = 0.05 cm (measured by a scale of least count 0.001 cm) and length is L = 110 cm (measured by a scale of least count 0.1 cm). A weight of 50N causes an extension of l = 0.125 cm (measured by a micrometer of least count 0.001 cm). Find maximum possible error in the values of Young's modulus. Screw gauge and meter scale are free from error. *(2004)*

Q.13 In plotting stress versus strain curves for two materials P and Q, a student by mistake puts strain on the y-axis and stress on the x-axis as shown in the figure. Then the correct statement(s) is(are) *(2015)*

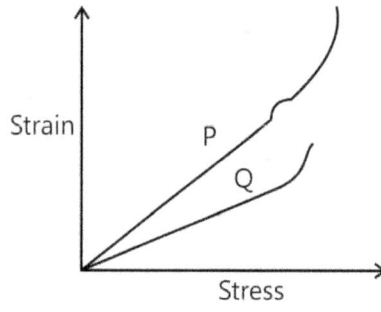

(A) P has more tensile strength than Q

(B) P is more ductile than Q

(C) P is more brittle than Q

(D) The Young's modulus of P is more than that of Q

Important Questions

JEE Main/Boards

Exercise 1

Q. 1 Q.3 Q.7

Q.10

Exercise 2

Q.4 Q.6 Q.7

Previous Years' Questions

Q.9 Q.11

JEE Advanced/Boards

Exercise 2

Q.1 Q.4

Previous Years' Questions

Q.1 Q.2 Q.3

Answer Key

JEE Main/ Boards

Exercise 1

Q.2 1: 8 **Q.3** (i) 2F (ii) 4F **Q.4** 2×10^7 N

Q.5 (a) 3.18×10^8 N m^{-2} (b) 1.59 mm (c) 0.16% **Q.6** 1.02×10^9 cm

Q.7 1.013×10^{11} **Q.8** 4.9×10^{10} **Q. 9** 2×10^6 Nm^{-2}

Q.10 2.4×10^{-5} J **Q.11** 1.5×10^{11} N/m^2.

Exercise 2

Single Correct Choice Type

Q.1 C **Q.2** A **Q.3** D **Q.4** D **Q.5** A

Q.6 D **Q.7** C

Previous Years' Questions

Q.1 A **Q.2** B **Q.3** D **Q.4** D

JEE Advanced/Boards

Exercise 1

Q.1 20 m/s   **Q.2** 1.728×10^8 N/m^2 **Q3** 2.4×10^{-5} J

Exercise 2

Q.1 B   **Q.2** C   **Q.3** A   **Q.4** A

Previous Years' Questions

Q.1 C   **Q.2** A   **Q.3** 1.09×10^{10} N/m^2 **Q.4** A, B

Solutions

JEE Main/Boards

Exercise 1

Sol 1: $\Delta L = \dfrac{fL}{Ay}$

If diameter is increased to twice then

(i) ΔL will decrease to ¼ value

(ii) $F = \dfrac{\Delta L}{L} A$

Maximum load capacity will decrease to ¼ of initial value.

Sol 2: $L_1 = \dfrac{L_2}{2}$

$d_1 = 2d_2 ;$ $A_1 = 4A_2$

$F_1 = F_2$

$\Delta L = \dfrac{FL}{Ay}$

$\dfrac{\Delta L_1}{\Delta L_2} = \dfrac{L_1 A_2}{A_1 L_2} = \dfrac{1}{2} \cdot \dfrac{1}{4} = \dfrac{1}{8}$

Sol 3: Breaking force for two parallel wires of this size

(i) $F' = F_1 + F_2 = F + F = 2F$

(ii) If thickness is double that means area is 4 times.

$F = \dfrac{\Delta L}{L} YA \Rightarrow F' = \dfrac{\Delta L}{L} Y \, 4A = 4F$

Sol 4: $F = y \dfrac{\Delta L}{L} A$

$\Delta L = 2L - L = L$

$F = 2 \times 10^{11} \dfrac{L}{L} 10^{-4} = 2 \times 10^7$ N

Sol 5: $r = 10 \times 10^{-3}$ m

$R = 10^{-2}$ m

$L = 1$ m

(a) Stress $= \dfrac{F}{A} = \dfrac{100 \times 10^3}{\pi(10^{-2})^2} = \dfrac{10^5}{\pi \times 10^{-4}} = \dfrac{10^9}{\pi}$

$= 3.18 \times 10^8$ N/m^2

(b) Elongation $= \Delta L = \dfrac{\text{stress} \times \text{length}}{y}$

$= \dfrac{3.18 \times 10^8 \times 1}{2 \times 10^{11}} = 1.59 \times 10^{-3}$ m $= 1.59$ mm

(c) % Strain $= \dfrac{\Delta L}{L} \times 100 = \dfrac{1.59 \times 10^{-3} \times 100}{1} = 0.159\%$

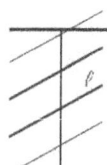

Sol 6: Stress $= 7.9 \times 10^7$ N/cm^2 $= 7.9 \times 10^{11}$ N/m^2

Stress $= \dfrac{F}{A} = \dfrac{\rho \ell A g}{A} = \rho \lambda g = 7.9 \times 10^{11}$

$\ell = \dfrac{7.9 \times 10^{11}}{7900 \times 10} = 10^7$ m $= 10^9$ cm

3.28

Sol 7: $\dfrac{\Delta V}{V} = -10^{-4}$

$P = 100 \times 10^5 \text{ N/m}^2$

$B = \dfrac{-P}{\Delta V / V} = \dfrac{1.013 \times 10^7}{10-4} = 1.013 \times 10^{11} \text{ Nm}^{-2}$

Sol 8: Pressure at 1 km depth = P

$= P_0 + 1000 \times 98 \times 1000$

$= 10^5 + 98 \times 10^7 = 99 \times 10^5 \text{ N/m}^2$

Bulk modulus $= \dfrac{-P}{\Delta V / V} = \dfrac{99 \times 10^5}{2 \times 10^{-4}}$

$= 4.9 \times 10^{10} \text{ Pa}$

Sol 9: Stream $= \dfrac{\Delta X}{L} = \dfrac{0.05}{10} = 5 \times 10^{-3}$

Modulus $= \dfrac{\text{stress}}{\text{strain}} = \dfrac{10^4}{5 \times 10^{-3}}$

$= \dfrac{10}{5} \times 10^6 = 2 \times 10^6 \text{ N/m}^2$

Sol 10: Strain $= \dfrac{5 \times 10}{10^{-4} \times 10^{11}} = \dfrac{5 \times 10}{10^7} = 5 \times 10^{-6}$

Increase in energy = work done

$\dfrac{1}{2} \times \dfrac{50}{10^{-4}} \times 5 \times 10^{-6} \times 10^{-4} \times 0.2$

$= \dfrac{250 \times 10^{-11}}{10^{-4}} = 250 \times 10^{-7} = 2.5 \times 10^{-5} \text{ J}$

Sol 11: (i) The initial length of the wire

$= L = 30 \text{m}$

The increase in length of the wire,

$l = 0.44 \times 10^{-2}$

Longitudinal stress

$= l / L = 1.633 \times 10^{-4}$.

(ii) The tension applied to the wire $= Mg = 5 \times 9.8 \text{ N}$

Area of cross section of the wire,

$A = 2 \text{ mm}^2 = 2 \times 10^{-6} \text{ m}^2$

∴ Longitudinal stress

$= Mg / A = \dfrac{5 \times 9.8}{2 \times 10^{-6}}$

$= 2.45 \times 10^7 \text{ N} / \text{m}^2$

(iii) Young's modulus $= \dfrac{\text{stress}}{\text{strain}}$

$\dfrac{2.45 \times 10^7}{1.633 \times 10^{-4}} = 1.5 \times 10^{11} \text{ N/m}^2$

Exercise 2

Sol 1: (C) Stress $= F/A = 10/(2 \times 10^{-6})$

$= 5 \times 10^6 \text{ N/m}^2$

Strain $= \dfrac{\text{Stress}}{Y} = \dfrac{5 \times 10^6}{2 \times 10^{11}}$

$= 2.5 \times 10^{-5}$

$l = L \times \text{strain} = 1 \times 2.5 \times 10^{-5}$

$l = 2.5 \times 10^{-5} \text{ m}$

Sol 2: (A) Volume = Mass / density

Area of cross-section = Volume/length

$= \dfrac{\text{mass}}{\text{density} \times \text{length}} = \dfrac{15.6 \times 10^{-3}}{7800 \times 2.5} = 8 \times 10^{-7} \text{ m}^2$

$Y = \dfrac{Fl}{A\Delta L} = \dfrac{8 \times 9.8 \times 2.5}{(8 \times 10^{-7}) \times 1.25 \times 10^{-3}}$

$Y = 1.96 \times 10^{11} \text{ N/m}^2$

Sol 3: (D) Work done on the wire

$W = \dfrac{1}{2} F \times l$

$= \dfrac{1}{2} \times \text{stress} \times \text{volume} \times \text{strain}$

$W = \dfrac{1}{2} \times Y \times \text{strain}^2 \times \text{volume}$

$W = \dfrac{1}{2} \times Y \times \dfrac{\Delta L^2}{L^2} \times AL = \dfrac{YA\Delta L^2}{2L}$

$W = \dfrac{2 \times 10^{11} \times 10^{-6} \times 10^{-6}}{2 \times 1} = 0.1 \text{ J}$

Sol 4: (D) $\dfrac{l_1}{l_2} = \dfrac{L_1 r_2^2}{L_2 r_1^2}$

$L_1 = L, L_2 = 2L, r_1 = 2R., r_2 = R$

∴ $\dfrac{l_1}{l_2} = \dfrac{L}{2L} \dfrac{R^2}{4R^2} = \dfrac{1}{8}$

Sol 5: (A) The changed density, $\rho' = \dfrac{\rho}{1 - \dfrac{dp}{B}}$

Substituting the value, we have

$$\rho' = \frac{11.4}{1 - \dfrac{2.0 \times 10^8}{8.0 \times 10^9}}$$

$\rho' = 11.69 \, g/cm^3 \approx 11.7 \, g/cm^3$

Sol 6: (D) Mass of the rod $= \dfrac{AL}{\rho}$ if A is its cross sectional area

Weight acts at the mid-point

$$\therefore \; Y = \frac{mg}{A} \times \frac{(L/2)}{\Delta L}$$

If L is the original length

$$\Rightarrow \Delta L = \frac{mgL}{2AY} = \frac{g\rho L^2}{2Y}$$

$$= \frac{9.8 \times 1.3}{120} = 10.6 \, cm$$

Sol 7: (C) If L = initial length of the rod, increase in length caused by temperature increase

$= L \, \alpha \, \theta$

If this expansion is prevented by a compressive force, then

$$\text{Strain} = \frac{L\alpha\theta}{L} \alpha\theta = 6 \times 10^{-5}$$

$\therefore$ Stress developed in the rod

$= Y \times \text{strain} = 12 \times 10^6 \, N/m^2$

$= 1.2 \times 10^7 \, N/m^2$

Previous Years' Questions

Sol 1: (A) $\Delta l = \dfrac{Fl}{AY} = \left(\dfrac{Fl}{\left(\dfrac{\pi d^2}{4} \right) Y} \right)$ or $(\Delta l) \propto \dfrac{1}{d^2}$

Now, $\dfrac{1}{d^2}$ is maximum in option (A).

Sol 2: (B) In isothermal process

pV = constant

$\therefore \; pdV + Vdp = 0$ or $\left(\dfrac{dp}{dV} \right) = -\left(\dfrac{p}{V} \right)$

$\therefore$ Bulk modulus,

$$B = -\left(\frac{dp}{dV/V} \right) = -\left(\frac{dp}{dV} \right) V$$

$$\therefore \; B = -\left[\left(-\frac{p}{V} \right) V \right] = p$$

$\therefore \; B = p$

Note: Adiabatic bulk modulus is given by $B = \gamma p$.

Sol 3: (D) From the definition of bulk modulus,

$$B = \frac{-dp}{(dV/V)}$$

Substituting the values, we have

$$B = \frac{(1.165 - 1.01) \times 10^5}{(10/100)} = 1.55 \times 10^5 \, Pa$$

Sol 4: (D)

Time period, $T = 2\pi \sqrt{\dfrac{\ell}{g}}$

When additional mass M is added to its bob

$$T_M = 2\pi \sqrt{\frac{\ell + \Delta\ell}{g}}$$

$$\Delta\ell = \frac{Mg\ell}{AY} \Rightarrow T_M = 2\pi \sqrt{\frac{\ell + \dfrac{Mg\ell}{AY}}{g}}$$

$$\left(\frac{T_M}{T} \right)^2 = 1 + \frac{Mg}{AY}$$

$$\frac{1}{Y} = \frac{A}{Mg} \left[\left(\frac{T_M}{T} \right)^2 - 1 \right]$$

JEE Advanced/Boards

Exercise 1

Sol 1: Equivalent force constant of rubber cord.

$$k = \frac{YA}{l} = \frac{(5.0 \times 10^8)(1.0 \times 10^{-6})}{(0.1)} = 5.0 \times 10^3 \, N/m$$

Now, from conservation of mechanical energy, elastic potential energy of cord

= Kinetic energy of missile

$$\therefore \frac{1}{2}k(\Delta l)^2 = \frac{1}{2}mv^2$$

$$\therefore v = \left(\sqrt{\frac{k}{m}}\right)\Delta l = \left(\sqrt{\frac{5.0\times10^3}{5.0\times10^{-3}}}\right)(12.0-10.0)\times10^{-2}$$

= 20 m/s

Note: Following assumptions have been made in this problem:

(i) k has been assumed constant, even though it depends on the length (l).

(ii) The whole of the elastic potential energy is converting into kinetic energy of missile.

Sol 2: Let the initial volume of the block be V and v the increase in volume when it is heated t_1 to t_2. Then

$$v = V \times \gamma \times (t_2 - t_1)$$

Where γ is the coefficient of volume expansion. The volume strain is therefore,

$$\frac{v}{V} = \gamma(t_2 - t_1)$$

The bulk modulus is

$$B = \frac{\text{change in pressure}}{\text{volume strain}}$$

$$B = \frac{P}{\gamma(t_2 - t_1)}$$

$$P = B\gamma(t_2 - t_1)$$

Given B = 3.6 × 10^{11} N/m^2

$\gamma = 3\alpha = 3 \times 8.0 \times 10^{-6}$

$= 24 \times 10^{-6} /°C$

$(t_2 - t_1) = 70 - 50 = 20°C$

$\therefore$ P $(3.6 \times 10^{11}) \times (24 \times 10^{-6}) \times 20$

$= 1.728 \times 10^8$ N/m^2

Sol 3: Work done in compressing the bar is given by

$$W = \frac{1}{2}Fl$$

Where F is the force applied on the bar and l is the compression in the length of the bar. By Hooke's law, the Young's modulus of the material of the bar is given by

$$Y = \frac{F/A}{l/L} = \frac{FL}{Al}$$

Where A is the area of cross-section of the bar and L is the initial length

$$\therefore l = \frac{FL}{AY}$$

Hence from equation (i), we have

$$W = \frac{F^2L}{2AY}$$

Here F = 5kg, wt=5 × 9.8 N, L=0.2 m

A = 1.0 cm^2 = 1.0 × 10^{-6} m^2 and

Y = 1.0 × 10^{-5} N/m^2

$$\therefore W = \frac{(5\times9.8)^2 \times 0.2}{2\times(1.0\times10^{-4})\times(1.0\times10^{11})}$$

$= 2.4 \times 10^{-5}$ J

This is the increase in energy of the bar.

Exercise 2

Sol 1: (B)

F = Y α ΔtA;   A = 2 × 10^{-6} m^2

Y = 2 × 11 N/m^2; α = 1.1 × 10^{-5}

T = 50 – 30 = 20°C

F = 2 × 10^{11}×1.1×10^{-5}×20 × 2 × 10^{-6} = 88 N

Sol 2: (C) Potential energy per unit volume = u

$$= \frac{1}{2}\times\text{stress}\times\text{strain}; \quad \text{But } Y = \frac{\text{stress}}{\text{strain}}$$

$\therefore$ stress = Y × strain = Y × S

$\therefore$ Potential energy per unit volume = u

$$= \frac{1}{2}\times(YS)S = \frac{1}{2}YS^2$$

Sol 3: (A)

$\because$ Compressibility $= \frac{1}{K} = \frac{\Delta V}{V\times\Delta P}$

$\Delta V = (V \times \Delta P) \times \frac{1}{K}$

$\Delta V = (100 \times 15 \times 10^6) \times 5 \times 10^{-10}$

$\Delta V = 0.75$ ml

Sol 4: (A) $\theta = \frac{r\phi}{L} = \frac{(2/1000)45°}{1} = 0.09°$

Previous Years' Questions

Sol 1: (C) Thermal stress $\sigma = Y \alpha \Delta\theta$ Given, $\sigma_1 = \sigma_2$

$\therefore Y_1 \alpha_1 \Delta\theta = Y_2 \alpha_2 \Delta\theta$

or $\dfrac{Y_1}{Y_2} = \dfrac{\alpha_2}{\alpha_1} = \dfrac{3}{2}$

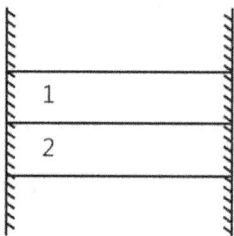

Sol 2: (A) $\Delta l = \left(\dfrac{l}{YA}\right).W$

i.e., graph is a straight line passing through origin (as shown in question also), the slope of which is $\dfrac{l}{YA}$.

$\therefore$ Slope $= \left(\dfrac{l}{YA}\right)$

$\therefore Y = \left(\dfrac{l}{YA}\right)\left(\dfrac{1}{\text{slope}}\right)$

$= \left(\dfrac{1.0}{10^{-6}}\right)\dfrac{(80-20)}{(4-1)\times10^{-4}}$

$= 2.0 \times 10^{11}$ N/m²

Sol 3: Young's modulus of elasticity is given by

$Y = \dfrac{\text{Stress}}{\text{Strain}} = \dfrac{F/A}{l/L} = \dfrac{FL}{lA} = \dfrac{FL}{l\left(\dfrac{\pi d^2}{4}\right)}$

Substituting the values, we get

$Y = \dfrac{50\times1.1\times4}{(1.25\times10^{-3})\times\pi\times(5.0\times10^{-4})^2}$

$= 2.24\times10^{11}$ N/m²

Now, $\dfrac{\Delta Y}{Y} = \dfrac{\Delta L}{L} + \dfrac{\Delta l}{l} + 2\dfrac{\Delta d}{d}$

$= \left(\dfrac{0.1}{110}\right)+\left(\dfrac{0.001}{0.125}\right)+2\left(\dfrac{0.001}{0.05}\right) = 0.0489$

$\Delta Y = (0.0489)\, Y$

$= (0.0489) \times (2.24 \times 10^{11})$ N/m² $= 1.09 \times 10^{10}$ N/m²

Sol 4: (A B)

$Y = \dfrac{\text{stress}}{\text{strain}}$

$\Rightarrow \dfrac{1}{Y} = \dfrac{\text{strain}}{\text{stress}} \Rightarrow \dfrac{1}{Y_P} > \dfrac{1}{Y_\theta} \Rightarrow Y_P < Y_Q$

3. FLUID MECHANICS

1. INTRODUCTION

Fluid is a collective term for liquid and gas. A fluid cannot sustain shear stress when at rest. We will study the dynamics of non-viscous, incompressible fluid. We will be learning about pressure variation, Archemides principle, equation of continity, Bernoulli's Theorem and its applications and surface tension, Stoke's Law and Terminal velocity of a spherical body.

2. DEFINITION OF A FLUID

A fluid is a substance that deforms continuously under the application of a shear (tangential) stress no matter how small the shear stress may be.

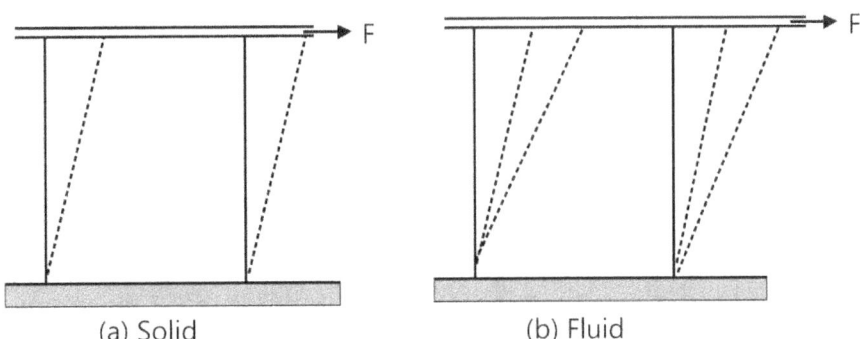

(a) Solid   (b) Fluid

Figure 9.1: Behavior of a solid and a fluid, under the action of a constant shear force.

3. FLUID STATICS

It refers to the state when there is no relative velocity between fluid elements. In this section we will learn some of the properties of fluid statics.

3.1 Density

The density ρ of a substance is defined as the mass per unit volume of a sample of the substance. If a small mass element Δm occupies a volume ΔV, the density is given by $\rho = \dfrac{\Delta m}{\Delta V}$

In general, the density of an object depends on position, so that $\rho = f(x, y, z)$

If the object is homogeneous, its physical parameters do not change with position throughout its volume. Thus for a homogeneous object of mass M and volume V, the density is defined as $\rho = \frac{M}{V}$

Thus SI units of density are kg m^{-3}.

NOMORECLASS CONCEPTS

Note: As pressure is increased, volume decreases and hence density will increase.

As the temperature of a liquid is increased, mass remains the same while the volume is increased.

3.2 Specific Gravity

The specific gravity of a substance is the ratio of its density to that of water at 4°C, which is 1000 kg/m^3. Specific gravity is a dimensionless quantity numerically equal to the density quoted in g/cm^3. For example, the specific gravity of mercury is 13.6, and the specific gravity of water at 100°C is 0.999. $RD = \frac{\text{Density of substance}}{\text{Density of water at } 4°C}$

Illustration 1: Find the density and specific gravity of gasoline if 51 g occupies 75 cm^3? **(JEE MAIN)**

Sol: Density is mass per unit volume, and specific gravity is the ratio of density of substance and density of water.

$$\text{Density} = \frac{\text{mass}}{\text{volume}} = \frac{0.051\text{kg}}{75\times10^{-6}\text{m}^3} = 680 \text{ kg/m}^3$$

$$\text{Sp. gr} = \frac{\text{density of gasoline}}{\text{density of water}} = \frac{680\text{kg}/\text{m}^3}{1000\text{kg}/\text{m}^3} = 0.68 \text{ or Sp. gravity} = \frac{\text{mass of 75 cm}^3\text{gasoline}}{\text{mass of 75 cm}^3\text{water}}$$

$$= \frac{51\text{g}}{75\text{g}} = 0.68$$

Illustration 2: The mass of a liter of milk is 1.032 kg. The butterfat that it contains has a density of 865 kg/m^3 when pure, and it constitutes 4 percent of the milk by volume. What is the density of the fat-free skimmed milk? **(JEE MAIN)**

Sol: Find the mass of butterfat present in the milk. Subtract this from total mass to get mass of fat-free milk. The density of fat-free milk is equal to its mass divided by its volume.

Volume of fat in 1000 cm^3 of milk = 4% × 1000 cm^3 = 40 cm^3

Mass of 40 cm^3 fat = $V\rho$ = $(40 \times 10^{-6}\ m^3)(865\ kg/m^3)$ = 0.0346 kg

$$\text{Density of skimmed milk} = \frac{\text{mass}}{\text{volume}} = \frac{(1.032-0.0346)\text{kg}}{(1000-40)\times10^{-6}\text{m}^3}$$

3.3 Pressure

The pressure exerted by a fluid is defined as the force per unit area at a point within the fluid. Consider an element of area ΔA as shown in the figure and an external force ΔF is acting normal to the surface. The average pressure in the fluid at the position of the element is given by $P_{av} = \frac{\Delta F}{\Delta A}$ [A normal force ΔF acts on a small cylindrical element of cross-section area ΔA.]

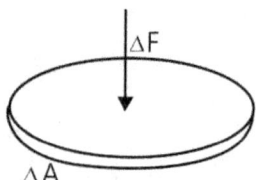

Figure 9.2

As $\Delta A \to 0$, the element reduces to a point, and thus, pressure at a point is defined as

$$p = \lim_{\Delta A \to 0} \frac{\Delta F}{\Delta A} = \frac{dF}{dA}$$

When the force is constant over the surface A, the above equation reduces to $p = \dfrac{F}{A}$

The SI unit of pressure is Nm^{-2} and is also called Pascal (Pa). The other common pressure units are atmosphere and bar.

1 atm = 1.01325×10^5 Pa; 1 bar = 1.00000×10^5 Pa; 1 atm = 1.01325 bar

3.3.1 Pressure Is Isotropic

Imagine a static fluid and consider a small cubic element of the fluid deep within the fluid as shown in the figure. Since this fluid element is in equilibrium therefore, forces acting on each lateral face of this element must also be equal in magnitude. Because the areas of each face are equal, therefore, the pressure on each face is equal in magnitude. Therefore the pressure on each of the lateral faces must also be the same. In the limit as the cube element to a point, the forces on top and bottom surfaces also become equal. Thus, the pressure exerted by a fluid at a point is the same in all directions – pressure is isotropic.

Note: Since the fluid cannot support a shear stress, the force exerted by a fluid pressure must also be perpendicular to the surface of the container that holds it.

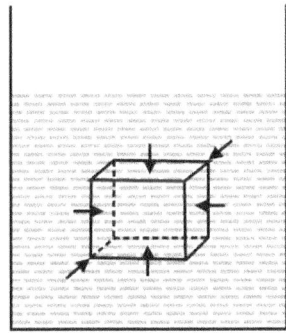

Figure 9.3: A small cubical element is in equilibrium inside a fluid

3.3.2 Atmospheric Pressure (P_0)

It is pressure of the earth's atmosphere. This changes with weather and elevation. Normal atmospheric pressure at sea level (an average value) is 1.013×10^5 Pa. Thus,

1 atm = 1.013×10^5 Pa=1.013 Bar

3.3.3 Absolute Pressure and Gauge Pressure

The excess pressure above atmospheric pressure is usually called gauge pressure and the total pressure is called absolute pressure. Thus, Gauge pressure = absolute pressure – atmospheric pressure. Aboslute pressure is always greater than or equal to zero. While gauge pressure can be negative also.

Illustration 3: Atmospheric pressure is about 1.01×10^5 Pa. How large a force does the atmosphere exert on a 2 cm^2 area on the top of your head? **(JEE MAIN)**

Sol: Force = Pressure × Area

Because p = F/A, where F is perpendicular to A, we have F = pA. Assuming that 2 cm^2 of your head is flat (nearly correct) and that the force due to the atmosphere is perpendicular to the surface (as it is), we have F = pA = $(1.01 \times 10^5$ N/m^2) $(2 \times 10^{-4} m^2) \approx 20N$

3.3.4 Variation of Pressure with Depth

Weight of a fluid element of mass Δm, $\Delta W = (\Delta m)g$. The force acting on the lower face of the element is pA and that on the upper face is $(p + \Delta p)A$. The figure (b) shows the free body diagram of the element. Applying the condition of equilibrium, we get, $pA - (p + \Delta p) A - (\Delta m)g = 0$

if ρ is the density of the fluid at the position of the element, then $\Delta m = \rho A(\Delta y)$

and pA − (p + Δp) A − ρgA(Δy) = 0

or $\dfrac{\Delta p}{\Delta y} = -\rho g$

In the limit Δy approaches to zero, $\dfrac{\Delta p}{\Delta y}$ becomes $\dfrac{dp}{dy} = -\rho g$. The above equation indicates that the slope of p versus y is negative. That is, the pressure p decreases with height y from the bottom of the fluid. In other words, the pressure p increases with depth h, i.e., $\dfrac{dp}{dh} = \rho g$

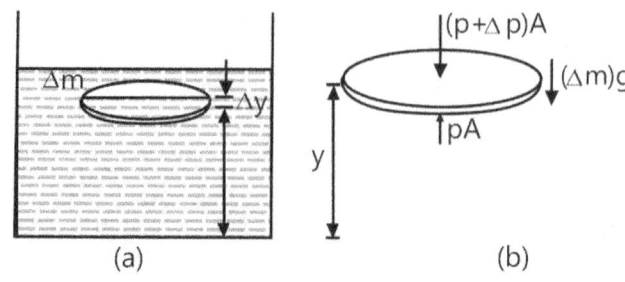

(a)   (b)

Figure 9.4

3.4 The Incompressible Fluid Model

For an incompressible fluid, the density ρ of the fluid remains constant throughout its volume. It is a good assumption for liquids. To find pressure at the point A in a fluid column as shown in the figure, is obtained by integrating the following equation:

$$dp = \rho g dh \text{ or } \int_{p_0}^{p} dp = \rho g \int_{0}^{h} dh \text{ or } p - p_0 = \rho gh \text{ or } p = p_0 + \rho gh \qquad ...(xvi)$$

where ρ is the density of the fluid, and p_0 is the atmospheric pressure at the free surface of the liquid.

Note: Further, the pressure is the same at any two points at the same level in the fluid. The shape of the container does not matter.

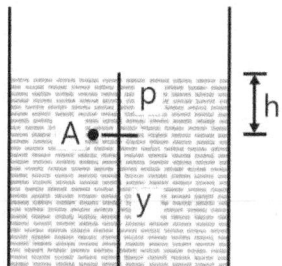

Figure 9.5: A point A is located in a fluid at a height from the bottom and at a deth h from the free surface

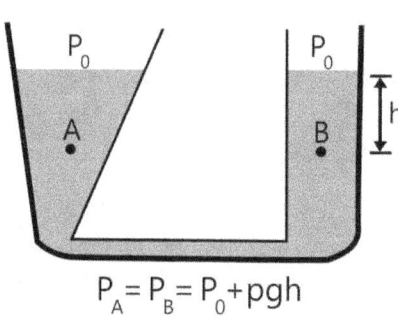

$$P_A = P_B = P_0 + \rho gh$$

Illustration 4: Find the absolute pressure and gauge pressure at point A, B and C as shown in the Fig. 9.6 (1 atm = 10^5 Pa) **(JEE MAIN)**

Sol: Gauge Pressure = ρgh, Absolute Pressure is sum of gauge pressure and atmospheric pressure.

$P_{atm} = 10^5$ Pa.

Absolute Pressure **A** -> $P_A + P_{atm} = r_1 gh_A = (800)(10)1 = 8$ kPa

$P'_A = P_A + P_{atm} = 108$ kPa

Gauge Pressure = 8 kPa.

B -> $P_B = \rho_1 g(2) + \rho_2 g(1.5)$

$P'_B = P_B + P_{atm} = 131$ kPa = $(800)(10)(2) + (10^3)(10)(1.5) = 131$ kPa

Gauge Pressure = 31 kPa.

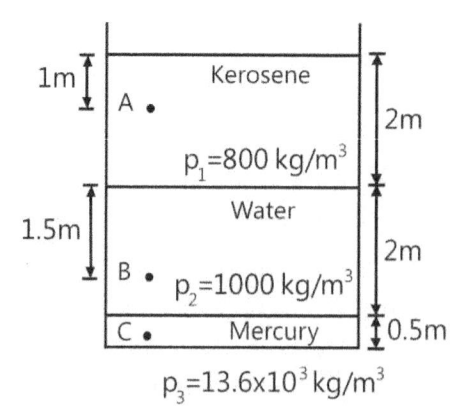

Figure 9.6

$C -> p_c = \rho_1 g(2) + \rho_2 g(2) + \rho_3 g(0.5)$

$\rho_C' = p_C + p_{atm} = 204\,kPa$

$= (800)(10)(2) + (10)^3(10)(2) + 1(13.6 \times 10^3)(10)(0.5) = 204\ kPa$

Gauge Pressure = 104 kPa.

Illustration 5: A glass full of water of a height of 10 cm has a bottom of area 10 cm², top of area 30 cm² and volume 1 litre. **(JEE ADVANCED)**

(a) Find the force exerted by the water on the bottom.

(b) Find the resultant force exerted by the side of the glass on the water.

(c) If the glass is covered by a jar and the air inside the jar is completely pumped out, what will be the answer to parts (a) and (b).

(d) If a glass of different shape is used provided the height, the bottom area and the volume are unchanged, will the answers to parts (a) and (b) change.

Take $g = 10\,m/s^2$, density of water $= 10^3\ kg/m^3$ and atmospheric pressure $= 1.01 \times 10^5\ N/m^2$.

Sol: Pressure at the bottom depends on the height of water in the container. Force = Pressure × Area. The force on water surface due to atmospheric pressure plus the weight of water are balanced by the force on water by the container bottom and its walls.

(a) Force exerted by the water on the bottom $F_1 = (P_0 + \rho gh)A_1$ ... (i)

Here, P_0 = atmospheric pressure = $1.01 \times 10^5\ N/m^2$; ρ = density of water = $10^3\ kg/m^3$

$g = 10\ m/s^2$, h = 10 cm = 0.1 m and A_1 = area of base 10 cm² $= 10^{-3}\ m^2$. Substituting in Eq. (i), we get $F_1 = (1.01 \times 10^5 + 10^3 \times 10 \times 0.1) \times 10^{-3}$ or $F_1 = 102\ N$ (downwards)

(b) Force exerted by atmosphere on water $F_2 = (P_0)A_2$

Here, A_2 = area of top = 30 cm² = $3 \times 10^{-3}\ m^2$; $F_2 = (1.01 \times 10^5)(3 \times 10^{-3}) = 303\ N$ (downwards)

Force exerted by bottom on the water $F_3 = -F_1$ or $F_3 = 102\ N$ (upwards)

Weight of water W = (volume)(density)(g) = $(10^{-3})(10^3)(10) = 10\ N$ (downwards)

Let F be the force exerted by side walls on the water (upwards). Then, from equilibrium of water

Net upward force = net downward force or $F + F_3 = F_2 + W$

$F - F_2 + W - F_3 = 303 + 10 - 102$ or F = 211 N (upwards)

(c) If the air inside of the Jar is completely pumped out,

$F_1 = (\rho gh)A_1$ (as $P_0 = 0$) = $(10^3)(10)(0.1)(10^{-3}) = 1\ N$ (downwards). In this case $F_2 = 0$ and $F_3 = 1\ N$ (upwards)
∴ $F = F_2 + W - F_3 = 0 + 10 - 1 = 9\ N$ (upwards)

(d) No, the answer will remain the same. Because the answers depend upon P_0, ρ, g, h , A_1 and A_2.

Illustration 6: Two vessels have the same base area but different shapes. The first vessel takes twice the volume of water that the second vessel requires to fill up to a particular common height. Is the force exerted by water on the base of the vessel the same in the two cases? If so, why do the vessels filled with water to the same height give different reading on a weighing scale? **(JEE MAIN)**

Sol: Force on the base of the vessel depends on the pressure on it, and pressure depends on the height of the liquid in the vessel. On the other hand the normal reaction from the surface on which the vessel is kept, depends on both the pressure at the base as well as the weight of the liquid in the vessel.

Pressure (and therefore force) on the two equal base areas are identical. But force is exerted by water on the sides of the vessels also, which has non-zero vertical component when the sides of the vessel are not perfectly normal to the base. This net vertical component of force by water on the side of the vessel is greater for the first vessel than the second. Hence, the vessels weigh different when the force on the base is the same in the two cases.

3.4.1 Pascal's Laws

According to the equation $p = p_0 + \rho gh$. Pressure at any depth h in a fluid may be increased by increasing the pressure p_0 at the surface. Pascal recognized a consequence of this fact that we now call Pascal's Law. A pressure applied to a confined fluid at rest is transmitted equally undiminished to every part of the fluid and the walls of the container.

This principle is used in a hydraulic jack or lift, as shown in the figure.

The pressure due to a small force F_1 applied to a piston of area A_1 is transmitted to the large piston of area A_2. The pressure at the two pistons is the same because they are at the same level.

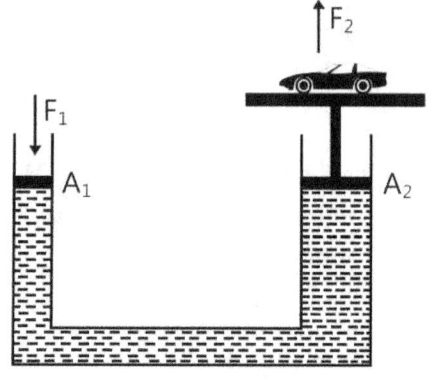

A hydraulic jack

Figure 9.7

$$p = \frac{F_1}{A_1} = \frac{F_2}{A_2} \quad \text{Or} \quad F_2 = \left(\frac{A_2}{A_1}\right) F_1.$$ Consequently, the force on the larger piston is large.

Thus, a small force F_1 acting on a small area A_1 results in a larger force F_2 acting on a larger area A_2.

NOMORECLASS CONCEPTS

Since energy is always conserved, $F_1 x_1 = F_2 x_2$ where x_1 and x_2 are the distances moved by the pistons.

Illustration 7: Find the pressure in the air column at which the piston remains in equilibrium. Assume the pistons to be massless and frictionless.

(JEE MAIN)

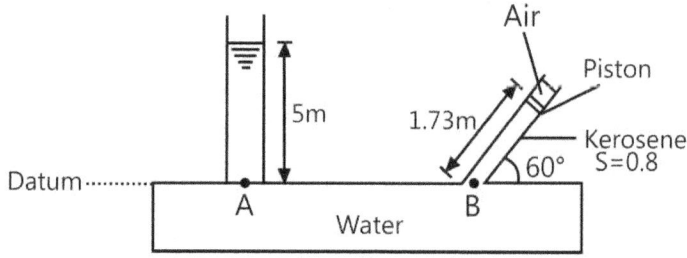

Figure 9.8

Sol: Apply Pascal's law at two points at equal height from a common datum.

Let p_a be the air pressure above the piston.

Applying Pascal's law at point A and B.

$$P_{atm} + r_w g(5) = p_a + r_k g(1.73)\frac{\sqrt{3}}{2} ; P_a = 138 \text{ kPa}$$

Illustration 8: A weighted piston confines a fluid of density ρ in a closed container, as shown in the figure. The combined weight of piston and container is W = 200 N, and the cross-sectional area of the piston is A = 8 cm². Find the total pressure at point B if the fluid is mercury and h = 25 cm (p_m = 13600 kgm⁻³). What would be an ordinary pressure gauge reading at B?
(JEE ADVANCED)

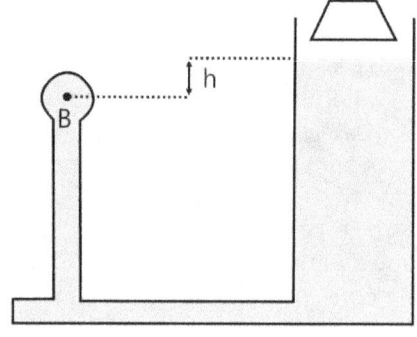

Figure 9.9

Sol: Pressure difference between two points at different heights is equal to ρgh, where h is difference in heights of two points. Apply Pascal's law at two points at different heights from a common datum.

4.6

Pascal's principle tells us about the pressure applied to the fluid by the piston and atmosphere. This added pressure is applied at all points within the fluid. Therefore the total pressure at B is composed of three parts: Pressure of atmosphere = 1.0×10^5 Pa

Pressure due to piston and weight = $\dfrac{W}{A} = \dfrac{200N}{8 \times 10^{-4} m^2}$ = 2.5×10^5 Pa

Pressure due to height h of fluid = $h\rho g$ = 0.33×10^5 Pa

In this case, the pressure of the fluid itself is relatively small. We have

Total pressure at B = 3.8×10^5 Pa = 383 kPa. The gauge pressure does not include atmospheric pressure. Therefore,

Gauge pressure at B = 280 kPa

Illustration 9: For the system shown in the figure, the cylinder on the left, at L, has a mass of 600 kg and a cross-sectional area of 800 cm². The piston on the right at S, has cross-sectional area 25 cm² and negligible weight. If the apparatus is filled with oil ($\rho=0.78$ g/cm³), find the force F required to hold the system in equilibrium as shown in figure.　　　**(JEE ADVANCED)**

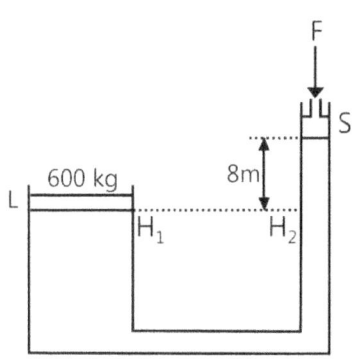

Sol: Apply Pascal's law at two points at different heights from a common datum.

The pressures at point H_1 and H_2 are equal because they are at the same level in the single connected fluid. Therefore, Pressure at H_1 = pressure at H_2 = (pressure due to F plus pressure due to liquid column above H_2)

$$\dfrac{(600)(9.8)N}{0.08m^2} = \dfrac{F}{25 \times 10^{-4} m^2} + (8m)(780\ kg/m^{-3})(9.8)$$

Figure 9.10

After solving, we get, F = 31 N

Illustration 10: As shown in the figure, as column of water 40 cm high supports 31 cm of an unknown fluid. What is the density of the unknown fluid?　　　**(JEE MAIN)**

Sol: Find the hydrostatic pressure at the bottom most point A due to both the water column and the unknown fluid column.

The pressure at point A due to the two fluids must be equal (or the one with the higher pressure would push lower pressure fluid away). Therefore, pressure due to water = pressure due to known fluid; $h_1 r_1 g = h_2 r_2 g$, from which $r_2 = \dfrac{h_1}{h_2} p_1 = \dfrac{40}{31} (1000\ kg/m^2) = 1290$ kg/m³

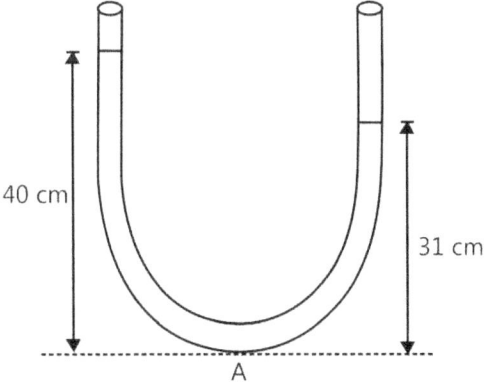

40 cm

31 cm

A

Figure 9.11

For gases, the constant density assumed in the compressible model is often not adequate. However, an alternative simplifying assumption can be made that the density is proportional to the pressure, i.e., $\rho = kp$. Let r_0 be the density of air at the earth's surface

where the pressure is atmospheric p_0, then $r_0 = kp_0$; After eliminating k, we get $\rho = \dfrac{p_0}{p_0} p$

Putting the value of ρ in equation $dp = -\rho g dy$ or $dp = -\left(\dfrac{p_0}{p^0}p\right)g dy$

On rearranging, we get $\displaystyle\int_{p_0}^{p} \dfrac{dp}{p} = -\dfrac{p_0}{p_0}g\int_0^h dy$ where p is the pressure at a height y = h above the earth's surface.

After integrating, we get $\ln\left|\dfrac{p}{p_0}\right| = -\dfrac{\rho_0}{p_0}gh$ or $p = p_0 p_0^{\frac{-p_0}{p_0}gh}$

Note: Instead of a linear decrease in pressure with increasing height as in the case of an incompressible fluid, in this case pressure decreases exponentially.

4. PRESSURE MEASURING DEVICES

4.1 Manometer

A manometer is a tube open at both ends and bent into the shape of a "U" and is partially filled with mercury. When one end of the tube is subjected to an unknown pressure p, the mercury level drops on that side of the tube and rises on the other so that the difference in mercury level is h as shown in the figure.

When we move down in a fluid, pressure increases with depth and when we move up the pressure decreases with height. When we move horizontally in a fluid, pressure remains constant. Therefore, $p + r_0gh_0 - r_mgh = p_0$ where p_0 is atmospheric pressure, and r_m is the density of the fluid inside the vessel.

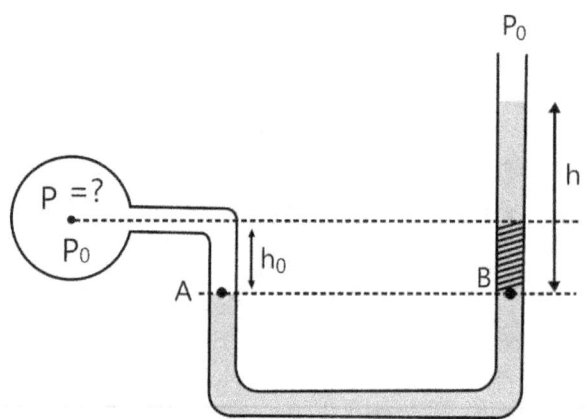

Figure 9.12: An U-shaped manometer tube connected to a vessel

4.2 The Mercury Barometer

It is a straight glass tube (closed at one end) completely filled with mercury and inserted into a dish which is also filled with mercury as shown in the figure. Atmospheric pressure supports the column of mercury in the tube to a height h. The pressure between the closed end of the tube and the column of mercury is zero, p = 0. Therefore, pressure at points A and B are equal and thus $p_0 = 0 + r_mgh$. Hence, $p_0 = (13.6 \times 10^3)(9.8)(0.76) = 1.01 \times 10^5$ Nm^{-2} for Pa.

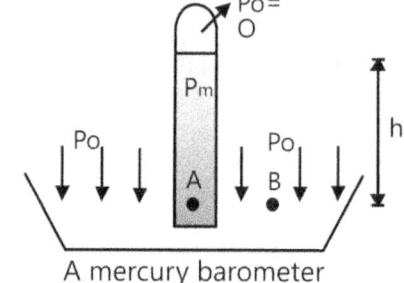

A mercury barometer

Figure 9.13

Illustration 11: What must be the length of a barometer tube used to measure atmospheric pressure if we are to use water instead of mercury? **(JEE MAIN)**

Sol: The length of the barometer tube will be inversely proportional to the density of fluid used in it.

We know that $p_0 = r_mgh_m = r_wgh_w$ where r_w and h_w are the density and height of the water column supporting the atmospheric pressure p_0.

$\therefore$ $h_w = \dfrac{\rho_m}{\rho_w}h_w$; Since $\dfrac{\rho_m}{\rho_w} = 13.6$; $h_w = 0.76$ m $= (13.6)(0.76) = 10.33$ m.

5. PRESSURE DIFFERENCE IN ACCELERATING FLUIDS

Consider a beaker filled with some liquid of density p accelerating upwards with an acceleration a_y along positive y-direction. Let us draw the free body diagram of a small element of fluid of area A and length dy as shown in figure. Equation of motion for this fluid element is, $PA - W - (P + dP)A = (mass)(a_y)$ or $-W - (dP) A = (A\rho\, dy)(a_y)$

or $(A\rho g\, dy) - (dP)A = (A\rho\, dy)(a_y)$ or $\dfrac{dP}{dy} = -\rho(g + a_y)$

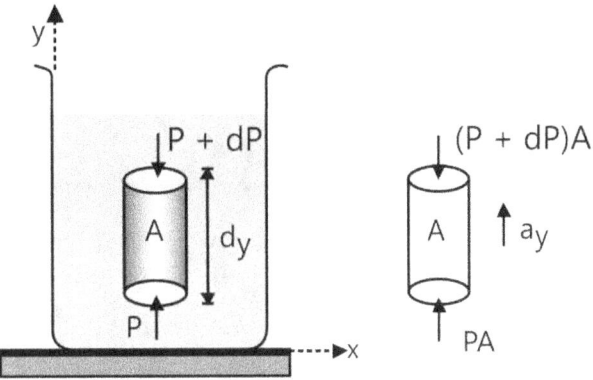

Figure 9.14

Similarly, if the beaker moves along positive x-direction with acceleration a_x, the equation of motion for the fluid element shown in the figure is, $PA - (P + dP) A$
$= (mass)(a_x)$

or $(dP)A = (A\rho\ dx)a_x$ Or $\dfrac{dP}{dx} = -\rho a_x$

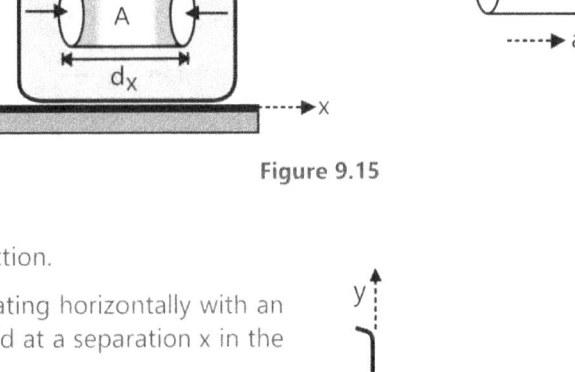

But suppose the beaker is accelerated and it has components of acceleration a_x and a_y in x and y directions respectively, then the pressure decreases along both x and y directions. The above equation

in that case reduces to,

Figure 9.15

$$\dfrac{dP}{dx} = -\rho a_x \qquad \text{and} \qquad \dfrac{dP}{dy} = -\rho(g + a_y) \qquad \dots (i)$$

For surface of a Liquid Accelerated in Horizontal Direction.

Consider a liquid placed in a beaker which is accelerating horizontally with an acceleration 'a'. Let A and B be two points in the liquid at a separation x in the same horizontal line. As we have seen in this case.

$\dfrac{dP}{dx} = -\rho a$ or dP = -ra dx. Integrating this with proper limits, we get

$P_A - P_B = pax \qquad \dots (ii)$

Further, $P_A = P_0 + \rho gh_1$ And $P_B = P_0 + \rho gh_2$

Substituting in Eq. (ii), we get $pg(h_1 - h_2) = pax$ ∴ $\dfrac{h_1 - h_2}{x}$

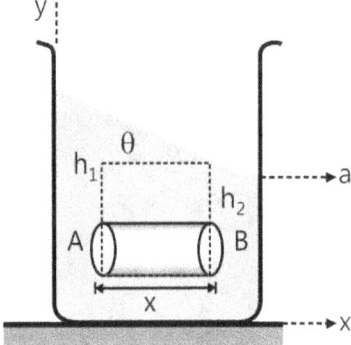

$$= \dfrac{a}{g} = \tan\theta \therefore \quad \boxed{\tan\theta = \dfrac{a}{g}}$$

Figure 9.16

Note: When a_y is not equal to zero then the angle of inclination is given by

$$\tan\theta = \dfrac{dy}{dx} = \left(\dfrac{\overline{(dp)}}{\left(\dfrac{dp}{dy} \right)} \right) = \dfrac{a_x}{g + a_y}$$

Illustration 12: A liquid of density ρ is in a bucket that spins with angular velocity ω as shown is the figure. Show that the pressure at a radial distance r from the axis is

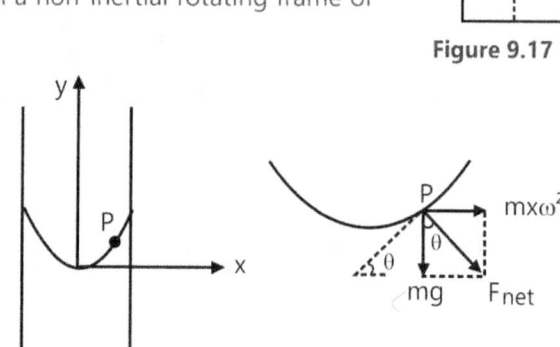

$P = P_0 + \dfrac{\rho\omega^2 r^2}{2}$ where P_0 is the atmospheric pressure.   **(JEE ADVANCED)**

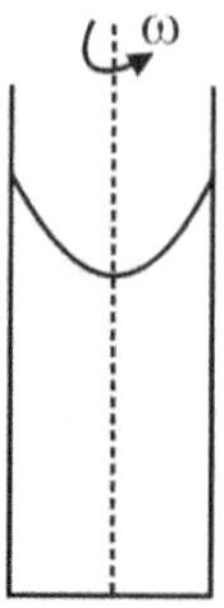

Sol: The net force on the liquid surface in equilibrium is always perpendicular to it as the liquid surface cannot sustain shear stress.

Consider a fluid particle P of mass m at coordinates (x, y). From a non-inertial rotating frame of reference, two forces are acting on it.

(i) Pseudo force $(mx\omega^2)$

(ii) Weight (mg) in the direction shown in figure.

Net force on it should be perpendicular to the free surface (in equilibrium). Hence,

$\tan\theta = \dfrac{mx\omega^2}{mg} = \dfrac{x\omega^2}{g}$ or $\dfrac{dy}{dx} = \dfrac{x\omega^2}{g}$

$\therefore \int_0^y dy = \int_0^x \dfrac{x\omega^2}{g}\cdot dx \therefore y = \dfrac{x^2\omega^2}{2g}$

Figure 9.18

This is the equation of the free surface of the liquid, which is a parabola.

As $x = r$, $y = \dfrac{r^2\omega^2}{2g} \therefore P(r) = P_0 + \rho g y$ or $\quad P(r) = P_0 + \dfrac{\rho\omega^2 r^2}{2}$

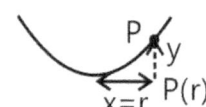

Figure 9.19

Hence proved.

Illustration 13: An open rectangular tank 5 m × 4 m × 3 m high containing water up to a height of 2 m is accelerated horizontally along the longer side.

(a) Determine the maximum acceleration that can be given without spilling the water.

(b) Calculate the percentage of water split over, if this acceleration is increased by 20%.

(c) If initially, the tank is closed at the top and is accelerated horizontally by 9 m/s², find the gauge pressure at the bottom of the front and rear walls of the tank. (Take g = 10 m/s²)   **(JEE MAIN)**

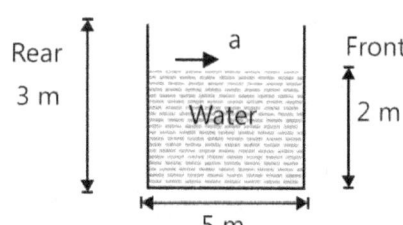

Figure 9.20

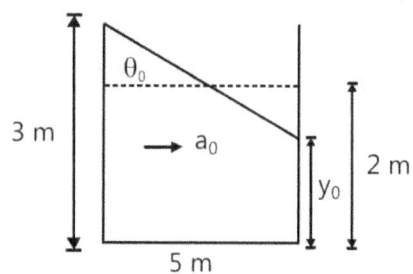

Figure 9.21

Sol: As the water column is accelerated towards right in horizontal direction, the free surface will not be horizontal but will be inclined at an angle with the θ horizontal, such that the left edge of the surface is at a higher level than the right edge. This is because the pressure at the left of water column will be more than the pressure at the right of it.

(a) Volume of water inside the tank remains constant

$\left(\dfrac{3+y_0}{2}\right) 5 \times 4 = 5 \times 2 \times 4$ or $y_0 = 1\text{m} \therefore \tan q_0 = \dfrac{3-1}{5} = 0.4$

Since, $\tan q_0 = \dfrac{a_0}{g}$, therefore $a_0 = 0.4\,g = 4\ \text{m/s}^2$

(b) When acceleration is increased by 20%

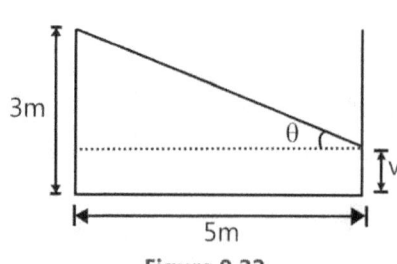

Figure 9.22

4.10

$a = 1.2 \, a_0 = 0.48 \, g \therefore \tan \theta = \dfrac{a}{g} = 0.48$

Now, $y = 3 - 5 \tan \theta = 3 - 5 \,(0.48) = 0.6 \, m$

Fraction of water split over $= \dfrac{4 \times 2 \times 5 - \dfrac{(3 + 0.6)}{2} \times 5 \times 4}{2 \times 5 \times 4} = 0.1$

Percentage of water split over $= 10\%$

(c) $a' = 0.9 \, g;\ \tan \theta' = \dfrac{a'}{g} = 0.9$

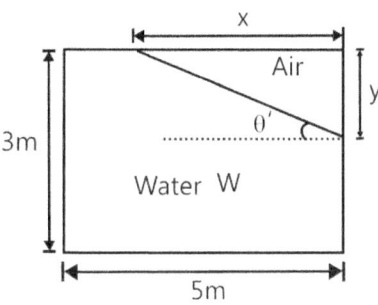

Figure 9.23

Volume of air remains constant $\rightarrow 4 \times \dfrac{1}{2} yx = (5)(1) \times 4 \Rightarrow$ Pressure does not change in the air.

Since $y = x \tan \theta' \therefore \dfrac{1}{2} x^2 \tan \theta' = 5$ or $\qquad x = 3.33 \, m;\ y = 3.0 \, m$

Gauge pressure at the bottom of the

(i) Front wall p_f = zero

(ii) Rear wall $p_r = (5 \tan \theta') rwg = 5(0.9)(10^3)(10) = 4.5 \times 10^4 \, Pa$

Illustration 14: A vertical U-tube with the two limbs 0.75 m apart with water and rotated about a vertical axis 0.5 m from the left limb, as shown in the figure. Determine the difference in elevation of the water levels in the two limbs, when the speed of rotation is 60 rpm.

(JEE MAIN)

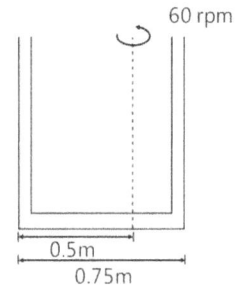

Figure 9.24

Sol: Each element of water in the tube is accelerated towards the axis. Along the horizontal part of the tube, the pressure will increase gradually as one moves radially away from the axis. The extra pressure provides the required centripetal acceleation.

Consider a small element of length dr at a distance r from the axis of rotation. Considering the equilibrium of this element.

$(p + dp) - p = r\omega^2 r \, dr \qquad$ or $dp = r\omega^2 r \, dr$

On integrating between 1 and 2

$p_1 - p_2 = \rho \omega^2 \displaystyle\int_{r_2}^{r_1} r \, dr = \dfrac{\rho \omega^2}{2} (r_1^2 - r_2^2)$

or $h_1 - h_2 = \dfrac{\omega^2}{2g} [r_1^2 - r_2^2] = \dfrac{(2\pi)^2}{2(10)} [(0.5)^2 - (0.25)^2] = 0.37 \, m.$

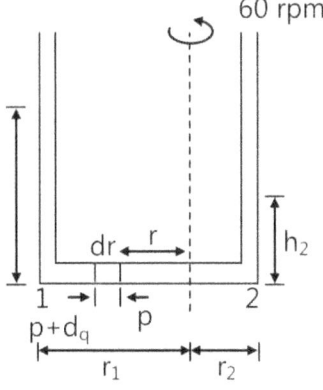

Figure 9.25

6. BUOYANCY

If a body is partially or wholly immersed in a fluid, it experiences an upward force due to the fluid surrounding it.

The phenomenon of force exerted by fluid on the body is called buoyancy and the force is called buoyant force. A body experiences buoyant force whether it floats or sinks, under its own weight or due to other forces applied on it.

Note: The buoyant force is due to the fact that the hydrostatic pressure at different depths is not the same. Buoyant force is independent of:

(a) Total volume and shape of the body.

(b) Density of the body.

6.1 Archimedes Principle

A body immersed in a fluid experiences an upward buoyant force equivalent to the weight of the fluid displaced by it. The proof of this principle is very simple. Imagine a body of arbitrary shape completely immersed in a liquid of density ρ. A body is being acted upon by the forces from all directions. Let us consider a vertical element of height h and cross-sectional area dA. The force acting on the upper surface of the element is F_1 (downward) and that on the lower surface is F_2 (upward). Since $F_2 > F_1$, therefore, the net upward force acting on the element is $dF = F_2 - F_1$. It can be easily seen that

$F_1 = (\rho g h_1) dA$ and $\qquad F_2 = (\rho g h_2) dA.$ $\qquad$ So $\quad dF = \rho g(h)\, dA$

Also, $\qquad h_2 - h_1 = h \qquad$ and $h(dA) = dV \qquad \therefore \quad$ The net upward force is $F = \int \rho g dV = \rho V g$

Hence, for the entire body, the buoyant force is the weight of the volume of the fluid displaced.

Note: Buoyant force acts on the centre of gravity of the displacement liquid. This point is called centre of Buoyancy.

NOMORECLASS CONCEPTS

The fluid exerts force on the immersed part of the body from all directions.

The net force experienced by every vertical element of the body is in the upward direction.

A uniform body floats in a liquid if density of the body is less than or equal to the density of the liquid and sinks if density of the uniform body is greater than that of the liquid.

6.1.1 Detailed Explanation

An object floats on water if it can displace a volume of water whose weight is greater than that of the object. If the density of the material is less than that of the liquid, it will float even if the material is a uniform solid, such as a block of wood floats on water surface. If the density of the material is greater than that of water, such as iron, the object can be made to float provided it is not a uniform solid. An iron built ship is an example to this case

Apparent weight of a body immersed in a liquid = $w - w_0$, where 'w' is the true weight of the body and w_0 is the apparent loss in weight of the body, when immersed in the liquid.

6.1.2 Buoyant Force in Accelerating Fluids

Suppose a body is dipped inside a liquid of density ρ_L placed in an elevator moving with acceleration $\vec{a}$. The buoyant force F in this case becomes, $F = V \rho_L g_{eff}$;

Here, $\qquad g_{eff} = |\vec{g} - \vec{a}|$

Illustration 15: An iceberg with a density of 920 kgm^{-3} floats on an ocean of density 1025 kgm^{-3}. What fraction of the iceberg is visible? **(JEE MAIN)**

Sol: The buoyant force on the iceberg will be equal to its weight. The buoyant force is equal to the weight of water displaced by the iceberg, i.e. the weight of volume of water equal to the volume of iceberg immersed.

Let V be the volume of the iceberg above the water surface, then the volume under inside is $V_0 - V$. Under floating conditions, the weight $(\rho_i V_0 g)$ of the iceberg is balanced by the buoyant force $\rho_w (V_0 - V) g$.

Thus, $\qquad \rho_i V_0 g = \rho_w (V_0 - V)g$

or $\qquad \rho_w V = (\rho_w - \rho_i) V_0$

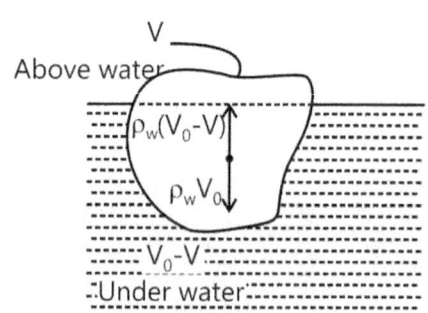

Figure 9.26

4.12

or $\dfrac{V}{V_0} = \dfrac{\rho_w - \rho_1}{\rho_w}$

Since, $r_w = 1025$ kg m^{-3} and $r_i = 920$ kg m^3, therefore, $\dfrac{V}{V_0} = \dfrac{1025 - 920}{1025} = 0.10$

Hence 10% of the total volume is visible.

Illustration 16: When a 2.5 kg crown is immersed in water, it has an apparent weight of 22 N. What is the density of the crown? **(JEE MAIN)**

Sol: Apply Archemides principle.

Let W = actual weight of the crown and W' = apparent weight of the crown

ρ = density of crown, ρ_0 = density of water. The buoyant force is given by $F_E = W - W'$ or

$\rho_0 Vg = W - W'$. Since $W = \rho Vg$, therefore, $V = \dfrac{W}{\rho g}$. Eliminating V from the above equation, we get

$\rho = \dfrac{\rho_0 W}{W - W'}$. Here W = 25 N; W' = 22 N; $\rho_0 = 10^3$ kg m^{-3}; $\rho = \dfrac{(10)^3 (25)}{25 - 22} = 9.3 \times 10^3$ kg m^{-3}.

Illustration 17: The tension in a string holding a solid block below the surface of a liquid (of density greater than that of solid) as shown in figure is T_0 when the system is at rest. What will be the tension in the string if the system has an upward acceleration a? **(JEE MAIN)**

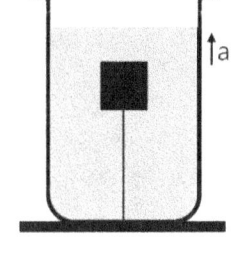

Figure 9.27

Sol: The weight and tension force on the block are balanced by the buoyant force on it. When the system is accelerated upwards, the effective value of g is increased.

Let m be the mass of block.

Initially for the equilibrium of block, $F = T_0 + mg$ ....(i)

Here, F is the up thrust on the block.

When the lift is accelerated upwards, g_{eff} becomes g + a instead of g.

Hence $F' = F\left(\dfrac{g + a}{g}\right)$ ...(ii)

From Newton's second law, $F' - T - mg = ma$ ...(iii)

Solving equations (i), (ii) and (iii), we get $T = T_0\left(1 + \dfrac{a}{g}\right)$

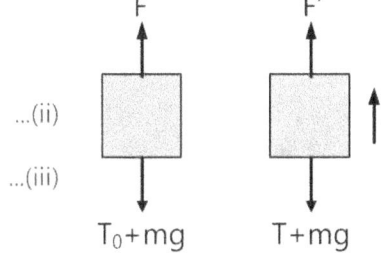

Figure 9.28

Illustration 18: An ice cube of side 1 cm is floating at the interface of kerosene and water in beaker of base area 10 cm^2. The level of kerosene is just covering the top surface of the ice cube.

(a) Find the depth of submergence in the kerosene and that in the water.

(b) Find the change in the total level of the liquid when the whole ice melts into water. **(JEE ADVANCED)**

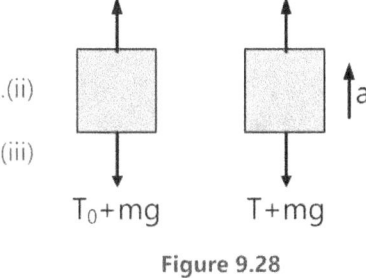

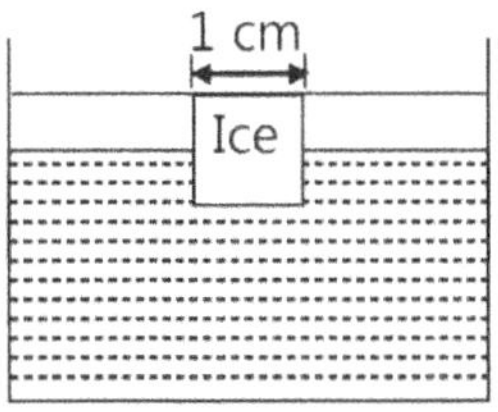

Figure 9.29

Sol: Apply Archemedes principle. Sum of the buoyant forces by kerosene and water will be equal to the weight of the ice cube.

(a) Condition of floating $0.8 \rho_w g h_k + \rho_w g h_w = 0.9 \rho_w g h$

or $0.8 h_k + h_w = (0.9)h$ ... (i)

Where h_k and h_w are the submerged depths of the ice in the kerosene and water, respectively.

Also $h_k + h_w = h$ ... (ii)

Here it is given that $h = 1$ cm

Solving equations (i) and (ii), we get

$h_k = 0.5$ cm, $h_w = 0.5$ cm

(b) $1 \text{ cm}^3 \xrightarrow[\text{Ice}]{\text{m heat}} 0.9 \text{ cm}^3 \text{ (water)}$

Fall in the level of kerosene $\Delta h_k = \dfrac{0.5}{A}$; Rise in the level of water $\Delta h_w = \dfrac{0.9 - 0.5}{A} = \dfrac{0.4}{A}$

Net fall in the overall level $\Delta h = \dfrac{0.1}{A} = \dfrac{0.1}{10} = 0.01$ cm $= 0.1$ mm.

6.2 Stability of a Floating Body

The stability of a floating body depends on the effective point of application of the buoyant force. The weight of the body acts at its centre of gravity. The buoyant force acts at the centre of gravity of the displaced liquid. This is called the centre of buoyancy. Under equilibrium condition, the centre of gravity G and the centre of buoyancy B lie along the vertical axis of the body as shown in the figure(s).

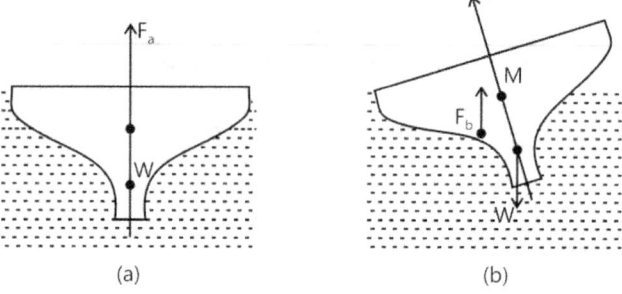

(a)   (b)

Figure 9.30

(a) The buoyant force acts at the centre of gravity of the displaced fluid.

(b) When the boat tilts, the line of action of the buoyant force intersects the axis of the boat at the metacentre M. In a stable boat, M is above the centre of gravity of the boat. When the body tilts to one side, the centre of buoyancy shifts relative to the centre of gravity as shown in the figure (b). The two forces act along different vertical lines. As a result, the buoyant force exerts a torque about the centre of gravity. The line of action of the buoyant force crosses the axis of the body at the point M, called metacentre. If G is below M, the torque will tend to restore the body to its equilibrium position. If G is above M, the torque will tend to rotate the body away from its equilibrium position and the body will be unstable.

Illustration 19: A wooden plank of length 1 m and uniform cross section is hinged at one end to the bottom of a tank as shown in the figure. The tank is filled with water up to a height of 0.5 m. The specific gravity of the plank is 0.5. Find the angle θ that the plank makes with the vertical in the equilibrium position. (Exclude the case $\theta = 0$)

(JEE ADVANCED)

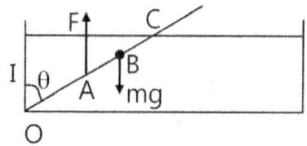

Figure 9.31

Sol: The net torque about the hinge due the weight of the plank and due to the buoyant force acting on the plank should be zero.

The forces acting on the plank are shown in the figure. The height of water level is 0.5m. The length of the plank is $1.0 = 2\ell$. We have OB $= \ell$. The buoyant force F acts through the mid-point of the dipped part OC of the plank.

We have OA $= \dfrac{OC}{2} = \dfrac{\ell}{2\cos\theta}$; Let the mass per unit length of the plank be ρ.

Its weight mg = $2\ell\rho g$; The mass of the part OC of the plank = $\left(\dfrac{\ell}{\cos\theta}\right)\rho$.

The mass of water displaced = $\dfrac{1}{0.5}\dfrac{1}{\cos\theta}\rho = \dfrac{2\ell\rho}{\cos\theta}$; The buoyant force F is, therefore, F = $\dfrac{2\ell\rho g}{\cos\theta}$.

Now, for equilibrium, the torque of mg about O should balance the torque of F about O.

So, mg (OB) sin θ = F(OA) sin θ or $(2\ell\rho)\,\ell = \left(\dfrac{2\ell\rho}{\cos\theta}\right)\left(\dfrac{\ell}{2\cos\theta}\right)$ or $\cos^2\theta = \dfrac{1}{2}$ or $\cos\theta = \dfrac{1}{\sqrt{2}}$, or $\theta = 45°$

6.3 Forces on Fluid Boundaries

Whenever a fluid comes in contact with solid boundaries, it exerts a force on it. Consider a rectangular vessel of base size l × b filled with water to a height H as shown in figure The force acting at the base of the container is given by F_b = p × (area of the base)

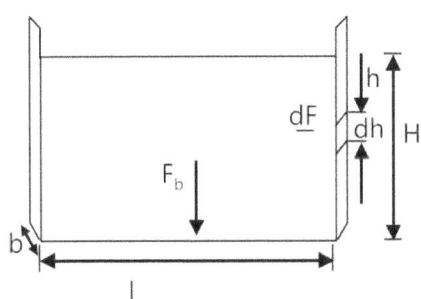

Pressure is same everywhere at the base and is equal to $\rho g H$. Therefore, $F_b = \rho g H(lb) = \rho\, glb\, H$ Since, $lbH = V$ (volume of the liquid) .Thus, $F_b = \rho g V$ = weight of the liquid inside the vessel.

Figure 9.32

A fluid contained in a vessel exerts forces on the boundaries. Unlike the base, the pressure on the vertical wall of the vessel is not uniform but increases linearly with depth from the free surface. Therefore, we have to perform the integration to calculate the total force on the wall. Consider a small rectangular element of width b and thickness dh at depth h from the free surface. The liquid pressure at this position is given by $p = \rho g h$. The force at the element is $dF = p(dbh) = \rho g b h\, dh$;

The total force is $F = \rho g b \displaystyle\int_O^H h\ dh = \dfrac{1}{2}\rho g b H^2$. The total force acting per unit width of the critical walls is $\dfrac{F}{b} = \dfrac{1}{2}\rho g H^2$

The point of application (the centre of force) of the total force from the free surface is given by $h_c = \dfrac{1}{F}\displaystyle\int_0^H h\, dF$

Where $\displaystyle\int_0^H h\, dF$ is the moment of force about the free surface.

Here $\displaystyle\int_0^H h\, dF = \int_0^H h(\rho\, g b h\, dh) = \rho g b\int_0^H h^2 dh = \dfrac{1}{3}\rho g H^3$;

Since F = $\dfrac{1}{2}\rho g b H^2$, therefore, $h_c = \dfrac{2}{3}H$

Illustration 20: Find the force acting per unit width on a plane wall inclined at an angle θ with the horizontal as shown in the figure.

(JEE MAIN)

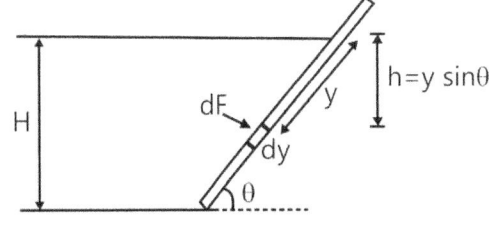

Sol: The pressure at each point on the wall will be different, depending on the height. Find pressure on a small element, and use the method of integration.

Figure 9.33

Consider a small element of thickness dy at a distance y measured along the wall from the free surface. There pressure at the position of the element is $p = \rho g h = \rho g y \sin\theta$. The force given by $dF = p(b\, dy) = \rho g b(y\, dy)\sin\theta$

The total force per unit width b is given by $\dfrac{F}{b} = \rho g \sin\theta . \displaystyle\int_0^{H/\sin\theta} y\, dy = \rho g \sin\theta\left[\dfrac{y^2}{2}\right]_0^{H/\sin\theta}$

Or $\dfrac{F}{b} = \dfrac{1}{2}\rho g \dfrac{H^2}{\sin\theta}$

Note: That the above formula reduces to $\dfrac{1}{2}\rho g H^2$ for a vertical wall ($\theta = 90°$)

6.4 Oscillations of a Fluid Column

The initial level of liquid in both the columns is the same. The area of cross-section of the tube is uniform. If the liquid is depressed by x in one limb, it will rise by x along the length of the tube is the other limb. Here, the restoring force is provided by the hydrostatic pressure difference.

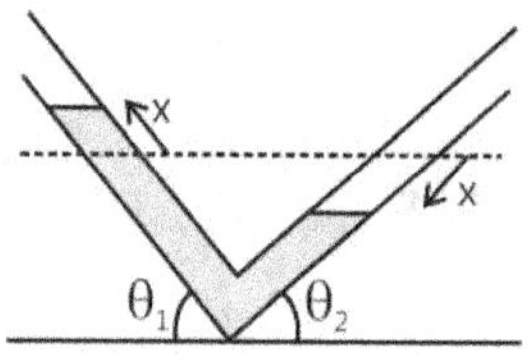

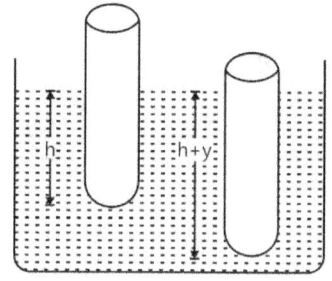

$\therefore \qquad F = -(\Delta P)A = -(h_1 + h_2)\rho g A = -\rho g A(\sin\theta_1 + \sin\theta_2)x$

suppose, m is the mass of the liquid in the tube. Then, $\quad ma = -\rho g A(\sin\theta_1 + \sin\theta_2)x$

Figure 9.34

Since, F or a is proportional to –x, the motion of the liquid column is simple harmonic in nature, time period of which is given by,

$$T = 2\pi\sqrt{\left|\dfrac{x}{a}\right|} \quad \text{or} \quad T = 2\pi\sqrt{\dfrac{m}{\rho g A(\sin\theta_1 + \sin\theta_2)}}$$

6.5 Oscillations of a Floating Cylinder

Consider a wooden cylinder of mass m and cross-sectional area A floating in a liquid of density ρ. At equilibrium, the cylinder is floating with a depth h submerged [See Fig. 8.35]. If the cylinder is pushed downwards by a small distance y and then released, it will move up and down with SHM. It is desired to find the time period and the frequency of oscillations.

According to the principle of flotation, the weight of the liquid displaced by the immersed part of the body is equal to the weight of the body. Therefore, at equilibrium,

Figure 9.35

Weight of cylinder = Weight of liquid displaced by the immersed part of cylinder

or $\quad mg = (\rho Ah)g \quad \therefore$ Mass of cylinder, $m = \rho Ah$

When the cylinder is pushed down to an additional distance y, the restoring force F (upward) equal to the weight of additional liquid displaced acts on the cylinder.

$\therefore$ Restoring force, F= - (weight of additional liquid displaced) or $F = -(\rho Ay)g$

The negative sign indicates that the restoring force acts opposite to the direction of the displacement.

Acceleration a of the cylinder is given by $a = \dfrac{F}{m} = \dfrac{-(\rho Ay)g}{\rho Ah} = -\left(\dfrac{g}{h}\right)y$ ... (i)

Since g/h is constant, $a \alpha - y$ Thus the acceleration a of the body (wooden cylinder) is directly proportional to the displacement y and its direction is opposite to the displacement. Therefore, motion of the cylinder is simple harmonic.

$\therefore$ Time period $T = 2\pi\sqrt{\dfrac{h}{g}}$ ... (ii)

$\therefore$ Frequency $f = \dfrac{1}{T} = \dfrac{1}{2\pi}\sqrt{\dfrac{g}{h}}$ ... (iii)

These very interesting results show that time period and frequency have the same form as that of simple pendulum. The submerged depth at equilibrium takes the place of the length of the pendulum.

7. FLUID DYNAMICS

In the order to describe the motion of a fluid, in principle, one might apply Newton's laws to a particle (a small volume element of fluid) and follow its progress in time. This is a difficult approach. Instead, we consider the properties of the fluid, such as velocity, pressure, at fixed points in space. In order to simplify the discussion we take several assumptions:

(i) The fluid is non viscous   (ii) The flow is steady

(iii) The flow is non rotational   (iv) The fluid is incompressible.

7.1 Equation of Continuity

It states that for streamlined motion of the liquid, the volume of liquid flowing per unit time is constant through different cross-sections of the container of the liquid. Thus, if v_1 and v_2 are velocities of fluid at respective points A and B of areas of cross-sections a_1 and a_2 and r_1 and r_2 be the densities respectively. Then the equation of continuity is given by $r_1 a_1 v_1 = r_2 a_2 v_2$ ... (i)

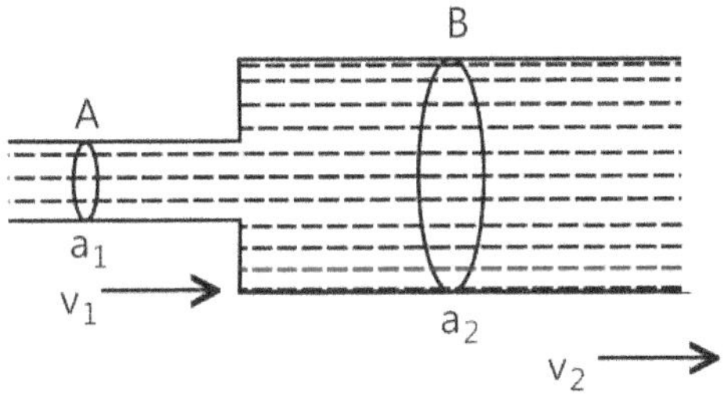

If the same liquid is flowing, then $\rho_1 = \rho_2$; then the equation (i) can be written

As $a_1 v_1 = a_2 v_2$ ...(ii)

$\Rightarrow$ av = constant $\Rightarrow$ $v \propto 1/a$

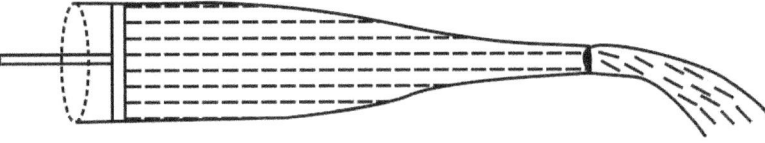

Figure 9.36

NOMORECLASS CONCEPTS

Equation of continunity repersents the law of conservation of mass of moving fluids.

$a_1 v_1 r_1 = a_2 v_2 r_2$ (General equation of continuity)

This equation is applicable to actual liquids or to other fluids which are not incompressible.

Illustration 21: Water is flowing through a horizontal tube of non-uniform cross-section. At a place, the radius of the tube is 1.0 cm and the velocity of water is 2 m/s. What will be the velocity of water, where the radius of the pipe is 2.0 cm? **(JEE MAIN)**

Sol: Apply the equation of continuity. Where area of cross-section is larger, the velocity of water is lesser and vice-versa.

Using equation of continuity, $A_1 v_1 = A_2 v_2$; $v_2 = \left(\dfrac{A_1}{A_2}\right) v_1$ or $v_2 = \left(\dfrac{\pi r_1^2}{\pi r_2^2}\right) v_1 = \left(\dfrac{r_1}{r_2}\right)^2 v_1$

Substituting the value, we get $v_2 = \left(\dfrac{1.0 \times 10^{-2}}{2.0 \times 10^{-2}}\right)$ or $v_2 = 0.5$ m/s

Illustration 22: Figure shows a liquid being pushed out of a tube by pressing a piston. The area of cross-section of the piston is 1.0 cm² and that of the tube at the outlet is 20 mm² If the piston is pushed at a speed of 2 cm·s⁻¹, what is the speed of the outgoing liquid?

Figure 9.37

Sol: Apply the equation of continuity. Where area of cross-section is larger, the velocity of liquid is lesser and vice-versa.

From the equation of continuity $A_1v_1 = A_2v_2$

or $(1.0 \text{ cm}^2)(2 \text{ cm s}^{-1}) = (20 \text{ mm}^2) v_2$

or $v_2 = \dfrac{1.0 \text{ cm}^2}{20 \text{ mm}^2} \times 2 \text{ cm s}^{-1}$

$= \dfrac{100 \text{ mm}^2}{20 \text{ mm}^2} \times 2 \text{ cm s}^{-1} = 10 \text{ cm s}^{-1}$

SHM of fluids in tubes:

Tubes form angles θ_1 and θ_2 with the horizontal.

$$T = 2\pi \sqrt{\dfrac{m}{\rho g A \left(\sin\theta_1 + \sin\theta_2 \right)}}$$

m is total mass of fluid in tubes, A is area of cross – section ρ is density of fluid.

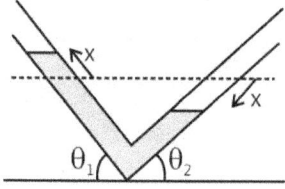

Figure 9.38

8. BERNOULLI'S THEOREM

When a non-viscous and an incompressible fluid flows in a streamlined motion from one place to another in a container, then the total energy of the fluid per unit volume is constant at every point of its path. Total energy = pressure energy + Kinetic energy + Potential energy

$= PV + \dfrac{1}{2}Mv^2 + Mgh$

Where P is pressure, V is volume, M is mass and h is height from a reference level.

$\therefore$ The total energy per unit volume $= P + \dfrac{1}{2}\rho v^2 + \rho gh$

Where ρ is density. Thus if a liquid of density ρ, pressure P_1 at a height h_1 which flows with velocity v_1 to another point in streamline motion where the liquid has pressure P_2, at height h_2 which flows with velocity v_2,

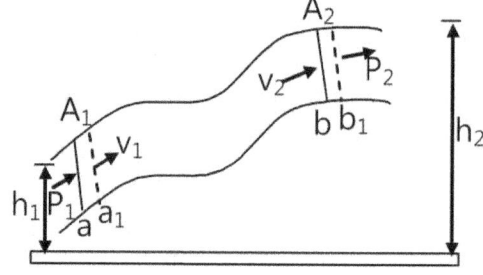

Figure 9.39

then $P_1 + \dfrac{1}{2}\rho v_1^2 + \rho gh_1 = P_2 + \dfrac{1}{2}\rho v_2^2 + \rho gh_2$

8.1 Derivations

8.1.1 Pressure Energy

If P is the pressure on the area A of a fluid, and the liquid moves through a distance due to this pressure, then Pressure energy of liquid = work done = force × displacement = PAl

The volume of the liquid is Al.

$\therefore$ Pressure energy per unit volume of liquid $= \dfrac{PAl}{Al} = P$

8.1.2 Kinetic Energy

If a liquid of mass m and volume V is flowing with velocity v, then the kinetic energy is $\dfrac{1}{2}mv^2$.

$\therefore$ Kinetic energy per unit volume of liquid. $= \frac{1}{2}\left(\frac{m}{V}\right)v^2 = \frac{1}{2}\rho v^2$. Here, ρ is the density of liquid.

8.1.3 Potential energy

If a liquid of mass m is at a height h from the reference line (h = 0), then its potential energy is mgh. $\therefore$ Potential energy per unit volume of the liquid $= \left(\frac{m}{v}\right)gh = \rho gh$

Thus, the Bernoulli's equation $P + \frac{1}{2}\rho v^2 + \rho gh = $ constant

This can also be written as: Sum of total energy per unit volume (pressure + kinetic + potential) is constant for an ideal fluid.

NOMORECLASS CONCEPTS

$\frac{P}{\rho g}$ is called the 'pressure head', $\frac{v^2}{2g}$ the velocity head and h the gravitational head.

Intresting takeaway is the SI unit of each of these is meter (m).

Illustration 23: Calculate the rate of flow of glycerin of density 1.25×10^3 kg/m^3 through the conical section of a pipe, if the radii of its ends are 0.1 m and 0.04 m and the pressure drop across its length is 10 N/m^2. **(JEE MAIN)**

Sol: Apply the equation of continuity. Where area of cross-section is larger, the velocity of fluid is lesser and vice-versa.

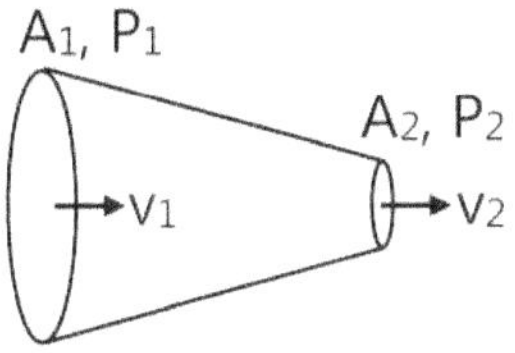

From continuity equation, $A_1 v_1 = A_2 v_2$

or $\dfrac{v_1}{v_2} = \dfrac{A_2}{A_1} = \dfrac{\pi r_2^2}{\pi r_1^2} = \left(\dfrac{r_2}{r_1}\right)^2 = \left(\dfrac{0.04}{0.1}\right)^2 = \dfrac{4}{25}$ ... (i)

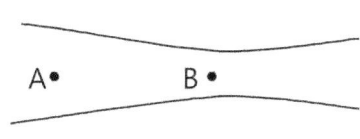

Figure 9.40

From Bernoulli's equation , $P_1 + \frac{1}{2}\rho v_1^2 = P_2 + \frac{1}{2}\rho v_2^2$

or $v_2^2 - v_1^2 = \dfrac{2\times 10}{1.25\times 10^3} = 1.6\times 10^{-2}$ m^2/s^2 ... (ii)

Solving equations (i) and (ii), we get $v_2 = 0.128$ m/s

$\therefore$ Rate of volume flow through the tube

$Q = A_2 v_2 = (\pi r_2^2)\, v_2 = \pi\,(0.04)^2 (0.128) = 6.43 \times 10^{-4}$ m^3/s

Illustration 24: Figure shows a liquid of density 1200 kg m^{-3} flowing steadily in a tube of varying cross section. The cross section at a point A is 1.0 cm^2 and that at B is 20 mm^2, the points A and B are in the same horizontal plane. The speed of the liquid at A is 10 cm s^{-1}. Calculate the difference in pressure at A and B.

(JEE ADVANCED)

Figure 9.41

Sol: Apply the equation of continuity. Where area of cross-section is larger, the velocity of fluid is lesser and vice-versa.

From equation of continuity. The speed v_2 at B is given by, $A_1v_1 = A_2v_2$

or $(1.0 \text{ cm}^2)(10 \text{ cm s}^{-1}) = (20 \text{ mm}^2)v_2$ or $v_2 = \dfrac{1.0\text{cm}^2}{20\text{mm}^2} \times 10\text{cm s}^{-1} = 50\text{cm s}^{-1}$

By Bernoulli equation, $P_1 + \rho g h_1 + \dfrac{1}{2}\rho v_1^2 = P_2 + \rho g h_2 + \dfrac{1}{2}\rho v_2^2$

Here $h_1 = h_2$.   Thus $P_1 - P_2 = \dfrac{1}{2}\rho v_2^2 - \dfrac{1}{2}\rho v_1^2 = \dfrac{1}{2} \times (1200 \text{ kg m}^{-2})(2500 \text{ cm}^2 \text{ s}^{-2} - 100 \text{ cm}^2 \text{ s}^{-2})$

$= 600 \text{ kg m}^{-3} \times 2400 \text{ cm}^2 \text{ s}^{-2} = 144 \text{ Pa}$

8.2 Application Based on Bernoulli's Equation

8.2.1 Venturimeter

Figure shows a venturimeter used to measure flow speed in a pipe of non-uniform cross-section. We apply Bernoulli's equation to the wide (point 1) and narrow (point 2) parts of the pipe, with $h_1 = h_2$

$$P_1 + \dfrac{1}{2}\rho v_1^2 = P_2 + \dfrac{1}{2}\rho v_2^2$$

From the continuity equation $v_2 = \dfrac{A_1 v_1}{A_2}$

Substituting and rearranging,

we get $P_1 - P_2 = \dfrac{1}{2}\rho v_1^2 \left(\dfrac{A_1^2}{A_2^2} - 1 \right)$ ...(i)

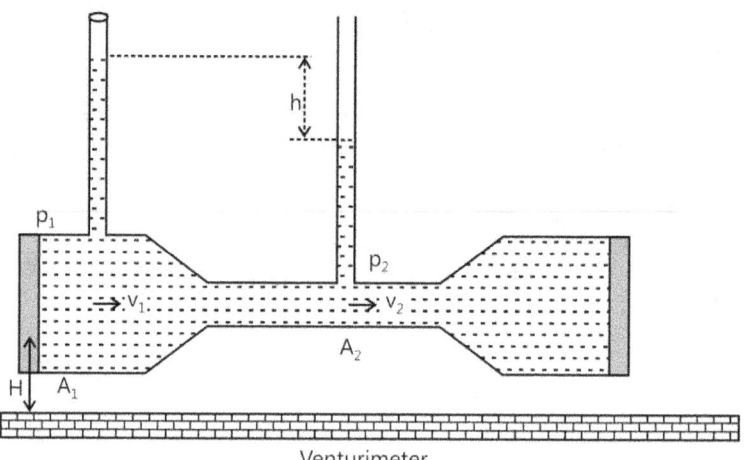

Venturimeter

Figure 9.42

The pressure difference is also equal to ρgh, where h is the difference in liquid level in the two tubes.

Substituting in equation (i), we get $v_1 = \sqrt{\dfrac{2gh}{\left(\dfrac{A_1}{A_2} \right)^2 - 1}}$

NOMORECLASS CONCEPTS

Because A_1 is greater than A_2, v_2 is greater than v_1 and hence the pressure P_2 is less than P_1.

The discharge or volume flow rate can be obtained as, $\dfrac{dV}{dt} = A_1 v_1 = A_1 \sqrt{\dfrac{2gh}{\left(\dfrac{A_1}{A_2} \right)^2 - 1}}$

9. TORRICELLI'S THEOREM

It states that the velocity of efflux of a liquid through an orifice is equal to that velocity which a body would attain in falling from a height from the free surface of a liquid to the orifice. If h is the height of the orifice below the free surface of a liquid and g is acceleration due to gravity, the velocity of efflux of liquid = $v = \sqrt{2gh}$. Total energy per unit volume of the liquid at the surface = KE + PE + Pressure energy = $0 + \rho gh + P_0$ ...(i)

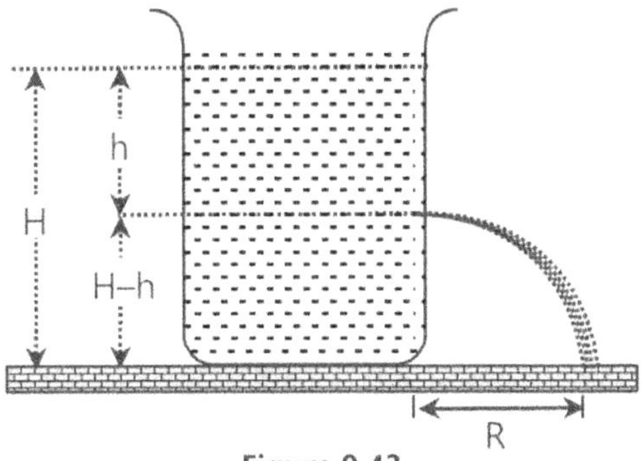

and total energy per unit volume at the orifice = KE + PE + Pressure energy = $\frac{1}{2}\rho v^2 + 0 + P_0$

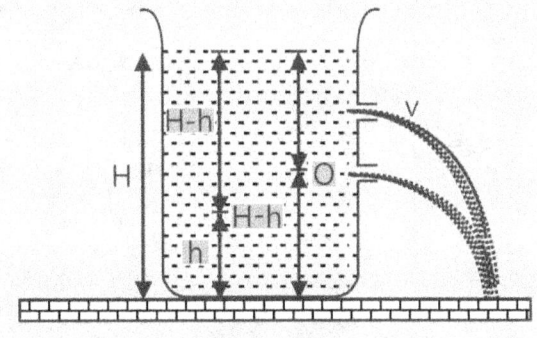

Figure 9.43

Since total energy of the liquid must remain constant in steady flow, in accordance with Bernoulli's equation,

we have $\rho gh + P_0 = \frac{1}{2}\rho v^2 + P_0$ or $v = \sqrt{2gh}$

Range = velocity × time ; $R = V_x \times$ time $= \sqrt{2gh} \times 1$

Now, $H - h = \frac{1}{2}gt^2 \Rightarrow t = \frac{\sqrt{2(H-h)}}{g}$. From equation (i),

$R = \sqrt{2gh} \times \frac{\sqrt{2(H-h)}}{g} = \sqrt{2h \times 2(H-h)} \times \sqrt{h(H-h)2}$

$\therefore \quad \boxed{R = 2\sqrt{h(H-h)}}$

Range is max. if $\frac{dR}{dh} = 0 \Rightarrow 2 \times \frac{H-2h}{2\sqrt{h(H-h)}} = 0 \Rightarrow H - 2h = 0 \Rightarrow \boxed{h = \frac{H}{2}}$

NOMORECLASS CONCEPTS

$R_h = R_{H-h}$

$R_h = 2\sqrt{h(H-h)}$

$R_{H-h} = 2\sqrt{h(H-h)}$

i.e. Range would be the same when the hole is at a height h

or at a height H – h from the ground or from the top of the beaker.

R is maximum at $h = \frac{H}{2}$ and $R_{max} = H$.

Figure 9.44

4.21

9.1 An Expression for the Force Experienced by the Vessel

The force experienced by the vessel from which liquid is coming out.

$F = \dfrac{dp}{dt}$ (Rate of change of momentum) $= \dfrac{d}{dt}(mv) = \dfrac{d}{dt}(\rho Avtv)$

$\boxed{F = \rho Av^2}$ Where ρ = It is the density of the liquid.

A = It is the area of hole through which liquid is coming out.

9.2 Time taken to Empty a Tank

Consider a tank filled with a liquid of density ρ up to a height H. A small hole of area of cross section a is made at the bottom of the tank. The area of cross-section of the tank is A.

Let at some instant of time the level of liquid in the tank be y. Velocity of efflux at this instant of time would be,
$v = \sqrt{2gy}$.

At this instant volume of liquid coming out of the hole per second is $\left(\dfrac{dV_1}{dt}\right)$.

Volume of liquid coming down in the tank per second is $\left(\dfrac{dV_2}{dt}\right)$.

$\dfrac{dV_1}{dt} = \dfrac{dV_2}{dt}$; $\therefore$ $av = A\left(-\dfrac{dy}{dt}\right)$ $\therefore a\sqrt{2gy} = A\left(-\dfrac{dy}{dt}\right)$ Or $\displaystyle\int_0^t dt = -\dfrac{A}{a\sqrt{2g}}\int_H^0 y^{-1/2}dy$

$\therefore t = \dfrac{2A}{a\sqrt{2g}}[\sqrt{y}]_0^H = \dfrac{A}{a}\sqrt{\dfrac{2H}{g}}$

Illustration 25: A tank is filled with a liquid up to a height H. A small hole is made at the bottom of this tank. Let t_1 be the time taken to empty first half of the tank and t_2 the time taken to empty rest half of the tank.

Then find $\dfrac{t_1}{t_2}$. **(JEE MAIN)**

Sol: This problem needs to be solved by method of integration.

Substituting the proper limit in equation (i), derived in the theory, we have

$\displaystyle\int_0^{t_1} dt = -\dfrac{A}{a\sqrt{2g}}\int_H^{H/2} y^{-1/2}dy$ Or $t_1 = \dfrac{2A}{a\sqrt{2g}}[\sqrt{y}]_{H/2}^H$ Or $= \dfrac{2A}{a\sqrt{2g}}\left[\sqrt{H} - \sqrt{\dfrac{H}{2}}\right]$

Or $t_1 = \dfrac{A}{a}\sqrt{\dfrac{H}{g}}(\sqrt{2} - 1)$ ...(ii)

Similarly $\displaystyle\int_0^{t_2} dt = -\dfrac{A}{a\sqrt{2g}}\int_{H/2}^0 y^{-1/2}dy$ Or $t_2 = \dfrac{A}{a}\sqrt{\dfrac{H}{g}}$ ... (iii)

From equations (ii) and (iii), we get $\dfrac{t_1}{t_2} = \sqrt{2} - 1$ Or $\dfrac{t_1}{t_2} = 0.414$

From here we see that $t_1 < t_2$. This is because inititally the pressure is high and the liquid comes out with greater speed.

10. VISCOSITY

When a liquid moves slowly and steadily on a horizontal surface, its layer in contact with the fixed surface is stationary and the velocity of the layers increase with the distance from the fixed surface.

Consider two layers CD and MN of a liquid at distances x and x + dx from the fixed surface AB having velocities v and v + dv respectively as shown in the figure. Here $\left(\dfrac{dv}{dx}\right)$ denotes the rate of change of velocity with distance and is known as velocity gradient. The tendency of the upper layer is to accelerate the motion and the lower layer tries to retard the motion of upper layer. The two layers together tend to destroy their relative motion as if there is some backward dragging force acting tangentially on the layers. To maintain the motion, an external force is applied to overcome this backward drag.

Hence the property of a liquid virtue of which it opposes the relative motion between its different layers is known as viscosity.

The viscous force is given by $F = -\eta A \dfrac{dv}{dx}$

Where η is a constant, called the coefficient of viscosity.

The SI unit of η is N-s/m². It is also called decapoise or Pascal second. Thus,

1 decapoise = N-s/m² = 1 Pa-s = 10 poise.

Dimensions of h are [ML⁻¹T⁻¹]

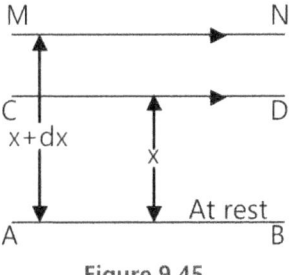

Figure 9.45

The negative sign in the above equation shows that the direction of viscous force F is opposite to the direction of relative velocity of the layer.

Viscous force depends upon the velocity gradient whereas the mechanical frictional force is independent of the velocity gradient.

10.1 Effect of Temperature

In case of liquids, coefficient of viscosity decreases with increase of temperature as the cohesive forces decrease with increase of temperature.

Illustration 26: A plate of area 2 m² is made to move horizontally with a speed of 2 m/s by applying a horizontal tangential force over the free surface of a liquid. The depth of the liquid is 1 m and the liquid in contact with the bed is stationary. Coefficient of viscosity of liquid is 0.01 poise. Find the tangential force needed to move the plate.

(JEE MAIN)

Sol: Apply the Newton's formula for the frictional force between two layers of a liquid.

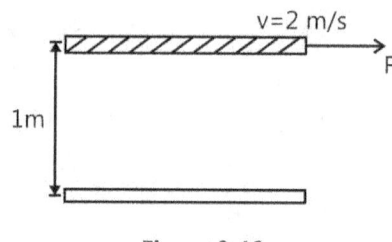

Velocity gradient $= \dfrac{\Delta v}{\Delta y} = \dfrac{2-0}{1-0} = 2\dfrac{m/s}{m}$

From Newton's law of viscous force,

$|F| = \eta A \dfrac{\Delta v}{\Delta y} = (0.01 \times 10^{-1})(2)(2) = 4 \times 10^{-3}$ N.

So, to keep the plate moving, a force of 4×10^{-3} N must be applied.

Figure 9.46

10.2 Stokes' Law and Terminal Velocity

Stokes established that the resistive force or F, due to the viscous drag, for a spherical body of radius r, moving with velocity V, in a medium of coefficient of viscosity η is given by

F = 6 pη rV

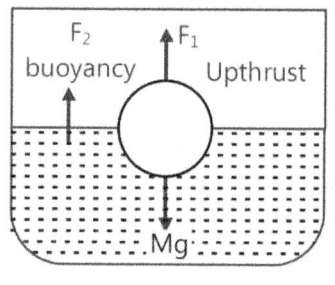

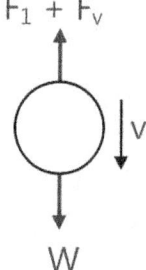

Figure 9.47

10.3.1 An Experiment for Terminal Velocity

Consider an established spherical body of radius r and density ρ falling freely from rest under gravity through a fluid of density σ and coefficient of viscosity η. When the body acquires the terminal velocity V

$W = F_t + 6\pi\eta rV$;

$$6\pi\eta rV = \frac{4}{3}\pi pr^3(\rho - \sigma)g \Rightarrow \boxed{V = \frac{2}{9}\frac{r^2(\rho - \sigma)g}{\eta}}$$

Note: From the above expression we can see that terminal velocity of a spherical body is directly proportional to the densities of the body and the fluid $(\rho - \sigma)$. If the density of the fluid is greater than that of the body (.i.e. $\sigma > \rho$), the terminal velocity is negative. This means that the body instead of falling, moves upward. This is why air bubbles rise up in water.

Illustration 28: Two spherical raindrops of equal size are falling vertically through air with a terminal velocity of 1 m/s. What would be the terminal speed if these two drops were to coalesce to form a large spherical drop?

(JEE MAIN)

Sol: Use the formula for terminal velocity for spherical body.

$v_T \propto r^2$. Let r be the radius of small rain drops and R the radius of large drop.

Equating the volume, we have $\dfrac{4}{3}\pi R^2 = 2\left(\dfrac{4}{3}\pi r^3\right)$

$\therefore \quad R = (2)^{1/3}.\, r \quad$ or $\quad \dfrac{R}{r} = (2)^{1/3} \setminus \quad \dfrac{v_T{}'}{v_T} = \left(\dfrac{R}{r}\right)^2 = (2)^{2/3}$

$\therefore \quad v_T{}' = (2)^{2/3}\, v_T = (2)^{2/3}\,(1.0)$ m/s = 1.587 m/s.

Illustration 29: An air bubble of diameter 2 mm rises steadily through a solution of density 1750 kg m⁻³ at the rate of 0.35 cm s⁻¹. Calculate the coefficient of viscosity of the solution. The density of air is negligible. **(JEE MAIN)**

Sol: As the air bubble rises with constant velocity, the net force on it is zero.

The force of buoyancy B is equal to the weight of the displaced liquid. Thus $B = \frac{4}{3} pr^3 \sigma g$.

This force is upward. The viscous force acting downward is $F = 6 \pi \eta r v$.

The weight of the air bubble may be neglected as the density of air is small. For uniform velocity

$$F = B \text{ or, } 6 \pi \eta r v = \frac{4}{3} \pi r^3 \sigma g \text{ or, } \eta = \frac{2r^2 \sigma g}{9v} = \frac{2 \times (1 \times 10^{-3}\,m)^2 \times (1750\,kg\,m^{-3})(9.8\,ms^{-2})}{9 \times (0.35 \times 10^{-2}\,ms^{-1})} \approx 11 \text{ poise}.$$

This appears to be a highly viscous liquid.

10.3 Stream Line Flow

When liquid flows in such a way that the velocity at a particular point is the same in magnitude as well as in direction. As shown in figure every molecule should have the same velocity at A, if it crossed from that point. Notice that the velocity at the point B will be different from that of A. But every molecule which reaches at the point B, gets the velocity of the point B.

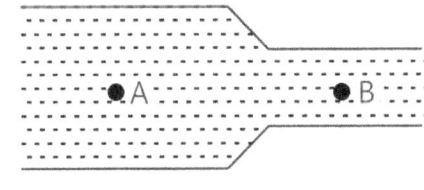

Figure 9.48

10.4 Turbulent Flow

When the motion of a particle at any point varies rapidly in magnitude and direction, the flow is said to be turbulent or beyond critical velocity. If the paths and velocities of particles change continuously and haphazardly, then the flow is called turbulent flow.

10.5 Critical Velocity and Reynolds Number

When a fluid flows in a tube with small velocity, the flow is steady. As the velocity is gradually increased, at one stage the flow becomes turbulent. The largest velocity which allows a steady flow is called the critical velocity.

Whether the flow will be steady or turbulent mainly depends on the density, velocity and the coefficient of viscosity of the fluid as well as the diameter of the tube through which the fluid is flowing. The quantity $N = \frac{\rho v D}{\eta}$ is called the Reynolds number and plays a key role in determining the nature of flow. It is found that if the Reynolds number is less than 2000, the flow is steady. If it is greater than 3000, the flow is turbulent. If it is between 2000 and 3000, the flow is unstable.

11. SURFACE TENSION

The properties of a surface are quite often marked different from the properties of the bulk material. A molecule well inside a body is surrounded by similar particles from all sides. But a molecule on the surface has particles of one type on one side and of a different type on the other side. Figure shows an example: A molecule of water well inside the bulk experiences force from water molecules from all sides, but a molecule at the surface interacts with air molecules from above and water molecules from below. This asymmetric force distribution is responsible for surface tension.

A surface layer is approximately 10-15 molecular diameters. The force between two molecules decreases as the separation between them increases. The force becomes

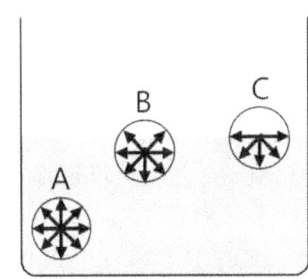

Figure 9.49

negligible if the separation exceeds 10-15 molecular diameters. Thus, if we go 10-15 molecular diameters deep, a molecule finds equal forces from all directions.

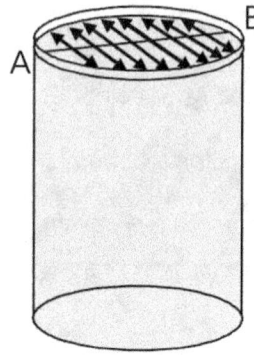

Imagine a line AB drawn on the surface of a liquid (figure). The line divides the surface in two parts, surface on one side and the surface on the other side of the line. Let us call them surface to the left of the line and surface to the right of the line. It is found that the two parts of the surface pull each other with a force proportional to the length of the line AB. These forces of pull are perpendicular to the line separating the two parts and are tangential to the surface. In this respect the surface of the liquid behave like a stretched rubber sheet. The rubber sheet which is stretched from all sides is in the state of tension. Any part of the sheet pulls the adjacent part towards itself.

Figure 9.50

Let F be the common magnitude of the forces exerted on each other by the two parts of the surface across a line of length ℓ. We define the surface tension T of the liquid as T = F/ℓ

The SI unit of surface tension is N/m.

Note: The surface tension of a particular liquid usually decreases as temperature increases. To wash clothing thoroughly, water must be forced through the tiny spaces between the fibers. This requires increasing the surface area of the water, which is difficult to do because of surface tension. Hence, hot water and soapy water is better for washing.

NOMORECLASS CONCEPTS

Surface tension acts over the free surface of a liquid only and not within the interior of the liquid.

Due to surface tension the insects can walk on liquid surface.

Illustration 30: Calculate the force required to take away a flat circular plate of radius 4 cm from the surface of water, surface tension of water being 75 dyne cm⁻¹. **(JEE MAIN)**

Sol: Force = Surface tension×length of the surface

Length of the surface = circumference of the circular plate = $2\pi r$ = (8π) cm

Required force = T × L = 72 × 8π = 1810 dyne.

12. SURFACE ENERGY

When the surface area of a liquid is increased, the molecules from the interior rise to the surface. This requires work against force of attraction of the molecules just below the surface. This work is stored in the form of potential energy. Thus, the molecules in the surface have some additional energy due to their position. This additional energy per unit area of the surface is called 'surface energy'. The surface energy is related to the surface tension as discussed below:

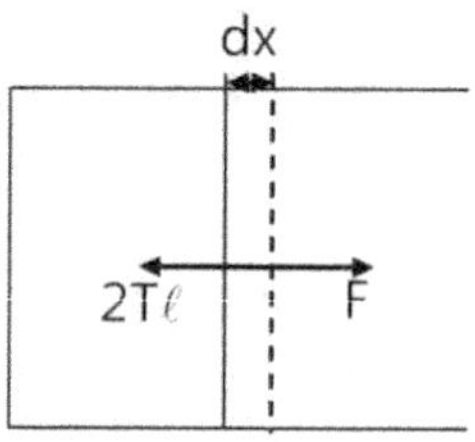

Figure 9.51

Let a liquid film be formed on a wire frame and a straight wire of length ℓ can slide on this wire frame as shown in figure. The film has two surfaces and both the surfaces are in contact with the sliding wire and hence, exert forces of surface tension on it. If T be the surface tension of the solution, each surface will pull the wire parallel to itself with a force Tℓ. Thus, net force on the wire due to both the surfaces is 2Tℓ. One has to apply an external force F equal and opposite to it to keep the wire in equilibrium. Thus, F = 2Tℓ

Now, suppose the wire is moved through a small distance dx, the work done by the force is,

$dW = F\,dx = (2T\ell)dx$

But $(2\ell)(dx)$ is the total increase in the area of both the surfaces of the film. Let it be dA. Then,

$dW = T\,da$ or $T = \dfrac{dW}{dA}$

Thus, the surface tension T can also be defined as the work done in increasing the surface area by unity. Further, since there is no change in kinetic energy, the work done by the external force is stored as the potential energy of the new surface.

$\therefore\ T = \dfrac{dU}{dA}$ (as $dW = dU$)

Thus, the surface tension of a liquid is equal to the surface energy per unit surface area.

Illustration 31: How much work will be done in increasing the diameter of a soap bubble from 2 cm to 5 cm? Surface tension of soap solution is 3.0×10^{-2} N/m.   **(JEE MAIN)**

Sol: Work done will be equal to the increase in the surface porential energy, which is surface tension multiplied by increase in area of surface of liquid.

Soap bubble has two surfaces. Hence, $W = T\,\Delta A$

Here, $\Delta A = 2[4\pi\{(2.5 \times 10^{-2})^2 - (1.0 \times 10^{-2})^2\}] = 1.32 \times 10^{-2}$ m²

$W = (3.0 \times 10^{-2})(1.32 \times 10^{-2})J = 3.96 \times 10^{-4}J$

Illustration 32: Calculate the energy released when 1000 small water drops each of same radius 10^{-7}m coalesce to form one large drop. The surface tension of water is 7.0×10^{-2} N/m.   **(JEE MAIN)**

Sol: Energy released will be equal to the loss in surface potential energy.

Let r be the radius of smaller drops and R of bigger one.

Equating the initial and final volumes, we have $\dfrac{4}{3}\pi R^3 = (1000)\left(\dfrac{4}{3}\pi r^3\right)$

R = 10r = (10)(10^{-7}) m = 10^{-6} m. Further, the water drops have only one free surface. Therefore,

$\Delta A = 4\pi R^2 - (1000)(4\pi r^2) = 4\pi[(10^{-6})^2 - (10^3)(10^{-7})^2] = -36\pi(10^{-12})$m²

Here, negative sign implies that surface area is decreasing. Hence, energy is released in the process.

$U = T[\Delta A] = (7 \times 10^{-2})(36\pi \times 10^{-12})J = 7.9 \times 10^{-12}J$

13. EXCESS PRESSURE

The pressure inside a liquid drop or a soap bubble must be in excess of the pressure outside the bubble drop because without such pressure difference, a drop or a bubble cannot be in stable equilibrium. Due to the surface tension, the drop or bubble has got the tendency to contract and disappear altogether. To balance this, there must be excess of pressure inside the bubble.

13.1 Excess Pressure Inside a Drop

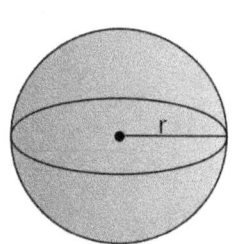
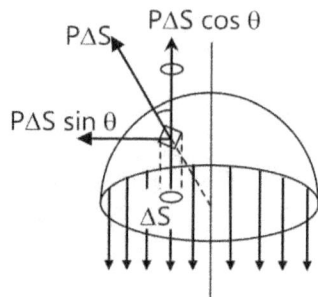

Figure 9.52

To obtain a relation between the excess of pressure and the surface tension, consider a water drop of radius r and surface tension T. Divide the drop into two halves by a horizontal passing through its centre as shown in figure and consider the equilibrium of one-half, say, the upper half. The force acting on it are:

(a) Force due to surface tension distributed along the circumference of the section.

(b) Outward thrust on elementary areas of it due to excess pressure.

Obviously, both the types of forces are distributed. The first type of distributed forces combine into a force of magnitude $2\pi r \times T$. To find the resultant of the other type of distributed forces, consider an elementary area ΔS of the surface. The outward thrust on $\Delta S = p\Delta S$ where p is the excess of the pressure inside the bubble. If this thrust makes an angle θ with the vertical, then it is equivalent to $\Delta Sp \cos \theta$ along the vertical and $\Delta Sp \sin \theta$ along the horizontal. The resolved component $\Delta Sp \sin \theta$ is infective as it is perpendicular to the resultant force due to surface tension. The resolved component $\Delta Sp \cos \theta$ is equal to balancing the force due to surface tension

The resultant outward thrust $= \Sigma \Delta Sp \cos \theta = p\Sigma \Delta S \cos \theta = p\Sigma \Delta S \cos \theta = p\Sigma \Delta S'$

where $\Delta S' = \Delta S \cos \theta$ = area of the projection of ΔS on the horizontal dividing plane

$= p \times \pi r^2 \ (\ell \ \Delta S' = \pi r^2)$

For equilibrium of the bubble we have $\pi r^2 \ p = 2\pi r \ T$ or $p = \dfrac{2T}{r}$

NOMORECLASS CONCEPTS

If we have an air bubble inside a liquid, a single surface is formed.

There is air on the concave side and liquid on the convex side.

The pressure in the concave side (that is in the air) is greater than

the pressure in the convex side (that is in the liquid) by an amount $\dfrac{2T}{R}$.

$\therefore \boxed{P_2 - P_1 = \dfrac{2T}{R}}$

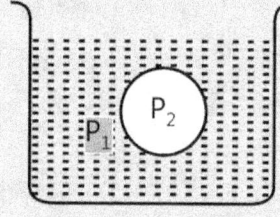

Figure 9.53

13.2 Excess Pressure Inside Soap Bubble

A soap bubble consists of two spherical surface films with a thin layer of liquid between them. $P' - P_1 = 2S/R$ where R is the radius of the bubble.

As the thickness of the bubble is small on a macroscopic scale, the difference in the radii of the two surfaces will be negligible.

Similarly, looking at the inner surface, the air is on the concave side of the surface, hence $P_2 - P' = 2S/R$. Adding the two equations, $P_2 - P_1 = 4S/R$

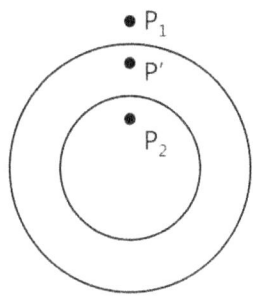

Illustration 33: What should be the pressure inside a small air bubble of 0.1 mm radius situated just below the water surface? Surface tension of water = 7.2×10^{-2} N/m and atmospheric pressure = 1.013×10^5 N/m².

(JEE MAIN)

Figure 9.54

Sol: Pressure inside the air bubble is larger than that outside it by amount 2T/R, where T is surface tension and R is its radius.

Surface tension of water T = 7.2×10^{-2} N/m; Radius of air bubble R = 0.1 mm = 10^{-4} m

The excess pressure inside the air bubble is given by, $\qquad P_2 - P_1 = \dfrac{2T}{R}$

∴ Pressure inside the air bubble, $P_2 = P_1 + \dfrac{2T}{R}$; Substituting the values, we have,

$$P_2 = (1.013 \times 10^5) + \dfrac{(2 \times 7.2 \times 10^{-2})}{10^{-4}} = 1.027 \times 10^5 \text{ N/m}^2$$

Illustration 34: A 0.02 cm liquid column balances the excess pressure inside a soap bubble of radius 7.5 mm. Determine the density of the liquid. Surface tension of soap solution = 0.03 Nm⁻¹. **(JEE MAIN)**

Sol: Pressure inside the soap bubble is larger than that outside it by amount 4T/R, where T is surface tension and R is its radius. Gauge pressure of liquid column is ρgh where symbols have the usual meaning.

The excess pressure inside a soap bubble is DP = 4S/R = $\dfrac{4 \times 0.03 \text{Nm}^{-1}}{7.5 \times 10^{-3} \text{m}}$ = 16 Nm⁻²

The pressure due to 0.02 cm of the liquid column is P = hρg = (0.02 × 10⁻² m) ρ (9.8 ms⁻²)

Thus, 16 N m⁻² = (0.02 × 10⁻² m) ρ (9.8 ms⁻²); ρ = 9.2 × 10³ kg m⁻³.

14. CAPILLARY ACTION

When a glass tube of very fine bore called a capillary tube is dipped in a liquid (like water), the liquid immediately rises into it due to the surface tension. The phenomenon of rise of a liquid in a narrow tube is known as capillarity.

Suppose that a capillary tube of radius r is dipped vertically in a liquid. The liquid surface meets the wall of the tube at some inclination θ called the angle of contact. Due to surface tension, a force, $\Delta \ell$ T acts on an element $\Delta \ell$ of the circle of contact along which the liquid surface meets the solid surface and it is tangential to the liquid surface at inclination θ to the wall of the tube. (The liquid on the wall of the tube exerts this force. The tube also exerts the same force on the liquid in the opposite direction.) Resolving this latter force along and perpendicular to the wall of the tube, we have $\Delta \ell$ Tcosθ along the tube vertically upwards and $\Delta \ell$ Tsinθ perpendicular to the wall. The latter component is ineffective. It simply comes the liquid against the wall of the tube. The vertical component $\Delta \ell$ Tcos θ pulls the liquid up the tube.

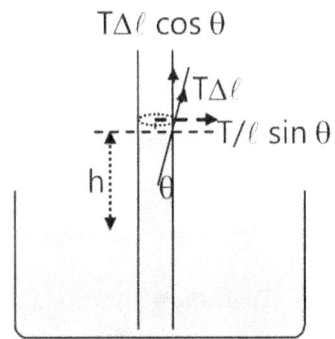

Figure 9.55

The total vertical upward force = $\Sigma \Delta \ell$ T cos θ = T cos θ$\Sigma \Delta \ell$ = T cosθ.2πr ($\ell \Sigma \Delta \ell$ = 2pr). Because of this upward pull liquid rises up in the capillary tube till it is balanced by the downward gravitational pull. If h is the height of the liquid column in the tube up to the bottom, the gravitational pull, i.e. weight of the liquid inside the tube is (πr²h + V)ρg, where V is the volume of the liquid in meniscus. For equilibrium of the liquid column in the tube 2πrT cos θ = (πr²h + V)ρg

If value of the liquid in meniscus is negligible then, $2\pi rT \cos\theta = (\pi r^2 h)\rho g$; $h = \dfrac{2T\cos\theta}{r\rho g}$

The small volume of the liquid above the horizontal plane through the lowest point of the meniscus can be calculated if θ is given or known. For pure water and glass $\theta = 0°$ and hence the meniscus is hemispherical.

∴   V = volume of the cylinder of height r – volume of hemisphere.

$$= \pi r^3 - \frac{1}{2}\frac{4\pi}{3}r^3 = \pi r^3 - \frac{2}{3}\pi r^3 = \frac{1}{3}\pi r^3$$

∴   For water and glass $2\pi rT = \left(\pi r^2 h + \dfrac{\pi r}{3}\right)^3 \rho g$

$$2T = r\left(h + \frac{r}{3}\right)\rho g \quad \Rightarrow \quad h = \frac{2T}{r\rho g} - \frac{r}{3}$$

For a given liquid and solid at a given place as ρ, T, θ and g are constant,   ∴   hr = constant

i.e. lesser t the radius of capillary greater will be the rise and vice-versa.

Illustration 36: A capillary tube of radius 0.20 mm is dipped vertically in water. Find the height of the water column raised in the tube. Surface tension of water = 0.075 N m^{-1} and density of water = 1000 kg m^{-3}. Take g = 10 m s^{-2}.   **(JEE MAIN)**

Sol: Use the formula for height of the liquid in the capillary.

We have, $h = \dfrac{2S\cos\theta}{r\rho g} = \dfrac{2\times 0.075\,\text{N m}^{-1}\times 1}{(0.20\times 10^{-3}\,\text{m})\times(1000\,\text{kg m}^{-3})(10\,\text{m s}^{-2})} = 0.075$ m = 7.5 cm.

PROBLEM SOLVING TACTICS

(a) Suppose two liquids of densities r_1 and r_2 having masses m_1 and m_2 are mixed together.

Then the density of the mixture will be $= \dfrac{(m_1 + m_2)}{\left(\dfrac{m_1}{\rho_1} + \dfrac{m_2}{\rho_2}\right)}$

If two liquids of densities r_1 and r_2 having volume V_1 and V_2 are mixed, then the density of the mixture will be

$\dfrac{\rho_1 V_1 + \rho_2 V_2}{V_1 + V_2}$.

(b) When solving questions on Bernoulli's always assume a reference level and calculate the heights from the reference level.

FORMULAE SHEET

Fluid Statics:

1. Density $= \dfrac{\text{mass}}{\text{volume}}$, S.I. units: kg/m³

2. Specific gravity / Relative density / Specific density $= \dfrac{\text{Ratio of its density}}{\text{Ratio of density of water at } 4°C}$,

   S.I. units: No units

3. If two liquids of volume V_1 and V_2 and densities d_1 and d_2 respectively are mixed then the density d of the

   mixture is $d = \dfrac{V_1 d_1 + V_2 d_2}{V_1 + V_2}$; If $V_1 = V_2$ then $d = \dfrac{d_1 + d_2}{2}$

4. If two liquids of densities d_1 and d_2 and masses m_1 and m_2 respectively are mixed together,

   then the density d of the mixture is $d = \dfrac{m_1 + m_2}{\dfrac{m_1}{d_1} + \dfrac{m_2}{d_2}}$; if $m_1 = m_2$ then $d = \dfrac{2 d_1 d_2}{d_1 + d_2}$

5. Pressure $= \dfrac{\text{Normal component of force}}{\text{Area on which force acts}} = \dfrac{f}{A}$, S.I. units: N/m², Pa

6. Pressure P acting at the bottom of an open fluid column of height h and density d is

   $= 1.013 \times 10^5$ Pa $= 1.013 \times 10^5$ Pa $= 1.013 \times 10^6$ dynes/cm² $= 76$ cm of Hg $= 760$ torr $= 1.013$ bars.

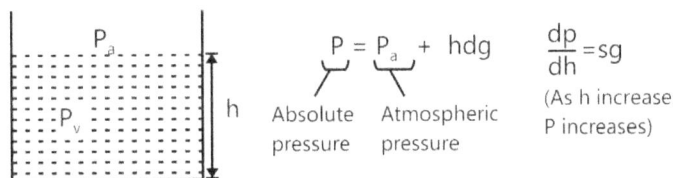

Figure 9.56

$P - P_a = hdg$

gauge pressure = absolute – atmospheric pressure.

7.

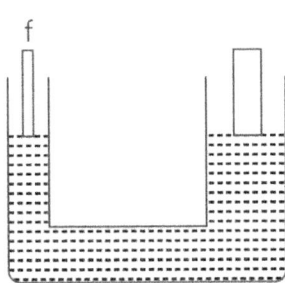

Figure 9.57

Area of smaller piston, a; area of larger piston, A, f is applied on the smaller piston

Force F developed on the larger piston $\dfrac{F}{A} = \dfrac{f}{a}$

$$\therefore \quad F = \frac{fA}{a}$$

8. Beaker is accelerated in horizontal direction

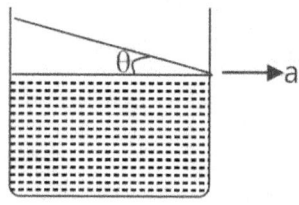

$$\tan\theta = \frac{a}{g}$$

a is the acceleration of the beaker in horizontal direction.

Figure 9.58

9. Beaker is accelerated and it has components of acceleration a_x, and a_y in x and y directions respectively.

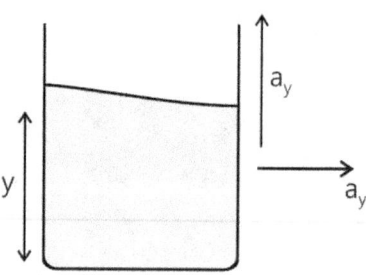

Figure 9.59

P increases with depth $\rightarrow \dfrac{dP}{dy} = p(g + ay)$

P is the density of the fluid.

ρ is the density of the fluid. $\dfrac{dP}{dx} = -pax$

10. Buoyant force $F = V_1 \rho_1 (\vec{g} - \vec{a})$

V_1 = immersed volume of liquid

ρ_1 = density of liquid

g = acceleration due to gravity

a = acceleration of body dipped inside liquid.

11. Body floats when Buoyant force balances the weight of the body.

$$\underset{\left(\substack{\text{Buoyant} \\ \text{force}}\right)}{V_i \rho_2 g} = \underset{\left(\substack{\text{Weight} \\ \text{of body}}\right)}{V_b \rho_b g}$$

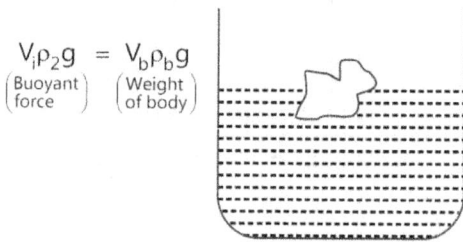

Figure 9.60

V_b, ρ_b = volume and density of body.

V_i = Volume of the immersed part of body.

ρ_2 = density of liquid.

Fraction of volume immersed $\dfrac{V_i}{V_b} = \dfrac{\rho_b}{\rho_2}$

% of volume immersed $\dfrac{V_i}{V_b} \times 100 = \dfrac{\rho_b}{\rho_2} \times 100$.

12. Apparent weight of a body inside a fluid is $W_{app} = W_{act}$ – Upthrust

$$W_{app} = V_b g\,(\rho_b - \rho_2)$$

V_b, δ_b = volume and density of body.

V_i = Volume of the immersed part of body.

ρ_2 = density of liquid.

13. General equation of continuity

$\rho_1 A_1 V_1 = \rho_2 A_2 V_2$ Generally $\rho_1 = \rho_2$ i.e., density is uniform.

A_1 & A_2 are area of cross-section at point P and Q.

V_1 & V_2 are velocities of the fluid at point P and Q.

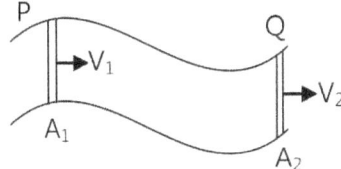

Figure 9.61

14. **Bernoulli's Equation**

$$P_1 + \rho g h_1 + \frac{1}{2}\rho V_1^2 = P_2 + \rho g h_2 + \frac{1}{2}\rho V_2^2$$

i.e.,

$$\underbrace{P}_{} + \underbrace{\rho g h}_{} + \underbrace{\frac{1}{2}\rho v^2}_{} = constant$$

Pressure   Height   Velocity at the point

at that point from the reference level

$$\underbrace{\frac{P}{\rho g}}_{} + \underbrace{\frac{V^2}{2g}}_{} + \underbrace{h}_{} = constant$$

Pressure   Velocity   gravitational head
head   head

15. Volumetric flow $Q = Av = \dfrac{dV}{dt}$ A – Area of cross section; v – Velocity; V– Volume

S.I. unit = $\dfrac{m^3}{S}$

16. **Torricelli Theorem:**

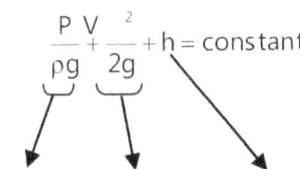

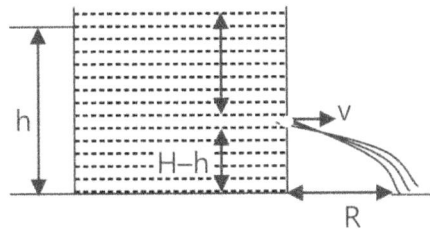

Figure 9.62

Range $R = 2\sqrt{h(H-h)}$

Range is maximum at $h = \dfrac{H}{2}$ and $R_{max} = H$

A_b – Area of orifice

A – Area of cross-section of the container.

Time taken to fall from H_1 to H_2 = $t \times \dfrac{A}{A_0}\sqrt{\dfrac{2}{g}}$

17. **Viscous Force** $F = \eta A \dfrac{dv}{dy}$

$\downarrow$

coefficient of viscosity

L – Length of pipe

P_1 and P_2 are pressure at two ends of pipe.

R – Radius of pipe.

When liquid is flowing through a tube, velocity of flow of a liquid at distance from the axis.

$V = \dfrac{P}{4\eta L}\left(r^2 - x^2\right)$. Velocity distribution curve is a parabola.

18. **Stoke's Law:** Formula for the viscous force on a sphere

$F = 6\pi\eta rv$ \qquad (η – coefficient of viscosity)

\qquad\qquad (r – radius of sphere)

\qquad\qquad (v – velocity of sphere)

$V_T = \dfrac{2}{9}r^2\dfrac{(\rho - \sigma)g}{\eta}$ \qquad (ρ – density of sphere)

\qquad\qquad (∞ – density of fluid)

19. **Surface Tension**

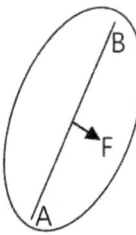

$T = \dfrac{F}{L}$

F is the total force acting on either side of AB.
L is length of AB.

Figure 9.63

20. **Surface Energy:** $dW = TdA$

Surface Tension $T = \dfrac{dV}{dA} = \dfrac{\text{Surface energy}}{\text{Area}}$

21. Pressure inside the soap bubble is P, then

$P - P_0 = \dfrac{4T}{R}$

22. Air Bubble Inside a Liquid

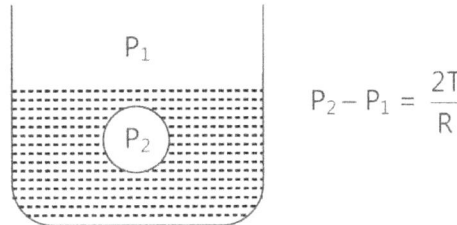

$$P_2 - P_1 = \frac{2T}{R}$$

Figure 9.64

R – radius of bubble

T – surface tension force

23. Capillary Rise

$$h = \frac{2T\cos\theta}{r\rho g}$$ r = is the radius of capillary tube

θ = angle of contact

Solved Examples

JEE Main/Boards

Example 1: For the arrangement shown in the figure. What is the density of oil?

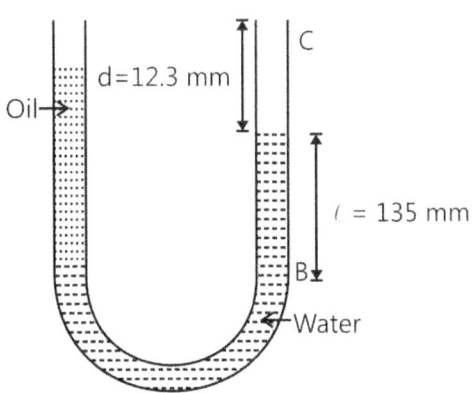

Sol: Pressure will be same at all points at the same height in the same liquid.

$$P_0 + \rho_w\, g l = P_0 + \rho_{oil}\, (l + d)g$$

$$\Rightarrow \rho_{oil} = \frac{\rho_w l}{l + d} = \frac{1000.(135)}{(135 + 12.3)} = 916\ kg / m^3$$

Example 2: A solid floats in a liquid of different material. Carry out an analysis to see whether the level of liquid in the container will rise or fall when the solid melts.

Sol: Level of liquid will rise or fall depending on the density of the solid.

Let M = Mass of the floating solid.

ρ_1 = density of liquid formed by the melting of the solid.

ρ_2 = density of the liquid in which the solid is floating. The mass of liquid displaced by the solid is M. Hence, the volume of liquid displaced is $\dfrac{M}{\rho_2}$. When the solid melts, the volume occupied by it is $\dfrac{M}{\rho_1}$. Hence, the level of liquid in container will rise or fall according as

$\dfrac{M}{\rho_2} - \dfrac{M}{\rho_1}$ is less than or greater than zero.

$\Rightarrow$ rises for $\rho_1 < \rho_2$

$\Rightarrow$ falls for $\rho_1 > \rho_2$

There will be no change in the level if the level if $\rho_1 = \rho_2$. In case of ice floating in water $\rho_1 = \rho_2$ and hence, the level of water remains unchanged when ice melts.

Example 3: An iron casting containing a number of cavities weighs 6000 N in air and 4000 N in water. What is the volume of the cavities in the casting? Density of iron is 7.87 g/cm³.

Take g = 9.8 m/s² and density of water = 10³ kg/m³.

Sol: Apply Archemides principal. The volume of iron without the cavity is easily found. The total volume is found from the upthrust. The difference in volumes is the volume of cavity.

Let v be the volume of cavities and V the volume of solid iron. Then,

$$V = \frac{mass}{density} = \left(\frac{6000/9.8}{7.87 \times 10^3}\right) = 0.078 m^3$$

Further, decrease in weight = upthrust

∴ $(6000 - 4000) = (V + v)\rho_w g$

or $2000 = (0.078 + v) \times 10^3 \times 9.8$

or $0.078 + v \approx 0.2$

∴ $v = 0.12\ m^3$

Example 4: A boat floating in a water tank is carrying a number of stones. If the stones were unloaded into water, what will happen to the water level?

Sol: When the stones are in boat they will displace more water as compared to the case when they are out of the boat and inside water.

Let weight of boat = W and weight of stone = w.

Assuming density of water = 1 g/cc

Volume of water displaced initially = $(w + W)/\rho_w$

Later, Volume displaced = $\left(\frac{W}{\rho_w} + \frac{w}{\rho}\right)$

(ρ = density of stones)

⇒ Water level comes down.

Example 5: A conical glass capillary tube A of length 0.1 m has diameters 10⁻³m and 5×10⁻⁴m at the ends. When it is just immersed in a liquid at 0°C with larger radius in constant contact with it, the liquid rises to 8×10⁻²m in the tube. In another cylindrical glass capillary tube B, when immersed in the same liquid at 0°C, the liquid rises to 6×10⁻²m height. The rise of liquid in tube B is only 5.5×10⁻²m when the liquid is at 50°C. Find the rate at which the surface tension changes with temperature considering the change to be linear. The density of liquid is (1/4) × 10⁴ kg/m³ and the angle of contact is zero. Effect of temp on the density of liquid and glass is negligible.

Sol: Use the formula for height of the liquid in the capillary.

Let r_1 and r_2 be radii of upper and lower ends of the conical capillary tube. The radius r at the meniscus is given by

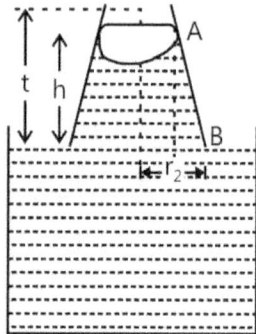

$$r = r_1 + (r_2 - r_1)\left(\frac{\ell - h}{\ell}\right)$$

$$= (2.5 \times 10^{-4}) + (2.5 \times 10^{-4})\left(\frac{0.1 - 0.08}{0.1}\right)$$

$$= 3.0 \times 10^{-4}\ m$$

The surface tension at 0°C is given by

$$T_0 = \frac{rh\rho g}{2}$$

$$= \frac{(3.0 \times 10^{-4})(8 \times 10^{-2})(1/4 \times 10^4) \times 9.8}{2} = 0.084$$

For tube B, N/m $\dfrac{T_0}{T_{50}} = \dfrac{h_0}{h_{50}} = \dfrac{6 \times 10^{-2}}{5.5 \times 10^{-2}} = \dfrac{12}{11}$

$$\Rightarrow T_0 = \frac{11}{12} \times T_0 = \frac{11}{12} \times 0.084 = 0.077\ N/m$$

Considering the change in the surface tension as linear, the change in surface tension with temp is given by

$$\alpha = \frac{T_{50} - T_0}{T_0 - T_{50}} = \frac{0.077 - 0.084}{0.084 \times 0.077} = -\frac{1}{60}k\ .$$

Negative sign shows that with rise in temp surface tension decreases.

Example 6: A piece of copper having an internal cavity weighs 264 gm in air and 221 gm when it is completely immersed in water. Find the volume of the cavity. The density of copper is 9.8 gm/cc.

Sol: Apply Archemides principal. The volume of copper without the cavity is easily found. The total volume is found from the upthrust. The difference in volumes is the volume of cavity.

Mass of copper in air = 264 gm

Mass of copper in water = 221 gm

Apparent loss of mass in water

= 264 – 221 = 43 gm

∴ Mass of water displaced by copper piece when completely immersed in water is equal to 43 gm.

$$\text{Volume of water displaced} = \frac{\text{mass of displaced}}{\text{density of water}}$$

$$= \frac{43}{1.0} = 43.0 \text{ cc}$$

∴ Volume of copper piece including volume of cavity = 43.0 cc. Volume of copper block only

$$= \frac{\text{mass}}{\text{density}} = \frac{264}{8.8} = 30.0 \text{cc}$$

Volume of cavity = 43.0 – 30.0 = 13.0 cc

Example 7: A cubical block of each side equal to 10 cm is made of steel of density 7.8 gm/cm³. It floats on mercury surface in a vessel with its sides vertical. The density of mercury is 13.6 gm/cm³.

(a) Find the length of the block above mercury surface.

(b) If water is poured on the surface of mercury, find the height of the water column when water just covers the top of the steel block.

Sol: Apply Archemides principal. The weight of the block will be equal to the weight of the liquid displaced.

(a) Volume of steel block

= (10)³ = 1000 cm³

Mass of steel block = 1000 × 7.8 = 7800 gm

Let ℓ_1 be the height of steel block above the surface of mercury. Height of block under mercury = 10 – ℓ_1. Weight of mercury displaced by block

= (10 – ℓ_1) × 100 × 13.6 × g gm

Archimedes' principle shows that upward thrust is equal to the weight of mercury displaced by block is equal to the weight of the block.

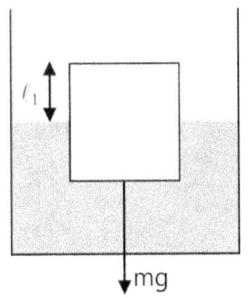

∴ (10– ℓ_1) × 100 × 13.6 × g = 7800 g

$$10 - \ell_1 = \frac{7800}{100 \times 13.6} = 5.74$$

∴ length of block above mercury surface

= 10 – 5.74 = 4.26 cm

(b) Let ℓ_2 be the height of water column above mercury surface so that water just covers the top of the steel block. The upward thrust due to mercury and water displaced is equal to the weight of the body

∴ weight of block = wt. of water displaced + wt. of mercury displaced

∴ 7800 g = ℓ_2 × 1000 × 1 × g

+ (10 – ℓ_2) × 100 × 13.6 × g

7800 = 100 ℓ_2 + 13600 – 1360 ℓ_2

1260 ℓ_2 = 13600 – 7800 = 5800

∴ Height of water column above mercury=

$$\ell_2 = \frac{5800}{1260} = 4.6 \text{ cm}$$

Example 8: A cubical block of wood of each side 10 cm long floats at the interface between oil and water with its lower surface 2 cm below the interface. The height of oil and water column is 10 cm each. The density of oil is 0.8 g cm⁻³.

(a) What is the mass of the block?

(b) What is the pressure at the lower side of surface of block?

Sol: Apply Archemides principle. The weight of the block will be equal to the weight of the liquid displaced.

(a) Buoyant force = (mass of liquid displaced) × g

= [10 × 10 × 8 × 0.8 + 10 × 10 × 1]g= 840 g

If m is mass of block

mg = 840 g   or   m = 840 gm

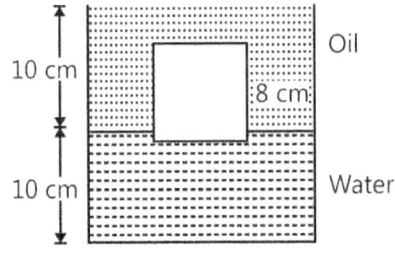

(b) Pressure at the lower surface of block

= pressure at any point on the same level.

$10 \times 0.8 \times g + 2 \times 1 \times g$

$= 10g = 10 \times 9.81 = 99.1$ Newton/meter^2

Example 9: A massless smooth piston forces water with a velocity of 8 m/s out of a tube shaped container with radii 4.0 cm and 1.0 cm respectively as shown in the figure. Assume that the water leaving the container enters air at 1 atmospheric pressure. Find

(a) The velocity of the piston

(b) Force F applied to the piston.

Sol: Apply Bernoulli's Theorem at two points, one near the piston and the other at the end of the tube.

(a) Let F be the force applied horizontally such that v_1 is the velocity of water in tube A of radius 4.0 cm and v_2 equal to 8 m/s is the velocity of water out of tube B of radius 1.0 cm.

Let p_0 be atmospheric pressure.

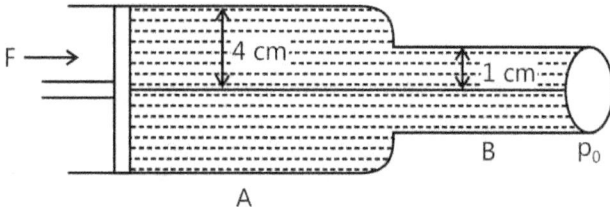

A

$\therefore$ At A, $v_1 = ?$, $r_1 = 4.0$ cm, $p_1 = p_0 + \dfrac{F}{a}$ Where a is area of cross-section of piston or tube A. At B, $v_2 = 8$ m/s, $r_2 = 1.0$ cm, $p_2 = p_0$

Bernoulli's theorem sat A and B gives,

$p_1 + \dfrac{1}{2}\rho v_1 + h\rho g = p_2 + \dfrac{1}{2}\rho v_2^2 + h\rho ga$

where ρ is density of water and h is height of axis of both tubes from ground level

$\therefore p_1 + \dfrac{1}{2}\rho v_1^2 = p_2 + \dfrac{1}{2}\rho v_2^2$

$\dfrac{F}{a} + p_0 = p_0 + \dfrac{1}{2}\rho(v_2^2 - v_1^2)$

$\dfrac{F}{a} = \dfrac{\rho}{2}(v_2^2 - v_1^2)$ ...(i)

Equation of continuity at A and B gives

$v_1 a_1 = v_2 a_2$

or $v_1 = v_2 \times \dfrac{\pi r_2^2}{\pi r_1^2} = 8 \times \left(\dfrac{1}{4}\right)^2 = 0.5 \, m/s$

(b) Equation (i) gives $F = \dfrac{a\rho}{2}(v_2^2 - v_1^2) =$

$\dfrac{\pi \times (4)^2 \times 1000}{2} \times \left[64 - \dfrac{1}{4}\right]$

$= \dfrac{1}{2} \times \dfrac{22}{7} \times 16 \times 1000 \times \dfrac{255}{4} = 160.3$N

Example 10: A horizontal tube has different cross-sections at two points A and B. The diameter at A is 4.0 cm and that at B is 2 cm/ The two manometer arms are fixed at A and B. When a liquid of density 800 kg/m^3 flows through the tube, the difference of pressure between the arms of two manometers is 8 cm. Calculate the rate of flow of tube liquid.

Sol: Apply Bernoulli's Theorem and equation of continuity.

From Bernoulli's principle:

$p_1 + \dfrac{1}{2}\rho v_1^2 = p_2 + \dfrac{1}{2}\rho v_2^2$

From the equation of continuity: $A_1 v = A_2 v_2$

pressure difference: $p_1 - p_2 = h\rho g$

These equations give $v_1 = A_2 \sqrt{\dfrac{2gh}{(A_1^2 - A_2^2)}}$

Rate of flow of volume

$V = A_1 v_1 = A_1 A_2 \sqrt{\dfrac{2gh}{(A_1^2 - A_2^2)}}$

$= \pi^2 (4 \times 10^{-4})(1 \times 10^{-2}) \sqrt{\dfrac{2 \times 9.8 \times 8 \times 10^{-2}}{(4\pi \times 10^{-4}) - (\pi \times 10^4)^2}}$

$= 4.06 \times 10^{-4}$ m^3/s

JEE Advanced/Boards

Example 1: Under isothermal condition two soap bubbles of radii a and b coalesce to form a single bubble of radius c. If the external pressure is p_0

show that surface tension $T = \dfrac{p_0(c^3 - a^3 - b^3)}{4(a^2 + b^2 + c^2)}$

Sol: Pressure inside the soap bubble is larger than that outside it by amount 4T/R, where T is surface tension and R is its radius.

As we know that for a soap bubble, the excess pressure

is $= \dfrac{4T}{r}$. External pressure is p_0

$\therefore p_a = p_0 + \dfrac{4T}{a}$ $\therefore p_b = p_0 + \dfrac{4T}{b}$ and

$p_c = p_0 + \dfrac{4T}{c}$

and $v_a = \dfrac{4}{3}\pi a^3$, $v_b = \dfrac{4}{3}\pi b^3$ & $v_c = \dfrac{4}{3}\pi c^3$ ....(i)

Applying conservation of mass

$$n_a + n_b = n_c$$

$\Rightarrow \dfrac{p_a v_a}{RT_a} + \dfrac{p_b v_b}{RT_b} = \dfrac{p_c v_c}{RT_c}$ $[\because pv = nRT \Rightarrow n = \dfrac{Pv}{RT}]$

Since, temp is constant

i.e. $T_a = T_b = T_c$, so the expression reduces to

$p_a v_a + p_b v_b = p_c v_c$

with the help of equation (i), we have

$$\left(p_0 + \dfrac{4T}{a}\right)\left(\dfrac{4}{3}\pi a^3\right) + \left(p_0 + \dfrac{4T}{b}\right)\left(\dfrac{4}{3}\pi b^3\right)$$

$$= \left(p_0 + \dfrac{4T}{c}\right)\left(\dfrac{4}{3}\pi c^3\right)$$

$$\Rightarrow 4T(a^2 + b^2 - c^2) = p_0(c^3 - a^3 - b^3)$$

$$\Rightarrow T = \dfrac{p_0(c^3 - a^3 - b^3)}{4(a^2 + b^2 - c^2)}$$

Example 2: Two identical cylindrical vessels with their bases at the same level contain a liquid of density ρ. The height of liquid in one vessel is h_1 and that in the other vessel is h_2. The areas of either base is A. What is the work done by gravity in equalizing the levels when the two vessels are connected.

Sol: Work done by gravity is equal to the loss in the gravitational potential energy.

The center of gravity of liquid column would be at height h_1 and h_2 respectively. A is area of cross-section.

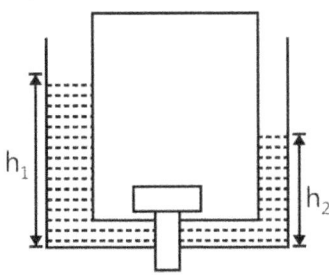

Total P.E. when they are not connected

$$Ah_1\rho g\left(\dfrac{h_1}{2}\right) + Ah_2\rho g\left(\dfrac{h_2}{2}\right) = A\rho g\left[\dfrac{h_1^2}{2} + \dfrac{h_2^2}{2}\right]$$

When the levels are equal, the potential energy is given as

$$= A\left(\dfrac{h_1 + h_2}{2}\right)\rho g\left(\dfrac{h_1 + h_2}{4}\right) + A\left(\dfrac{h_1 + h_2}{2}\right)\rho g\left(\dfrac{h_1 + h_2}{4}\right)$$

$$= 2A\rho g\dfrac{(h_1 + h_2)^2}{2 \times 4} = A\rho g\dfrac{(h_1 + h_2)^2}{4}$$

The change in potential energy

$$= \dfrac{A\rho g}{2}\left[\dfrac{(h_1 + h_2)^2}{2} - (h_1^2 - h_2^2)\right]$$

$$= \dfrac{A\rho g}{2}\left[\dfrac{h_1^2 + h_2^2 - 2h_1^2 - 2h_2^2 + 2h_1 h_2}{2}\right]$$

$$= \dfrac{A\rho g}{2}\left[\dfrac{-(h_1^2 + h_2^2 - 2h_1 h_2)}{2}\right]$$

Work done due to gravity $= -A\rho g\left[\dfrac{h_1 - h_2}{2}\right]^2$

The negative sign shows that the work is done by the gravitational field on the liquid.

Example 3: A container of large uniform cross-section area A resting on a horizontal surface holds two immiscible, non-viscous and incompressible liquids of densities d and 2d, each of height H/2 as shown in figure. The lower density liquid is open to the atmosphere having pressure P_0.

(a) A homogeneous solid cylinder of length L (L < H/2) and cross-section area A/5 is immersed such that, it floats with its axis vertical at the liquid-liquid interface with length L/4 in the dense liquid.

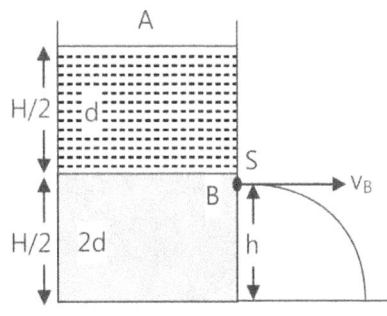

Determine:

(i) The density D of the solid.

(ii) The total pressure at the bottom of the container.

(b) The cylinder is removed and the original arrangement is restored. A tiny hole of area S(S<<A) is punched on the vertical side of the container at a height h(h<H/2).

Determine:

(i) The initial speed of efflux of liquid at the hole.

(ii) The horizontal distance x travelled by the liquid initially.

(iii) The height h_m at which the hole should be punched so that the liquid travels the maximum distance x_m initially. Also calculate x_m: (Neglect the air-resistance in these calculations)

Sol: Apply the principles of hydrostatic pressure, Archemedes and Bernoulli's Theorem.

(A) (i) As per Archimedes' principle, the buoyant force on a body is equal to the weight of the fluid displaced by the body.

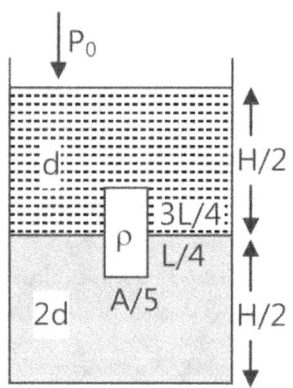

Weight of solid cylinder $= L \times \dfrac{A}{5} \times D \times g = F \downarrow$

$F \uparrow =$ Buoyant force = weight of liquid displaced

$= \dfrac{L}{4} \times \dfrac{A}{5} \times 2dg + \dfrac{3L}{4} \times \dfrac{A}{5} \times d \times g$

Equating: $L \times \dfrac{A}{5} \times D \times g$

$= \dfrac{L}{4} \times \dfrac{A}{5} \times 2dg + \dfrac{3L}{4} \times \dfrac{A}{5} \times d \times g$

$D = \dfrac{d}{2} + \dfrac{3d}{4} = \dfrac{2d+3d}{4} = \dfrac{5d}{4}$

(ii) Pressure at the bottom of the cylinder

$= P_{atmosphere} + P_{dense\ liquid} + P_{light\ liquid}$

Pressure due to liquid $= \dfrac{Force}{Area}$

$= \dfrac{1}{A}\left[Adg\left(\dfrac{H}{2}\right) + A(2d)g\left(\dfrac{H}{2}\right) \right] = dg\left(\dfrac{3H}{2}\right)$

Pressure due to buoyancy reaction

$= \dfrac{Buoyancy\ reaction\ force}{area} \qquad = \left(\dfrac{A}{5}\right)\dfrac{LDg}{A}$

$= \dfrac{A}{5} \times L \times \dfrac{5d}{4} \times \dfrac{1}{A} \times g = \dfrac{Lgd}{4}$

$\therefore$ Total pressure $= P_0 + dg\left(\dfrac{3H}{2}\right) + \dfrac{Ldg}{4}$

$= P_0 + dg\left[\dfrac{3H}{2} + \dfrac{L}{4}\right]$

(b) (i) Let v_A and v_B be velocity of fluids at points A and B.

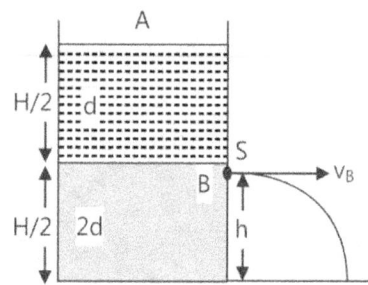

$Av_A = Sv_B$

$\because \quad v_A = \left(\dfrac{S}{A}\right)v_B \simeq 0 \qquad (\ell\ A>>H/2)$

Bernoulli's Equation: $p + \dfrac{1}{2}\rho v_2 + \rho g h = constant$

At A, $P_0 + \dfrac{1}{2}dv_A^2 + dg\dfrac{H}{2} + 2d(g)\left(\dfrac{H}{2}\right) = constant$ or

$P_0 + \dfrac{3}{2}dgH = constant \ (\because\ V_A = 0)$

At point B,
$P_0 + \dfrac{1}{2}dv_A^2 + \dfrac{1}{2}(2d)v_B^2 + 2dgh = cosntant$

or $\quad P_0 + dv_B^2 + 2dgh = constant$

Equating: $P_0 + dv_B^2 + 2dgh = P_0 + \dfrac{3}{2}dgH$

$dv_B^2 + 2dgh = \dfrac{3}{2}dgH$

$v_B^2 = g\left[\dfrac{3}{2}H - 2h\right]; \ v_B = \sqrt{g\left(\dfrac{3}{2}H - 2h\right)}$

(ii) Time t taken by liquid to fall through height h under g with zero initial velocity. $t = \sqrt{\dfrac{2h}{g}}$

Horizontal distance

$$x = v_B t = \sqrt{\dfrac{2h}{g}} \times \sqrt{g\left(\dfrac{3}{2}H - 2h\right)}$$

$$\sqrt{h(3H - 4h)} = 2 \times \sqrt{h} \times \sqrt{\dfrac{3H}{4} - h}$$

(iii) To find height h at which x is max, $\dfrac{dx}{dh} = 0$.

$$\dfrac{d}{dh}\left[3Hh - 4h^2\right]^{1/2} = 0 \ ; \ \dfrac{d}{dh}\left[h(3H - 4h)\right]^{1/2} = 0$$

$$\dfrac{d}{dh}\left[2 \times h\sqrt{\dfrac{3H}{4} - h}\right] = 0.$$

$$2 \times \dfrac{1}{2h}\left(\dfrac{3H}{4} - h\right)^{1/2} + 2\sqrt{h} \times \dfrac{1}{2}\left(\dfrac{3H}{4} - h\right)^{-1/2}(-1) = 0$$

$$\dfrac{1}{h}\left(\dfrac{3H}{4} - h\right)^{1/2} = \dfrac{\sqrt{h}}{\left[\dfrac{3H}{4} - h\right]^{1/2}}$$

or $\dfrac{3H}{4} - h = h$ or $h = \dfrac{3}{8}H$

$$\therefore \quad x_m = 2 \times \sqrt{\dfrac{3H}{8}\left(\dfrac{3H}{4} - \dfrac{3H}{8}\right)^{1/2}}$$

$$= 2 \times \sqrt{\dfrac{3H}{8}} \times \sqrt{\dfrac{3H}{8}} = \dfrac{3H}{4}$$

Example 4: A tube of length ℓ and radius R carries a steady flow of liquid whose density is ρ and viscosity η. The velocity v of flow is given by $V = V_0\left(1 - \dfrac{r^2}{R^2}\right)$, where r is the distance of flowing fluid from the axis. Find

(a) Volume of fluid, flowing across the section of the tube, in unit time.

(b) Kinetic energy of the fluid within the volume of the tube.

(c) The frictional force exerted on the tube by the fluid, and

(d) The difference of pressures at the ends of the tube.

Sol: The cross section of tube can be thought of made-up of elementary rings of infinitesimal thickness. Find the volume flow rate and kinetic energy of one ring. Use the method of integration to find the flow rate and energy for the tube.

(a) Let us consider a cylindrical section at a distance of r and having thickness dr. The volume of fluid flowing through this section per second. $dv = (2\pi r dr)v_0\left(1 - \dfrac{r^2}{R^2}\right)$

So, the volume of fluid flowing across the section of the tube in unit time.

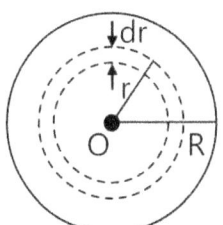

$$v = \int_0^R (2\pi r dr)v_0\left(1 - \dfrac{r^2}{R^2}\right) = 2\pi v_0 \int_0^R r\left(1 - \dfrac{r^2}{R^2}\right) dr$$

$$= 2\pi v_0\left[\dfrac{r^2}{2} - \dfrac{r^4}{4R^2}\right]_0^R = 2\pi v_0\left(\dfrac{R^2}{4}\right)$$

(b) The kinetic energy of the fluid within the volume element of thickness dr

$$\dfrac{1}{2}(dm)v^2 = \dfrac{1}{2}(2\pi r dr \ell)\rho v_0^2\left(1 - \dfrac{r^2}{R^2}\right)^2$$

So, the K.E. of fluid within the tube

$$= \dfrac{1}{2}(2\pi \ell)\rho v_0^2\int_0^R\left(1 - \dfrac{r^2}{R^2}\right)^2 r dr$$

Integrating, we get

$$\text{K.E.} = \pi r \ell \ell v_0^2\left(\dfrac{R^2}{6}\right)\ell$$

(c) The viscous drag exerts a force on the tube

$$F = \eta A\left(\dfrac{dv}{dx}\right)_{r=R}$$

Hence $\left(\dfrac{dv}{dr}\right)_{r=R} = v_0\left(-\dfrac{2r}{R^2}\right)_{r=R} = \dfrac{-2v_0}{R}$

$\therefore F = -\eta(2\pi R \ell)(-2v_0 \ell R) = 4\pi\eta h \ell v_0$

(d) The pressure difference ΔP is given by

$$\Delta P = P_2 - P_1 = P$$

Where $P_1 = O$ and $P_2 = P$

As we know that $P = \dfrac{\text{Force}(F)}{\text{area of section of tube}}$

$$P = \dfrac{F}{\pi R^2} = \dfrac{4\pi \eta \ell v_0}{\pi R^2}$$

$$P = \dfrac{4\eta \ell v_0}{R^2}$$

Example 5: A fresh water reservoir is 10 m deep. A horizontal pipe 4.0 cm in diameter passes through the reservoir 6.0 m below the water surface as shown in figure. A plug secures the pipe opening.

(a) Find the friction force between the plug and pipe wall.

(b) The plug is removed. What volume of water flows out of the pipe in 1 h? Assume area of reservoir to be too large.

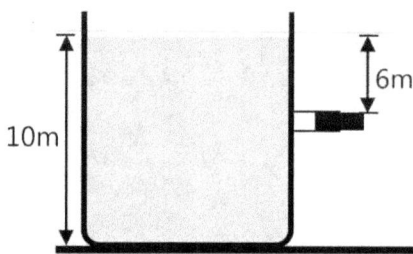

Sol: Force of friction will balance the force due to pressure difference on the plug. Use the formula for velocity of efflux for part (b)

(a) Force of friction

= pressure difference on the sides of the plug × area of cross section of the plug

= $(\rho g h)$ A = $(10)^3 (9.8)(6.0)(\pi)(2 \times 10^{-2})^2$

= 73.9 N

(b) Assuming the area of the reservoir to be too large.

Velocity of efflux $v = \sqrt{2gh} = \text{constant}$

$\therefore \quad v = \sqrt{2 \times 9.8 \times 6} = 10.84\,\text{m/s}$

Volume of water coming out per sec,

$$\dfrac{dV}{dt} = Av = \pi(2 \times 10^{-2})^2 (10.84) = 1.36 \times 10^{-2}\,\text{m}^3/\text{s}$$

$\therefore$ The volume of water flowing through the pipe in 1 h.

$$V = \left(\dfrac{dV}{dt}\right)t = (1.36 \times 10^{-2})(3600) = 49.96\,\text{m}^3$$

Example 6: The U-tube acts as a water siphon. The bend in the tube is 1m above the water surface. The tube outlet is 7 m below the water surface. The water issues from the bottom of the siphon as a free jet at atmospheric pressure. Determine the speed of the free jet and the minimum absolute pressure of the water in the bend. Given atmospheric pressure = 1.01×10^5 N/m².

g = 9.8 m/s² and density of water = 10^3 kg/m³.

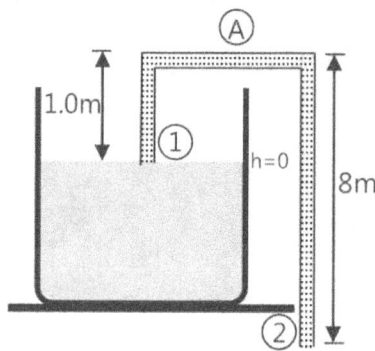

Sol: Apply Bernoulli's Theorem at points 1, A and 2.

(a) Applying Bernoulli's equation between point (1) and (2)

$$P_1 + \dfrac{1}{2}\rho v_1^2 + \rho h_1 + P_2 + \dfrac{1}{2}\rho v_2^2 + \rho g h_2$$

Since, area of reservoir >> area of pipe

$v_1 \approx 0$, also $P_1 = P_2$ = atmospheric pressure

So, $v_1 = \sqrt{2g(h_1 - h_2)} = \sqrt{2 \times 9.8 \times 7} = 11.7$ m/s

(b) The minimum pressure in the bend will be at A. Therefore, applying Bernoulli's equation between (1) and (A)

$$P_1 + \dfrac{1}{2}\rho v_1^2 + \rho g h_1 = P_A + \dfrac{1}{2}\rho v_A^2 + \rho g h_A$$

Again, $v_1 \approx 0$ and from conservation of mass $v_A = v_2$;

$$P_A = P_1 + \rho g(h_1 - h_A) - \dfrac{1}{2}\rho v_2^2$$

Therefore, substituting the values, we have

$P_A = (1.01 \times 10^5) + (1000)(9.8)(-1)$

$-\dfrac{1}{2} \times (1000)(11.7)^2 2 = 2.27 \times 10^4$ N/m²

Example 7: Two separate air bubbles (radii 0.004 m and 0.002 m) formed of the same liquid (surface tension 0.07 N/m) come together to form a double bubble. Find the radius and the sense of curvature of the internal film surface common to both the bubbles.

Sol: Pressure inside the soap bubble is larger than that outside it by amount 4T/R, where T is surface tension and R is its radius.

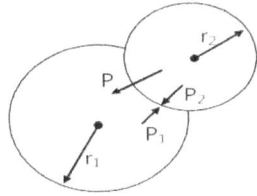

For the two bubbles,

$$P_1 = P_0 + \frac{4T}{r_1};$$

$$P_2 = P_0 + \frac{4T}{r_2}, r_2 < r_1$$

$$\therefore \quad P_2 > P_1$$

i.e. pressure inside the smaller bubble will be more. The excess pressure

$$P = P_2 - P_1 = 4T\left(\frac{r_1 - r_2}{r_1 r_2}\right) \qquad ...(i)$$

This excess pressure acts from concave to convex side, the interface will be concave towards smaller bubble and convex towards larger bubble. Let R be the radius of interface then,

$$P = \frac{4T}{R} \qquad ....(ii)$$

From equations (i) and (ii)

$$R = \frac{r_1 r_2}{r_1 - r_2} = \frac{(0.004)(0.002)}{(0.004 - 0.002)} = 0.004 \text{ m}$$

Example 8: A cylindrical tank of base area A has a small hole of area 'a' at the bottom. At time t = 0, a tap starts to supply water into the tank at a constant rate α m³/s.

(a) What is the maximum level of water h_{max} in the tank?

(b) Find the time when level of water becomes h(< h_{max}).

Sol: The height of water level will increase till the rate of inflow is greater than the rate of outflow. Use method of integration to find the time taken by water level to reach height h.

(a) Level will be maximum level when

Rate of inflow of water = rate of outflow of water

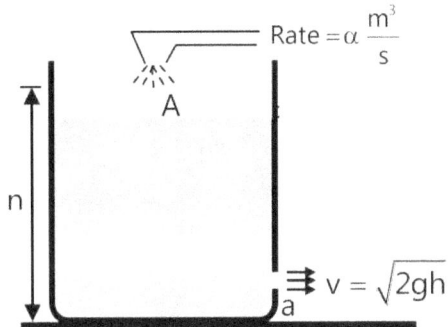

i.e., $\alpha = av$ or $\alpha = a\sqrt{2gh_{max}}$

$$\Rightarrow \quad h_{max} = \frac{\alpha^2}{2ga^2}$$

(b) Let at time t, the level of water be h. Then,

$$A\left(\frac{dh}{dt}\right) = \alpha - a\sqrt{2gh} \quad \text{or} \quad \int_0^h \frac{dh}{\alpha - a\sqrt{2gh}} = \int_0^t \frac{dt}{A}$$

Solving this, we get

$$t = \frac{A}{ag}\left[\frac{\alpha}{a}\ln\left\{\frac{\alpha - a\sqrt{2gh}}{\alpha}\right\} - \sqrt{2gh}\right]$$

Example 9: Under isothermal condition, two soap bubbles of radii r_1 and r_2 coalesce to form a single bubble of radius r. The external pressure is P_0. Find the surface tension of the soap in terms of the given parameters.

Sol: Pressure inside the soap bubble is larger than that outside it by amount 4T/R, where T is surface tension and R is its radius. Use ideal gas equation and the condition that he total number of moles of air is conserved.

As mass of the air is conserved,

$$\therefore \quad n_1 + n_2 = n \qquad \text{(as PV = nRT)}$$

$$\therefore \quad \frac{P_1V_1}{RT_1} + \frac{P_2V_2}{RT_2} = \frac{PV}{RT}$$

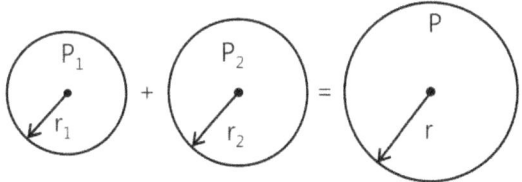

Although not given in the question, but we will have to assume that temperature of A and B are the same.

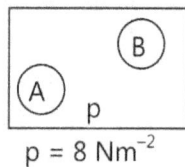

$$p = 8\ Nm^{-2}$$

$$\frac{n_B}{n_A} = \frac{p_B V_B / RT}{p_A V_A / RT} = \frac{p_B V_B}{p_A V_A}$$

$$= \frac{(p + 4s/r_A) \times 4/3\pi(r_A)^3}{(p + 4s/r_B) \times 4/3\pi(r_B)^3}$$

(s = surface tension)

Substituting the values, we get $\dfrac{n_B}{n_A} = 6$

Example 10: A thin rod of length L and area of cross section S is pivoted at its lowest point P inside a stationary, homogeneous and non-viscous liquid as shown in the figure. The rod is free to rotate in a vertical plane about a horizontal axis passing through P. The density d_1 of the material of the rod is smaller than the density d_2 of the liquid. The rod is displaced by a small angle. From its equilibrium position and then released, show that the motion of the rod is simple harmonic and determine its angular frequency in terms of the given parameters.

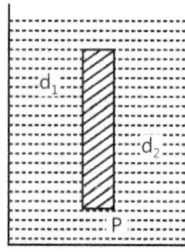

Sol: Use the restoring torque method to find the angular frequency.

Consider the rod be displaced through an angle θ. The different forces on the rod are shown in the figure.

Weight of rod acting downward $= S\,L\,d_1\,g = mg$

Buoyant force acting upwards $= S\,L\,d_2\,g$

Net thrust acting on the rod upwards; $F = S\,L\,(d_2 - d_1)g$

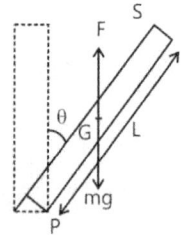

Restoring torque $\tau = F \times \dfrac{L}{2}\sin\theta = SL(d_2 - d_1)g\dfrac{L}{2}\sin\theta$

$\sin\theta \approx \theta$ (θ is small)

$$\therefore\ \tau = \frac{1}{2}SL^2(d_2 - d_1)g\,\theta$$

$$\tau = I\alpha = \left(\frac{ML^2}{3}\right)\frac{d^2\theta}{dt^2} = \left(\frac{SLd_1 \times L^2}{3}\right)\frac{d^2\theta}{dt^2}$$

$$\therefore\ \frac{d^2\theta}{dt} = \frac{3}{SL^3 d_1} \times \frac{1}{2}SL^2(d_2 - d_1)g\,\theta$$

or $\dfrac{d^2\theta}{dt} = \dfrac{3g}{2L}\left(\dfrac{d_2 - d_1}{d_1}\right)\theta$; so motion is S.H.M;

comparing with differential equation of S.H.M.

$$\frac{d^2\theta}{dt^2} + \omega^2\theta = 0;\ \ \omega = \sqrt{\frac{3g}{2L}\left(\frac{d_2 - d_1}{d_1}\right)};$$

Time period, $T = \dfrac{2\pi}{\omega} = 2\pi\sqrt{\dfrac{2Ld_1}{3g(d_2 - d_1)}}$

Example 11: Two non-viscous, incompressible and immiscible liquids of density ρ and 1.5ρ are poured into two limbs of a circular tube of radius R and small cross-section kept fixed in a vertical plane as shown in the figure.

Each liquid occupies one fourth the circumference of the tube.

(a) Find the angle that the radius vector to the interface makes with the vertical in the equilibrium position.

(b) If the whole liquid is given a small displacement from its equilibrium position, show that the resulting oscillations are simple harmonic. Find the time period of these oscillations.

Sol: Use the restoring torque method to find the angular frequency.

(a) Density of liquid column BC = 1.5 ρ ;

Density of liquid column CD = ρ

Pressure at A due to liquid column BA = ρ AB

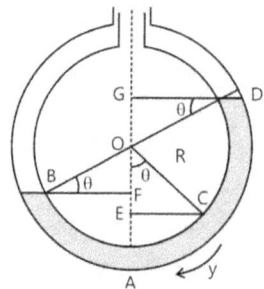

4.44

$= AF \times 1.5\rho \times g = (AO - OF)1.5\rho g \times g\rho$

$= (R - R\sin\theta)1.5g\rho$

Pressure at A due to liquid column $AD = \rho\,AD$

$= AE \times 1.5\rho \times g + EG\rho g$

$\therefore \rho AD - (AO - OE)1.5\rho g + (EO + OG)\rho g$

$-(R - R\cos\theta)1.5\rho g + R(\cos\theta + \sin\theta)\rho g$

In equilibrium $P_{AB} = P_{AD}$

$R(1 - \sin\theta)1.5\rho g = R(1 - \cos\theta)1.5\rho g + R(\cos\theta + \sin\theta)\rho g$

$\tan\theta = \dfrac{0.5}{2.5} = \dfrac{1}{5}.$ or $\tan^{-1}\left(\dfrac{1}{5}\right)$

(b) If a is area of cross – section,

length of each column $= \dfrac{2\pi R}{4} = \dfrac{\pi R}{2}$

Volume of each column $= \dfrac{\pi R a}{2}$

Mass of column $BC = \dfrac{\pi R a}{2} \times 1.5\rho$

Mass of column $CD = \dfrac{\pi R a}{2} \times p$

M.I. of whole liquid about $O = \left(\dfrac{\pi R a p}{2}\right)(1.5 + 1)R^2$

or $I = \dfrac{2.5\pi R^3 a p}{2}$

Let y be small displacement toward left and θ be the angular displacement,

$\theta = \dfrac{y}{R}$ or $y = R\theta$, Angular acceleration $= \dfrac{d^2\theta}{dt^2}$,

Torque about $A = I\dfrac{d^2\theta}{dt^2} = \dfrac{2.5\pi R^3 a p}{2}\left(\dfrac{d^2\theta}{dt^2}\right)$

Restoring torque due to displaced liquid.

$\tau_{rest} = -\left[ay \times 1.5pg + ay\,pg\right] \times R\cos\theta$

$= -2.5\,ay\,pg \times R\cos\theta = -2.5\,apgR^2\cos\theta.\theta$

[$R\cos\theta$ is perpendicular distance of gravitational force from axis of rotation]

Equating $\left(\dfrac{2.5\pi R^3 ap}{2}\right)\dfrac{d^2\theta}{dt^2} = -\left(2.5apgR^2\cos\theta\right)\theta$

$\dfrac{d^2\theta}{dt^2} = -\left(\dfrac{2g\cos\theta}{\pi R}\right)\theta = -\omega^2\theta$

As $\dfrac{2g\cos\theta}{\pi R}$ Acceleration is proportional to angular displacement and is directed towards mean position, the liquid undergoes SHM

$T = \dfrac{2\pi}{\omega} = 2\pi \times \sqrt{\dfrac{\pi R}{2g\cos\theta}}$

As $\tan\theta = \dfrac{1}{5}.\cos\theta = \dfrac{5}{\sqrt{26}}.$

$T = 2\pi\sqrt{\dfrac{\pi R}{2 \times g \times \dfrac{5}{\sqrt{26}}}} = 2(\pi)^2\sqrt[3]{\dfrac{R}{\left(\dfrac{10g}{\sqrt{26}}\right)}}$

JEE Main/Boards

Exercise 1

Q.1 If water in one flask and castor oil in other are violently shaken and kept on a table, then which one will come to rest earlier?

Q.2 What is the acceleration of a body falling through a viscous medium after terminal velocity is reached?

Q.3 The liquid is flowing steadily through a tube of varying diameter. How are the velocity of liquid flow (V) in any portion and the diameter (D) of the tube in that portion related?

Q.4 How does the viscosity of gases depend upon temperature?

Q.5 Explain the effect of (i) density (ii) temperature and (iii) pressure on the viscosity of liquids and gases.

Q.6 Two equal drops of water falling through air with a steady velocity v. If the drops coalesced, what will be the new steady velocity?

Q.7 What is the viscous force on a drop of liquid of radius 0.2 mm moving with a constant velocity 4 cm s^{-1} through a medium of viscosity 1.8×10^{-1} Nm^{-2} s.

Q.8 Eight rain drops of radius 1 mm each falling downwards with a terminal velocity of 5 cm s^{-1} coalesce to form a bigger drop. Find the terminal velocity of bigger drop.

Q.9 The flow rate of water from a tap of diameter 1.25 cm is 0.48 L/min. The coefficient of viscosity of water is 10^{-3} Pa-s. After sometime, the flow rate is increased to 3 L/min. The coefficient of viscosity of water is 10^{-3} Pa-s. Characterize the flow.

Q.10 A block of wood is floating in a lake? What is apparent weight of the floating block?

Q.11 A block of wood is floating in a lake. What is apparent weight of the floating block?

Q.12 A body floats in a liquid contained in a beaker. The whole system shown in the figure falls freely under gravity. What is the up thrust on the body due to the liquid?

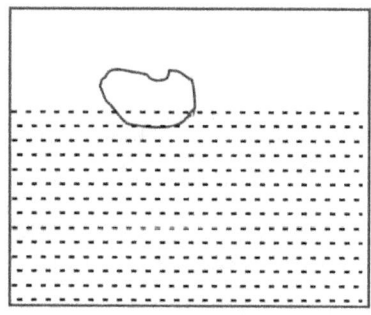

Q.13 A force of 60 N is applied on a nail, where tip has an area of cross-section of 0.0001 cm^2. Find the pressure on the tip.

Q.14 If the water pressure gauge shows the pressure at ground floor to be 270 kPa, how high would water rise in the pipes of a building?

Q.15 A metal cube is 5 cm side and relative density 9, suspended by a thread is completely immersed in a liquid of density 1.2×10^3 kg m^{-3}. Find the tension in the thread.

Q.16 A boat having a length of 3 m and breadth 2 m is floating on a lake. The boat sinks by one cm, when a man gets on it. What is the mass of the man?

Q.17 Calculate the force required to take away a flat plate of radius 5 cm from the surface of water. Given surface tension of water = 72×10^{-3} Nm^{-1}.

Q.18 A square wire frame of side 10 cm is dipped in a liquid of surface tension 28×10^{-3} Nm^{-1}. On taking out, a membrane is formed. What is the force acting on the surface of wire frame?

Q.19 The air pressure inside a soap bubble of diameter 3.5 mm is 8 mm of water above the atmosphere. Calculate the surface tension of soap solution.

Q.20 What should be the radius of the capillary tube so that water will rise to a height of 8 cm in it? Surface tension of water 70×10^{-3} Nm^{-1}.

Exercise 2

Single Correct Choice Type

Q.1 The area of cross-section of the wider tube shown in figure is 800 cm^2. If a mass of 12 kg is placed on the massless piston, the difference in heights h in the level of water in the two tubes is:

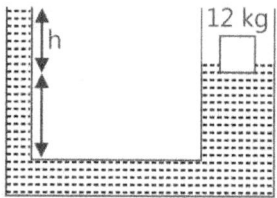

(A) 10 cm   (B) 6 cm   (C) 15 cm   (D) 2 cm

Q.2 Two cubes of size 1.0 m side, one of relative density 0.60 and another of relative density = 1.15 are connected by weightless wire and placed in a large tank of water. Under equilibrium the lighter cube will project above the water surface to a height of:

(A) 50 cm   (B) 25 cm   (C) 10 cm   (D) Zero

Q.3 A cuboidal piece of wood has dimensions a, b and c. Its relative density is d. It is floating in a large body of water such that side a is vertical. It is pushed down a bit and released. The time period of SHM executed by it is:

(A) $2\pi\sqrt{\dfrac{abc}{g}}$

(B) $2\pi\sqrt{\dfrac{h}{da}}$

(C) $2\pi\sqrt{\dfrac{bc}{dg}}$

(D) $2\pi\sqrt{\dfrac{da}{g}}$

Q.4 The frequency of a sonometer wire is f, but when the weights producing the tensions are completely immersed in water the frequency becomes f/2 and on immersing the weights in a certain liquid the frequency becomes f/3. The specific gravity of the liquid is:

(A) $\frac{4}{3}$ (B) $\frac{16}{9}$ (C) $\frac{15}{12}$ (D) $\frac{32}{27}$

Q.5 A small ball of relative density 0.8 falls into water from a height of 2m. The depth to which the ball will sink is (neglect viscous forces):

(A) 8 m   (B) 2 m   (C) 6 m   (D) 4 m

Q.6 A hollow sphere of mass M and radius r is immersed in a tank of water (density ρ_w). The sphere would float if it were set free. The sphere is tied to the bottom of the tank by two wires which makes angle 45° with the horizontal as shown in figure. The tension T_1 in the wire is:

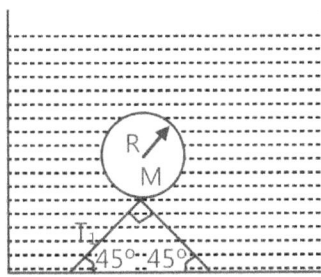

(A) $\dfrac{\frac{4}{3}\pi R^3 \rho_w g - Mg}{\sqrt{2}}$ (B) $\frac{2}{3}\pi R^3 \rho_w g - Mg$

(C) $\dfrac{\frac{4}{3}\pi R^3 \rho_w g - Mg}{2}$ (D) $\frac{4}{3}\pi R^3 \rho_w g - Mg$

Q.7 A large tank is filled with water to a height H. A small hole is made at the base of the tank. It takes T_1 times to decrease the height of water to H η, ($\eta > 1$) and it takes T_2 time to take out the rest of water. If $T_1 = T_2$, then the value of η is:

(A) 2   (B) 3   (C) 4   (D) 2.2

Q.8 In the case of a fluid, Bernoulli's theorem exes the application of the principle of conservation of:

(A) Linear momentum   (B) Energy

(C) Mass   (D) Angular momentum

Q.9 Fountains usually seen in gardens are generated by a wide pipe with an enclosure at one end having many small holes. Consider one such fountain which is produced by a pipe of internal diameter 2 cm in which water flows at a rate 3 ms⁻¹. The enclosure has 100 holes each of diameter 0.05 cm. The velocity of water coming out of the holes is (in ms⁻¹):

(A) 0.48   (B) 96   (C) 24   (D) 48

Q.10 A vertical tank open at the top, is filled with a liquid and rests on a smooth horizontal surface. A small hole is opened at the centre of one side of the tank. The area of cross-section of the tank is N times the area of the hole, where N is a large number. Neglect mass of the tank itself. The initial acceleration of the tank is:

(A) $\dfrac{g}{2N}$ (B) $\dfrac{g}{\sqrt{2N}}$

(C) $\dfrac{g}{N}$ (D) $\dfrac{g}{2\sqrt{N}}$

Q.11 Two water pipes P and Q having diameters 2×10^{-2} m and 4×10^{-2} m, respectively, are joined in series with the main supply line of water. The velocity of water flowing in pipe P is:

(A) 4 times that of Q   (B) 2 times that of Q

(C) 1/2 times that of Q   (D) 1/4 times that of Q

Q.12 A rectangular tank is placed on a horizontal ground and is filled with water to a height H above the base. A small hole is made on one vertical side at a depth D below the level of the water in the tank. The distance x from the bottom of the tank at which the water jet from the tank will hit the ground is:

(A) $2\sqrt{D(H-D)}$ (B) $2\sqrt{DH}$

(C) $2\sqrt{D(H+D)}$ (D) $\frac{1}{2}\sqrt{DH}$

Q.13 A horizontal pipe line carries water in a streamline flow. At a point along the tube where the cross-sectional area is 10^{-2} m², the water velocity is 2 ms⁻¹ and the pressure is 8000 Pa . The pressure of water at another point where the cross-sectional area is 0.5×10^{-2} m² is:

(A) 4000 Pa   (B) 1000 Pa

(C) 2000 Pa   (D) 3000 Pa

Q.14 Which of the following is not an assumption for an ideal fluid flow for which Bernoulli's principle is valid:

(A) Steady flow   (B) Incompressible

(C) Viscous   (D) Irrotational

Q.15 A solid metallic sphere of radius r is allowed to fall freely through air. If the frictional resistance due to air is proportional to the cross-sectional area and to the square of the velocity, then the terminal velocity of the sphere is proportional to which of the following?

(A) r^2 (B) r   (C) $r^{3/2}$ (D) $r^{1/2}$

Q.16 If two soap bubbles of different radii are connected by a tube.

(A) Air flows from the bigger bubble to the smaller bubble till the sizes become equal

(B) Air flows from bigger bubble to the smaller bubble till the sizes are interchanged

(C) Air flows from the smaller bubble to the bigger

(D) There is no flow of air

Q.17 A long capillary of radius r is initially just vertically completely immerged inside a liquid of angle of contact 0°. If the tube is slowly raised, then relation between radius of curvature of meniscus inside the capillary tube and displacement (h) of tube can be represented by:

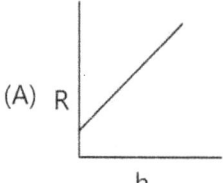 (A) R

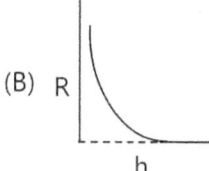 (B) R

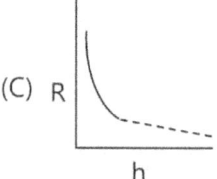 (C) R

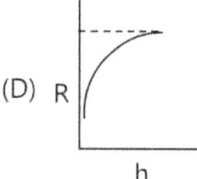 (D) R

Q.18 Figure shows a siphon. Choose the wrong statement:

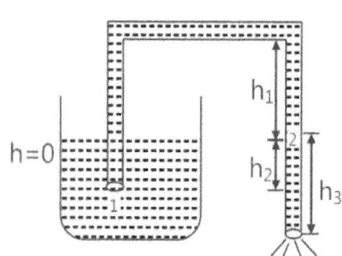

(A) Siphon works when $h_3 > 0$

(B) Pressure at point 2 is $P_2 = p_0 - \rho g h_3$

(C) Pressure at point 3 is P_0

(D) None of the above

Q.19 A steady flow of water passes along horizontal tube from a wide section X to the narrower section Y, see figure. Manometers are placed at P and Q of the sections. Which of the statements A,B,C,D is most correct?

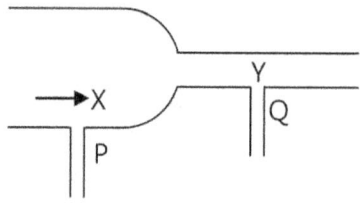

(A) water velocity at X is greater than at Y

(B) the manometer at P shows lower pressure than at Q

(C) kinetic energy per m^3 of water at X = kinetic energy per m^3 at Y

(D) the manometer at P shows greater pressure than at Y

Previous Years' Questions

Q.1 A metal ball immersed in alcohol weighs W_1 at 0° C and W_2 at 50°C. The coefficient of cubical expansion of the metal is less than that of the alcohol. Assuming that the density of the metal is large compared to that of alcohol, it can be shown that: *(1980)*

(A) $W_1 > W_2$ (B) $W_1 = W_2$

(C) $W_1 < W_2$ (D) All of these

Q.2 A vessel containing water is given a constant acceleration a towards the right along a straight horizontal path. Which of the following diagrams represent the surface of the liquid? *(1981)*

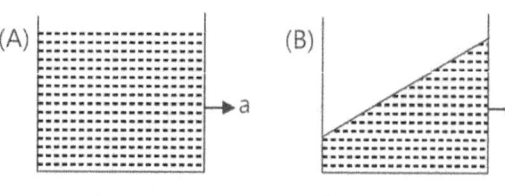

 (D) None of these

Q.3 A body floats in a liquid contained in a beaker. The whole system as shown in figure falls freely under gravity. The upthrust on the body due to the liquid is: *(1982)*

(A) Zero

(B) Equal to the weight of the liquid displaced

(C) Equal to the weight of the body in air

(D) Equal to the weight of the immersed position of the body

Q.4 A U-tube of uniform cross-section is partially filled with a liquid I. Another liquid II which does not mix with liquid I is poured into one side. It is found that the liquid levels of the two sides of the tube are the same, while the level of liquid I has risen by 2 cm. If specific gravity of liquid I is 1.1, the specific gravity of liquid I is 1.1, the specific gravity of liquid II must be: *(1983)*

(A) 1.12   (B) 1.1   (C) 1.05   (D) 1.0

Q.5 A homogeneous solid cylinder of length L. Cross-sectional area A/5 is immersed such that it floats with its axis vertical at the liquid-liquid interface with length L/4 in the denser liquid as shown in the fig. The lower density liquid is open to atmosphere having pressure p_0. Then density D of solid is given by: *(1995)*

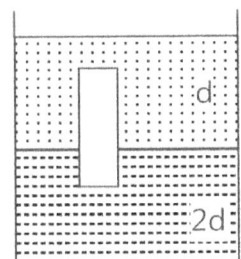

(A) $\frac{5}{4}d$ (B) $\frac{4}{5}d$ (C) $4d$ (D) $\frac{d}{5}$

Q.6 Water from a tap emerges vertically downwards with an initial speed of 1.0 m/s. The cross-section area of the tap is $10^{-4} m^2$. Assume that the pressure is constant throughout the steam of water and that the flow is steady, the cross-sectional area of stream 0.15 m below the tap is: *(1998)*

(A) $5.0 \times 10^{-4} m^2$ (B) $1.0 \times 10^{-4} m^2$

(C) $5.0 \times 10^{-5} m^2$ (D) $2.0 \times 10^{-4} m^2$

Q.7 A large open tank has two holes in the wall. One is a square hole of side L at a depth y from the top and the other is a circular hole of radius R at a depth 4y from the top. When the tank is completely filled with water the quantities of water flowing out per second from both the holes are the same. Then R is equal to *(2000)*

(A) $L / \sqrt{2\pi}$ (B) $2\pi L$

(C) L   (D) L/2p

Q.8 A wooden block, with a coin placed on its top, floats in water as shown in fig. The distance l and h are shown there. After some time the coin falls into the water. Then: *(2002)*

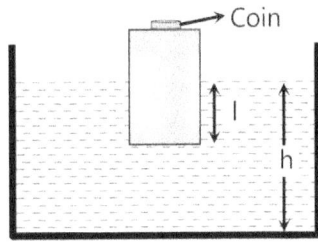

(A) *l* Decreases and h increases

(B) Increases and h decreases

(C) Both *l* and h increase

(D) Both *l* and h decrease

Q.9 Water is filled in a cylindrical container to a height of 3 m. The ratio of the cross-sectional area of the orifice and the beaker is 0.1. The square of the speed of the liquid coming out from the orifice is (g = 10 m/s²) *(2005)*

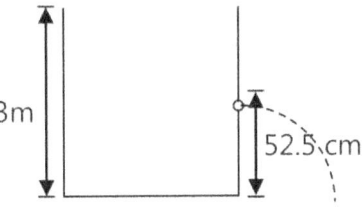

(A) $50 \ m^2/s^2$ (B) $50.5 \ m^2/s^2$

(C) $51 \ m^2/s^2$ (D) $52 \ m^2/s^2$

Q.10 A glass tube of uniform internal radius (r) has a valve separating the two identical ends. Initially, the valve is in a tightly closed position. End 1 has a hemispherical soap bubble of radius r. End 2 has sub-hemispherical soap bubble as shown in figure. *(2008)*

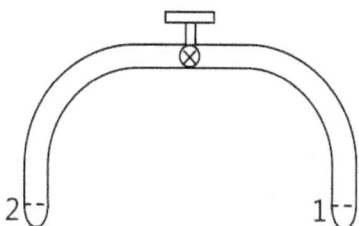

Just after opening the valve:

(A) air from end 1 flow towards end 2. No change in the volume of the soap bubbles.

(B) air from end 1 flows towards end 2. Volume of the soap bubble at end 1 decreases

(C) no change occurs

(D) air from end 2 flows towards end 1. Volume of the soap bubble at end 1 increases

Q.11 A uniform cylinder of length L and mass M having cross-sectional area A is suspended, with its length vertical, from a fixed point by a massless spring, such that it is half-submerged in a liquid of density p at equilibrium position. When the cylinder is given a small downward push and released it starts oscillating vertically with a small amplitude. If the force constant of the spring is k, the frequency of oscillation of the cylinder is *(1990)*

(A) $\dfrac{1}{2\pi}\left(\dfrac{k - A\rho g}{M}\right)^{1-2}$ (B) $\dfrac{1}{2\pi}\left(\dfrac{k + A\rho g}{M}\right)^{1/2}$

(C) $\dfrac{1}{2\pi}\left(\dfrac{k + \rho g L^2}{M}\right)^{1/2}$ (D) $\dfrac{1}{2\pi}\left(\dfrac{k + A\rho g}{A\rho g}\right)^{1/2}$

Q.12 A thin liquid film formed between a U-shaped wire and a light slider supports a weight of 1.5×10^{-2} N (see figure). The length of the slider is 30 cm and its weight negligible. The surface tension of the liquid film is *(2012)*

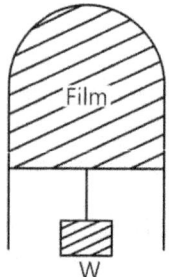

(A) 0.0125 Nm⁻¹   (B) 0.1 Nm⁻¹

(C) 0.05 Nm⁻¹   (D) 0.025 Nm⁻¹

Q.13 A uniform cylinder of length L and mass M having cross-sectional area A is suspended, with its length vertical, from a fixed point by a massless spring, such that it is half submerged in a liquid of density σ at equilibrium position. The extension x_0 of the spring when it is in equilibrium is: *(2013)*

(A) $\dfrac{Mg}{k}\left(1 - \dfrac{LA\sigma}{M}\right)$ (B) $\dfrac{Mg}{k}\left(1 - \dfrac{LA\sigma}{2M}\right)$

(C) $\dfrac{Mg}{k}\left(1 + \dfrac{LA\sigma}{M}\right)$ (D) $\dfrac{Mg}{k}$

(Here k is spring constant)

Q.14 Assume that a drop of liquid evaporates by decrease in its surface energy, so that its temperature remains unchanged. What should be the minimum radius of the drop for this to be possible? The surface tension is T, density of liquid is ρ and L is its latent heat of vaporization. *(2013)*

(A) $\sqrt{T / \rho L}$ (B) $T / \rho L$

(C) $2T / \rho L$ (D) $\rho L / T$

Q.15 An open glass tube is immersed in mercury in such a way that a length of 8 cm extends above the mercury level. The open end of the tube is then closed and sealed and the tube is raised vertically up by additional 46cm. What will be length of the air column above mercury in the tube now? (Atmospheric pressure = 76 cm of Hg) *(2014)*

(A) 38 cm   (B) 6 cm   (C) 16 cm   (D) 22 cm

Q.16 On heating water, bubbles being formed at the bottom of the vessel detach and rise. Take the bubbles to be spheres of radius R and making a circular contact of radius r with the bottom of the vessel. If r < < R, and the surface tension of water is T, value of r just before bubbles detach is: (density of water is ρ_w) *(2014)*

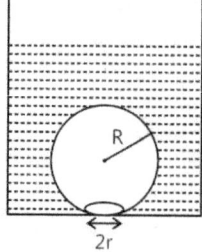

(A) $R^2\sqrt{\dfrac{\rho_w g}{T}}$ (B) $R^2\sqrt{\dfrac{3\rho_w g}{T}}$

(C) $R^2\sqrt{\dfrac{\rho_w g}{3T}}$ (D) None of these

4.50

Exercise 1

Q.1 A piston of mass M = 3 kg and radius R=4 cm has a hole into which a thin pipe of radius r = 1 cm is inserted. The piston can enter a cylinder tightly and without friction, and initially it is at the bottom of the cylinder. 750 gm of water is now poured into the pipe so that the piston and pipe are lifted up as shown. Find the height H of water in the cylinder and height h of water in pipe.

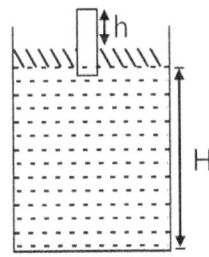

Q.2 A solid ball of density half of that of water falls freely under gravity from a height of 19.6m and then enters the water. Upto what depth will the ball go? How much time will it take to come again to the water surface? Neglect air resistance & velocity effects in water.

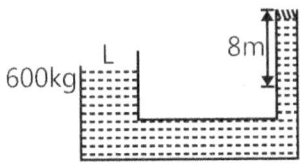

Q.3 For the system shown in the figure, the cylinder on left at L has a mass of 600 kg and a cross sectional area of 800 cm². The piston on the right, at S, has cross sectional area 25 cm² and negligible weight. If the apparatus is filled with oil (ρ = 0.75 gm/cm³). Find the force F required to hold the system in equilibrium.

Q.4 (a) A spherical tank of 1.2 m radius is half filled with oil of relative density 0.9. If the tank is given a horizontal acceleration of 10 m/s², calculate the inclination of the oil surface to horizontal and maximum gauge pressure on the tank.

(b) The volume of an air bubble is doubled as it rises from the bottom of a lake to its surface. If the atmospheric pressure is H m of mercury & the density of mercury is n times that of lake water, find the depth of the lake.

Q.5 A test tube of thin walls has lead shots in it at its bottom and the system floats vertically in water, sinking by a length l = 10 cm. A liquid of density less than that of water, is poured into the tube till the levels inside and outside the tube are even. If the tube now sinks to a length l = 40 cm, the specific gravity of the liquid is ....

Q.6 A large tank is filled with two liquids of specific gravities 2σ and σ. Two holes are made on the wall of the tank as shown. Find the ratio of distances from O of the points on the ground where the jets from holes A and B strike.

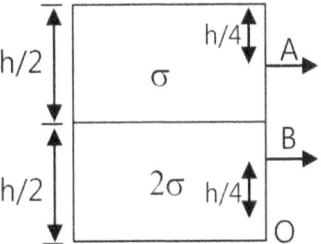

Q.7 A jet of water having velocity = 10 m/s and stream cross-section = 2 cm² hits a plate perpendicularly, with the water splashing out parallel to plate. Find the force that the plate experiences.

Q.8 A laminar stream is flowing vertically down from a tap of cross-section area 1 cm². At a distance 10 cm below the tap, the cross-section area of the stream has reduced to 1/2m² Find the volumetric flow rate of water from the tap.

Q.9 A cylindrical vessel open at the top is 20 cm high and 10 cm in diameter. A circular hole whose cross-sectional area is 1 cm² is cut at the centre of the bottom of the vessel. Water flows from a tube above it into the vessel at the rate 100 cm³ s⁻¹. Find the height of water in the vessel under steady state.

Q.10 Calculate the rate of flow of glycerin of density 1.25×10^3 kg/m³ through the conical section of a 0.1m and 0.04m and the pressure drop across its length is 10N/m².

Q.11 A ball is given velocity v_0 (greater than the terminal velocity v_T) in downward direction inside a highly viscous liquid placed inside a large container. The height of liquid in the container is H. The ball attains the terminal velocity just before striking at the bottom of the container. Draw graph between velocity of the ball and distance moved by the ball before getting terminal velocity.

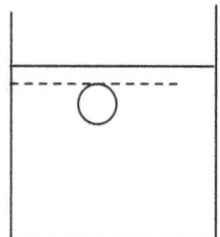

Q.12 A spherical ball of radius 1×10^{-4}m and density 10^4 kg/m^3 falls freely under gravity through a distance h before entering a tank of water. If after entering the water the velocity of the ball does not change, find h. The viscosity of water is 9.8×10^{-6} N-s/m^2.

Q.13 Two arms of a U-tube have unequal diameters d_1 = 10 mm and d_2 = 1.0cm. If water (surface tension 7×10^{-2}N/m) is poured into the tube held in the vertical position, find the difference of level of water in the U-tube. Assume the angle of contact to be zero.

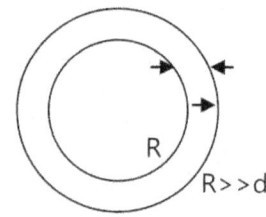

Q.14 A soap bubble has radius R and thickness d(<<R) as shown. It collapses into a spherical drop. Find the ratio of excess pressure in the drop to the excess pressure inside the bubble.

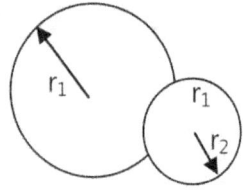

Q.15 Two soap bubbles with radii r_1 and r_2 ($r_1 > r_2$) come in contact. Their common surface has a radius curvature r.

Q.16 Place a glass beaker, partially filled with water, in a sink. The beaker has mass 390 gm and an interior volume of 500 cm^3. You now start to fill the sink with water and you find, by experiment, that if the beaker is less than half full, it will float; but if it is more than half full, it remains on the bottom of the sink as the water

rises to its rim. What is the density of the material of which the beaker is made?

Q.17 A level controller is shown in the figure. It consists of a thin circular plug of diameter 10 cm and a cylindrical float of diameter 20 cm tied together with a light rigid rod of length 10 cm. The plug fits in snugly in a drain hole at the bottom of the tank which opens into the atmosphere. As water fills up and the level reaches height h, the plug opens. Find h. Determine the level of water in the tank when the plug closes again. The float has a mass 3kg and the plug may be assumed as massless.

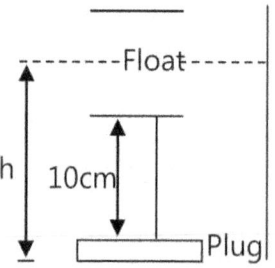

Q.18 A cylindrical rod of length l=2m and density $\frac{\rho}{2}$ floats vertically in a liquid of density ρ as shown in fig. (a)

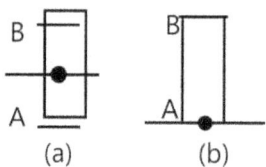

(a) Show that it performs SHM when pulled slightly up & released & find its time period. Neglect change in liquid level.

(b) Find the time taken by the rod to completely immerse when released from position shown in figure (b). Assume that it remains vertical throughout its motion.

(take g = π^2m/s^2)

Q.19 A thin rod of length L and area of cross-section S is pivoted at its lowest point P inside a stationary, homogeneous & non-viscous liquid (Figure). The rod is free to rotate in a vertical plane about a horizontal axis passing through P. the density d_1 of the material of the rod is smaller than the entity d_2 of the liquid. The rod is displaced by a small angle θ from its equilibrium position and then released. Show that the motion of the rod is simple harmonic and determine its angular frequency in terms of the given parameters.

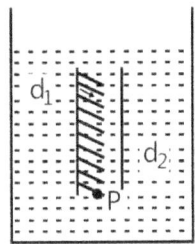

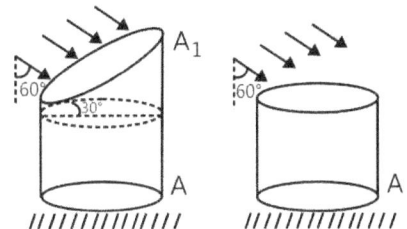

Q.20 A hollow cone floats with its axis vertical up to one-third liquid of its height in a liquid of relative density ρ is filled in it up to one-third of its height, the cone floats up to half its vertical height. The height of the cone is 0.10 m and the radius of the circular base is 0.05m. Find the specific gravity ρ.

Q.21 In the figure shown, the heavy cylinder (radius R) resting on a smooth surface separates two liquids of densities 2ρ and 3ρ. Find the height 'h' for the equilibrium of cylinder.

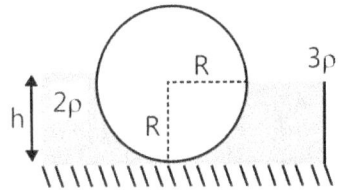

Q.22 The vertical limbs of a U shaped tube are filled with a liquid of density ρ up to a height h on each side. The horizontal portion of the U tube having length 2h contains a liquid of density 2ρ. The U tube is moved horizontally with an acceleration g/2 parallel to the horizontal arm. Find the difference in heights in liquid levels in the two vertical limbs, at steady state.

Q.23 A wooden stick of length l and radius R and density ρ has a small metal piece of mass m (of negligible volume) attached to its one end. Find the minimum value for the mass m (in terms of given parameters) that would make the stick float vertically in equilibrium in a liquid of density σ(>ρ).

Q.24 A vertical cylindrical container of base area A and upper cross-section area A_1 making angle 30° with the horizontal placed in an open rainy field as shown near another cylindrical container having same base area A. Find the ratio of rates of collection of water in the two containers.

Q.25 A siphon has a uniform circular base of diameter $\frac{8}{\pi}$ cm with its crest A 1.8 m above water level as in figure. Find

(a) Velocity of flow.

(b) Discharge rate of the flow in m³/sec.

(c) Absolute pressure at the crest level A.

[Use $P_0 = 10^5$ N/m² & g = 10m/s²]

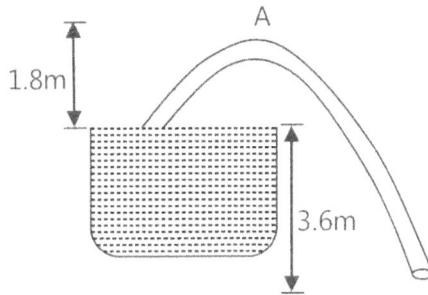

Q.26 Two very large open tanks A and F both contain the same liquid. A horizontal pipe BCD, having a constriction at C leads out of the bottom of tank A, and a vertical pipe E opens into the constriction at C and dips into the liquid in tank F. Assume streamline flow and no viscosity. If the cross section at C is one half that at D and if D is at a distance h_1 below the level of liquid in A, to what height h_2 (in terms of h_1) will liquid rise in pipe E?

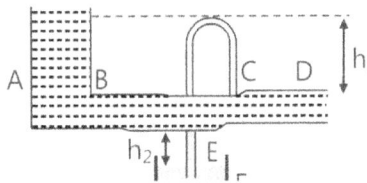

Q.27 A cube with mass 'm' completely wet by water floats on the surface of water. Each side of the cube is 'a'. What is the distance h between the lower face of cube and the surface of the water as $ρ_w$. Take angle of contact as zero.

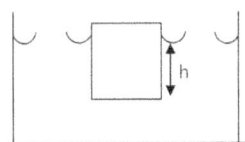

Exercise 2

Single Correct Choice Type

Q.1 A bucket contains water filled up to a height = 15cm. The bucket is tied to a rope which is passed on a frictionless light pulley and the other end of the rope is tied to a weight of mass which is half of that of the (bucket + water). The water pressure above atmosphere at the bottom is:

(A) 0.5 kPa

(B) 1 kPa

(C) 5 kPa

(D) None of these

Q.2 A cone of radius R and height H, is hanging inside a liquid of density ρ by means of a string as shown in the figure. The force, due to the liquid acting on the slant surface of the cone is (Neglect atmosphere pressure)

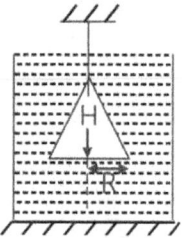

(A) prgHR²

(B) pρHR²

(C) 4/3 prgHR²

(D) 2/3 prgHR²

Q.3 An open cubical tank was initially fully filled with water. When the tank was accelerated on a horizontal plane along one of its side, it was found that one third of volume of water spilled out. The acceleration was:

(A) g/3

(B) 2g/3

(C) 3g/2

(D) None

Q.4 Some liquid is filled in a cylindrical vessel of radius R. Let F₁ be the force applied by the liquid on the bottom of the cylinder. Now the same liquid is poured into a vessel of uniform square cross-section of side R. Let F₂ be the force applied by the liquid on the bottom of this new vessel. (Neglect atmosphere pressure). Then:

(A) $F_1 = \pi F_2$

(B) $F_1 = F_2/p$

(C) $F_1 = \sqrt{\pi}F_2$

(D) $F_1 = F_2$

Q.5 A heavy hollow cone of radius R and height h is placed on a horizontal table surface, with its base on the table. The whole volume inside the cone is filled with water of density ρ. The circular rim of the cone's base has a water tight seal with the table's surface and

the top apex of the cone has a small hole. Neglecting atmospheric pressure, the total upward force exerted by water on the cone is:

(A) (2/3)pR²hrg

(B) (1/3)pR²hrg

(C) pR²hrg

(D) None

Q.6 A slender homogeneous rod of length 2L floats partly immersed in water, being supported by a string fastened to one of its ends, as shown. The specific gravity of the rod is 0.75. The length of rod that extends out of water is:

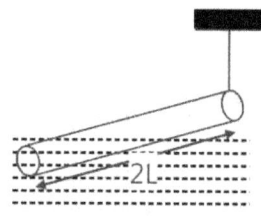

(A) L²

(B) L²/2

(C) L²/4

(D) 3L²/4

Q.7 A dumbbell is placed in water of density ρ. It is observed that by attaching a mass m to the rod, the dumbbell floats with the rod horizontal on the surface of water and each sphere exactly half submerged as shown in the figure. The volume of the mass m is negligible. The value of length ℓ is:

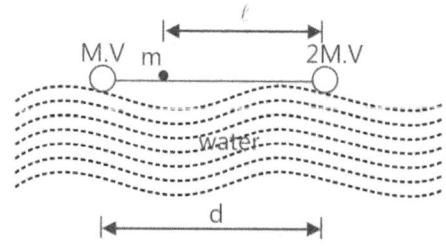

(A) $\dfrac{d(V\rho - 3M)}{2(V\rho - 2M)}$

(B) $\dfrac{d(V\rho - 2M)}{2(V\rho - 3M)}$

(C) $\dfrac{d(V\rho + 2M)}{2(V\rho - 3M)}$

(D) $\dfrac{d(V\rho - 2M)}{2(V\rho + 3M)}$

Q.8 A small wooden ball of density ρ is immersed in water of density σ to depth h and then released. The height H above the surface of water up to which the ball will jump out of water is:

(A) $\dfrac{\sigma h}{\rho}$

(B) $\left(\dfrac{\sigma}{\rho} - 1\right)h$

(C) h

(D) Zero

4.54

Q.9 A sphere of radius R and made of material of relative density σ has a concentric cavity of radius r. It just floats when placed in a tank full of water. The value of the ratio R/r will be:

(A) $\left(\dfrac{\sigma}{\sigma-1}\right)^{1/3}$

(B) $\left(\dfrac{\sigma-1}{\sigma}\right)^{1/3}$

(C) $\left(\dfrac{\sigma+1}{\sigma}\right)^{1/3}$

(D) $\left(\dfrac{\sigma-1}{\sigma+1}\right)^{1/3}$

Q.10 A fire hydrant delivers water of density ρ at a volume rate L. The water travels vertically upward through the hydrant and then does 90° turn to emerge horizontally at speed V. The pipe and nozzle have uniform cross-section throughout. The force exerted by the water on the corner of the hydrant is:

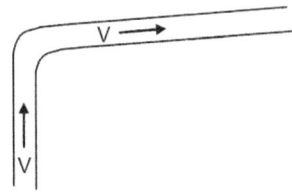

(A) ρVL

(B) Zero

(C) 2ρVL

(D) $\sqrt{2}$ ρVL

Q.11 A cylindrical vessel filled with water up to height of H stands on a horizontal plane. The side wall of the vessel has a plugged circular hole touching the bottom. The coefficient of friction between the bottom of vessel and plane is μ and total mass of water plus vessel is M. What should be the minimum diameter of the hole so that the vessel begins to move on the floor if plug is removed (here density of water is ρ)

(A) $\sqrt{\dfrac{2\mu M}{\pi\rho H}}$

(B) $\sqrt{\dfrac{\mu M}{2\pi\rho H}}$

(C) $\sqrt{\dfrac{\mu M}{\rho H}}$

(D) None

Q.12 A Newtonian fluid fills the clearance between a shaft and a sleeve. When a force of 800N is applied to shift, parallel to the sleeve, the shaft attains of 1.5 cm/sec. If a force of 2.4 kN is applied instead, the shaft would move with a speed of

(A) 1.5 cm/sec

(B) 13.5 cm/sec

(C) 4.5 cm/sec

(D) None

Q.13 A cubical block of side 'a' and density 'ρ' slides over a fixed inclined plane with constant velocity 'v'. There is a thin film of viscous fluid of thickness 't' between the plane and the block. Then the coefficient of viscosity of the thin film will be:

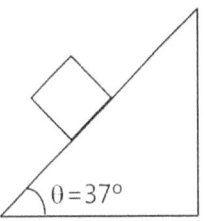

(A) $\dfrac{3\rho\,a\,g\,t}{5v}$

(B) $\dfrac{4\rho\,a\,g\,t}{5v}$

(C) $\dfrac{\rho\,a\,g\,t}{5v}$

(D) None of these

Q.14 Which of the following graphs best represent the motion of a raindrop?

(A)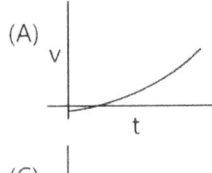

(B)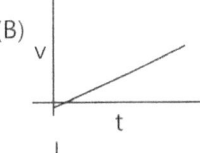

(C)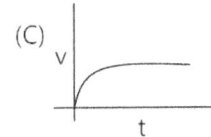

(D)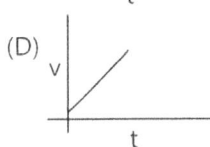

Q.15 Which of the following is the incorrect graph for a sphere falling in a viscous liquid? (Given at t = 0, velocity v = 0 and displacement x = 0)

(A)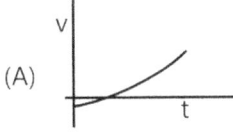

(B)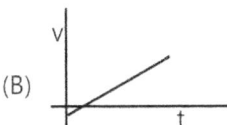

(C)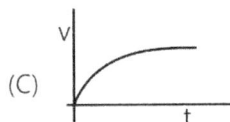

(D)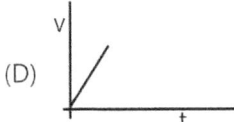

Q.16 A container, whose bottom has round holes with diameter 0.1 mm is filled with water. The maximum height in cm up to which water can be filled without leakage will be what?

Surface tension=75×10^{-3}N/m and g=10 m/s²:

(A) 20 cm   (B) 40 cm   (C) 30 cm   (D) 60 cm

Q.17 A liquid is filled in a spherical container of radius R till a height h. At this position the liquid surface at the edges is also horizontal. The contact angle is:

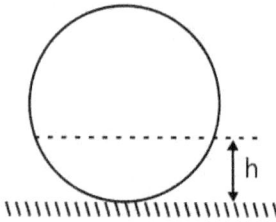

(A) 0

(B) $\cos^{-1}\left(\dfrac{R-h}{R}\right)$

(C) $\cos^{-1}\left(\dfrac{h-R}{R}\right)$

(D) $\sin^{-1}\left(\dfrac{R-h}{R}\right)$

Q.18 The vessel shown in the figure has two sections. The lower part is a rectangular vessel with area of cross-section A and height h. The upper part is a conical vessel of height h with base area 'A' and top area 'a' and the walls of the vessel are inclined at an angle 30° with the vertical. A liquid of density ρ fills both the sections up to a height 2h. Neglecting atmospheric pressure,

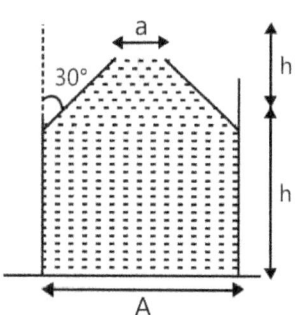

(A) The force F exerted by the liquid on the base of the vessel is $2h\rho g\dfrac{(A+a)}{2}$

(B) The pressure P at the base of the vessel is $2h\rho g\dfrac{A}{a}$

(C) The weight of the liquid W is greater than the force exerted by the liquid on the base.

(D) The walls of the vessel exert a downward force (F-W) on the liquid.

Multiple Correct Choice Type

Q.19 A cubical block of wood of edge 10 cm and mass 0.92 kg floats on a tank of water with oil of relative density 0.6 to a depth of 4 cm above water. When the block attains equilibrium with four of its side edges vertical,

(A) 1 cm of it will be above the force of oil.

(B) 5 cm of it will be under water.

(C) 2 cm of it will be above the common surface of oil and water.

(D) 8 cm of it will be under water.

Q.20 Water coming out of a horizontal tube at a speed v strikes normally a vertically wall close to the mouth of the tube and falls down vertically after impact. When is the speed of water increased to 2v.

(A) the thrust exerted by the water on the wall will be doubled.

(B) the thrust exerted by the water on the wall will be four times

(C) the energy lost per second by water striking the wall will also be four times

(D) the energy lost per second by water striking the wall be increased eight times.

Q.21 A beaker filled with water is accelerated a m/s² in +x direction. The surface of water shall make on angle:

(A) $\tan^{-1}(a/g)$ backwards

(B) $\tan^{-s}$ draw of $(g/a)^1$

(C) $\cot^{-1}(g/a)$ backwards

(D) $\cot^{-1}(a/g)$ backwards

Q.22 The spring balance A read 2 kg with a block m suspended from it. A balance B reads 5 kg when a beaker with liquid is put on the pan of the balance. The two balances are now so arranged that the hanging mass is inside the liquid in the beaker as shown in the figure in this situation:

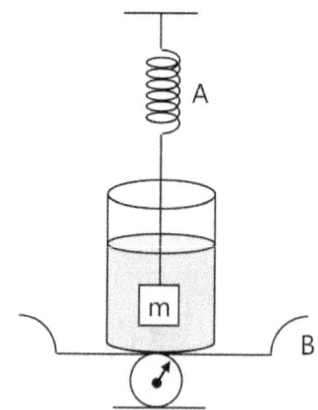

(A) The balance A will read more than 2kg

(B) The balance B will read more than 5 kg

(C) The balance A will read less than 2 kg and B will read more than 5 kg.

(D) The balance A and B will read 2 kg and 5 kg respectively

Q.23 When an air bubble rises from the bottom of a deep lake to a point just below the water surface, the pressure of air inside the bubble:

(A) Is greater than the pressure outside it

(B) Is less than the pressure outside it

(C) Increases as the bubble moves up

(D) Decreases as the bubble moves up

Q.24 A tank is filled up to a height h with a liquid and is placed on a platform of height h at a distance of y from the free surface of the liquid. Then

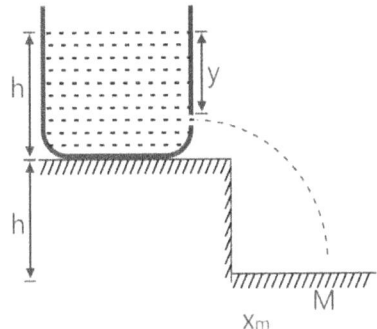

(A) $x_m = 2h$
(B) $x_m = 1.5 h$
(C) $y = h$
(D) $y = 0.75 h$

Assertion Reasoning Type

(A) Statement-I is true, statement-II is true and Statement-II is the correct explanation for statement-I

(B) Statement-I is true, statement-II is true and statement-II is NOT the correct explanation for statement-I

(C) Statement-I is true, statement-II is false.

(D) Statement-I is false, statement-II is true

Q.25 Statement-I: A helium filled balloon does not rise indefinitely in air but halts after a certain height.

Statement-II: Viscosity opposes the motion of balloon.

Q.26 Statement-I: A partly filled test tube is floating in a liquid as shown. The tube will remain as if its atmosphere pressure changes.

Statement-II: The buoyant force on a submerged object is independent of atmospheric pressure.

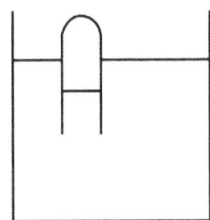

Q.27 Statement-I: Submarine sailors are advised that they should not be allowed to rest on floor of the ocean.

Statement-II: The force exerted by a liquid on a submerged body may be downwards.

Q.28 Statement-I: When a body floats such that it's parts are immersed into two immiscible liquids then force exerted by liquid-1 is of magnitude $r_1 v_1 g$.

Statement-II: Total Bouyant force $r_1 v_1 g + r_2 v_2 g$.

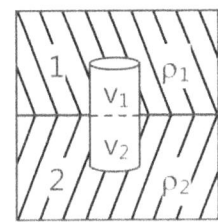

Q.29 Statement-I: When temperature rises the coefficient of viscosity of gases decreases.

Statement-II: Gases behave more like ideal gases at higher temperature.

Q.30 Statement-I: The free surface of a liquid at rest with respect to stationary container is always normal to the $\vec{g}_{eff}$.

Statement-II: Liquids at rest cannot have shear stress.

Previous Years' Questions

Q.1 A hemispherical portion of radius R is removed from the bottom of a cylinder of radius R. The volume of the remaining cylinder is V and mass M. It is suspended by a string in a liquid of density ρ, where it stays vertical. The upper surface of the cylinder is at a depth h below the liquid surface. The force on the bottom of the cylinder by the liquid is *(2001)*

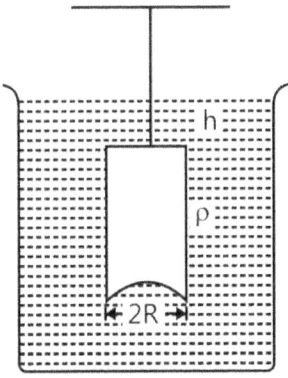

4.57

(A) Mg   (B) Mg – Vρg

(C) Mg + ρR²hρR   (D) ρg(V + pR²h)

Q.2 When a block of iron floats in mercury at 0°C, fraction k_1 of its volume is submerged, while at the temperature 60°C, a fraction k_2 is seen to be submerged. If the coefficient of volume expansion of iron is γ_{Fe} and that of mercury is γ_{Hg}, then the ratio k_1/k_2 can be expressed as **(2001)**

(A) $\dfrac{1+60\gamma_{Fe}}{1+60\gamma_{Hg}}$

(B) $\dfrac{1-60\gamma_{Fe}}{1+60\gamma_{Hg}}$

(C) $\dfrac{1+60\gamma_{Fe}}{1-60\gamma_{Hg}}$

(D) $\dfrac{1+60\gamma_{Hg}}{1+60\gamma_{Fe}}$

Q.3 Water is filled up to a height h in a beaker of radius R as shown in the figure. The density of water is ρ, the surface tension of water is T and the atmospheric pressure is p_0. Consider a vertical section ABCD of the water column through a diameter of the beaker. The force on water on one side of this section by water on the other side of this section has magnitude **(2007)**

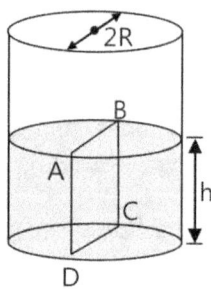

(A) $|2p_0Rh + \pi R^2\rho gh - 2RT|$

(B) $|2p_0Rh + R\rho gh^2 - 2RT|$

(C) $|p_0\pi R^2 + R\rho gh^2 - 2RT|$

(D) $|p_0\pi R^2 + R\rho gh^2 - 2RT|$

Paragraph 1: (Q.4 - Q.6)

A wooden cylinder of diameter 4r, height h and density ρ/3 is kept on a hole of diameter 2r of a tank, filled with liquid of density ρ as shown in the figure.

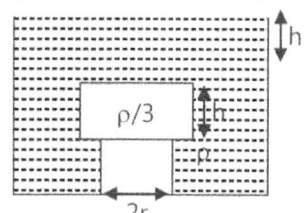

Q.4 Now level of the liquid starts decreasing slowly. When the level of liquid is at a height h_1 above the cylinder the block starts moving up. At what value of h_1, will the block rise? **(2005)**

(A) 4h/9   (B) 5h/9

(C) 5h/3   (D) Remains same

Q.5 The block in the above question is maintained at the position by external means and the level of liquid is lowered. The height h_2 when this external force reduces to zero is: **(2006)**

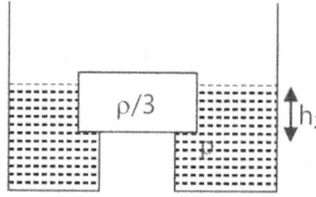

(A) $\dfrac{4h}{9}$ (B) $\dfrac{5h}{9}$ (C) Remains same   (D) $\dfrac{2h}{3}$

Q.6 If height h_2 of water level is further decreased, then: **(2006)**

(A) cylinder will not move up and remains at its original position

(B) for $h_2 = h/3$, cylinder again starts moving up

(C) for $h_2 = h/4$, cylinder again starts moving up

(D) $h_2 = h/5$, cylinder again starts moving up

Paragraph 2: (Q.7 - Q.9)

When a liquid medicine of density ρ is to be put in the eye, it is done with the help of a dropper. As the bulb on top of the dropper is pressed, a drop forms at the opening of the dropper. We wish to estimate the size of the drop.

We first assume that the drop formed at the opening is spherical because that requires minimum increase in its surface energy. To determine the size, we calculate the net vertical force due to the surface tension T when the radius of the drop is R. When this force becomes smaller than the weight of the drop, the drop gets detached from the dropper. **(2010)**

Q.7 If the radius of the opening of the dropper is r, the vertical force due to the surface tension on the drop of radius R (assuming r<<R) is

(A) $2\pi rT$ (B) $2\pi RT$ (C) $\dfrac{2\pi r^2T}{R}$ (D) $\dfrac{2\pi R^2T}{r}$

Q.8 If $r = 5\times10^{-4}$m, $r = 10^3$ kg m^{-3}, g$= 10$ms^{-2}, T $=0.11$ Nm^{-1}, the radius of the drop when it detaches from the dropper is approximately:

(A) 1.4×10^{-3} m

(B) 3.3×10^{-3} m

(C) 2.0×10^{-3} m

(D) 4.1×10^{-3} m

Q.9 After the drop detaches, its surface energy is:

(A) 1.4×10^{-6} J

(B) 2.7×10^{-6} J

(C) 5.4×10^{-6} J

(D) 9.1×10^{-9} J

Q.10 The spring A reads 2 kg with a block m suspended from it. A balance reads 5 kg when a beaker with liquid is put on the pan of the balance. The two balances are now so arranged that the hanging mass is inside the liquid in the beaker as shown in the figure. In this situation: **(1985)**

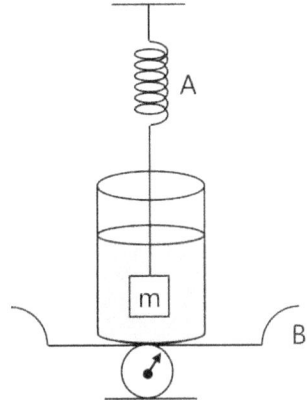

(A) The balance A will read more than 2 kg

(B) The balance A will read more than 5 kg

(C) The balance A will read less than 2 kg and B will read more than 5 kg

(D) The balances A and B will read 2 kg and 5 kg respectively.

Q.11 Two solid spheres A and B of equal volumes but of different densities d_A and d_B are connected by a string. They are fully immersed in a fluid of density d_F. They get arranged into the equilibrium state as shown in the figure with a tension in the string. The arrangement is possible only if **(2011)**

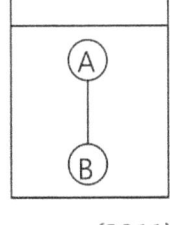

(A) $d_A < d_F$

(B) $d_B > d_F$

(C) $d_A > d_F$

(D) $d_A + d_B = 2d_F$

Q.12 A thin uniform cylindrical shell, closed at both ends, is partially filled with water. It is floating vertically in water in half-submerged state. If ρ_c is the relative density of the material of the shell with respect to water, then the correct statement is that the shell is – **(2012)**

(A) More than half-filled if ρ_c is les sthan 0.5

(B) More than half-filled if ρ_c is less than 0.5

(C) Half-filled if ρ_c is more than 0.5

(D) Less than half – filled if ρ_c is less than 0.5

Q.13 A solid sphere of radius R and density ρ is attached to one end of a mass-less spring of force constant k. The other end of the spring is connected to another solid sphere of radius R and density 3ρ. The complete arrangement is placed in a liquid of density 2ρ and is allowed to reach equilibrium. The correct statement(s) is (are) **(2013)**

(A) the net elongation of the spring is $\dfrac{4\pi R^3 \rho g}{3k}$

(B) the net elongation of the spring is $\dfrac{8\pi R^3 \rho g}{3k}$

(C) the light sphere is partially submerged.

(D) the light sphere is completely submerged.

Paragraph for Questions 14 and 15

A spray gun is shown in the figure where a piston pushes air out of a nozzle. A thin tube of uniform cross section is connected to the nozzle. The other end of the tube is in a small liquid container. As the piston pushes air through the nozzle, the liquid from the container rises into the nozzle and is sprayed out. For the spray gun shown, the radii of the piston and the nozzle are 20 mm and 1 mm respectively. The upper end of the container is open to the atmosphere.

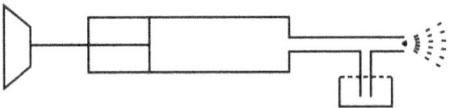

Q.14 If the piston is pushed at a speed of 5 mms^{-1}, the air comes out of the nozzle with a speed of **(2014)**

(A) 0.1 ms^{-1}

(B) 1 ms^{-1}

(C) 2 ms^{-1}

(D) 8 ms^{-1}

Q.15 If the density of air is ρ_a and that of the liquid ρ_ℓ, for a given piston speed the rate (volume per unit time) at which the liquid is sprayed will be proportional to
(2014)

(A) $\sqrt{\dfrac{\rho_a}{\rho_\ell}}$

(B) $\sqrt{\rho_a \rho_\ell}$

(C) $\sqrt{\dfrac{\rho_\ell}{\rho_a}}$

(D) ρ_ℓ

Q.16 A person in a lift is holding a water jar, which has a small hole at the lower end of its side. When the lift is at rest, the water jet coming out of the hole hits the floor of the lift at a distance d of 1.2 m from the person.

In the following, state of the lift's motion is given in List I and the distance where the water jethits the floor of the lift is given in List II. Match the statements from List I with those in List II and select the correct answer using the code given below the lists. *(2014)*

	List I		List II
1.	Lift is accelerating vertically up.	(p)	d=1.2 m
2.	Lift is accelerating vertically down with an acceleration less than the gravitational acceleration.	(q)	d > 1.2 m
3.	Lift is moving vertically up with constant speed.	(r)	d < 1.2 m
4.	Lift is falling freely.	(s)	No water leaks out of the jar

Code:

(A) 1 - q, 2 - r, 3 - q, 4-s

(B) 1 - q, 2 - r, 3 - p, 4 - s

(C) 1 - p, 2 - p, 3 - p, 4 - s

(D) 1 - q, 2 - r, 3 - p, 4 - p

Q.17 Two spheres P and Q of equal radii have densities ρ_1 and ρ_2, respectively. The spheres are connected by a massless string and placed in liquids L_1 and L_2 of densities σ_1 and σ_2 and viscosities η_1 and η_2, respectively. They float in equilibrium with the sphere P in L_1 and L_2 has terminal velocity $\vec{V}_P$ and Q alone in L_1 has terminal velocity $\vec{V}_Q$, then *(2015)*

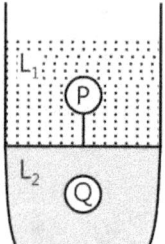

(A) $\dfrac{\left|\vec{V}_P\right|}{\left|\vec{V}_Q\right|} = \dfrac{\eta_1}{\eta_2}$

(B) $\dfrac{\left|\vec{V}_P\right|}{\left|\vec{V}_Q\right|} = \dfrac{\eta_2}{\eta_1}$

(C) $\vec{V}_P \cdot \vec{V}_Q > 0$

(D) $\vec{V}_P \cdot \vec{V}_Q < 0$

Q.18 A spherical body of radius R consists of a fluid of constant density and is in equilibrium under its own gravity. If P(r) is the pressure at r(r < R), then the correct option(s) is(are) *(2015)*

(A) $P(r = 0) = 0$

(B) $\dfrac{P(r = 3R/4)}{P(r = 2R/3)} = \dfrac{63}{80}$

(C) $\dfrac{P(r = 3R/5)}{P(r = 2R/5)} = \dfrac{16}{21}$

(D) $\dfrac{P(r = R/2)}{P(r = R/3)} = \dfrac{20}{27}$

Q.19 Consider two solid spheres P and Q each of density 8 gm cm^{-3} and diameters 1cm and 0.5cm, respectively. Sphere P is dropped into a liquid of density 0.8 gm cm^{-3} and viscosity $\eta = 3$ poiseulles. Sphere Q is dropped into a liquid of density 1.6 gm cm^{-3} and viscosity $\eta = 2$ poiseulles. The ratio of the terminal velocities of P and Q is *(2016)*

Important Questions

JEE Main/Boards

Exercise 1

Q. 7	Q.9	Q.15
Q.16	Q.20	

Exercise 2

Q. 1	Q.7	Q.9
Q.13	Q.17	

Previous Years' Questions

Q.8	Q.9	Q.10

JEE Advanced/Boards

Exercise 1

Q.3	Q.6	Q.9
Q.17		

Exercise 2

Q.1	Q.4	Q.10
Q.11	Q.19	Q.22

Previous Years' Questions

Q.7	Q.8	Q.9

Answer Key

JEE Main/Boards

Exercise 1

Q.2 Zero	**Q.6** $(2)^{2/3}\, v_T$	**Q.7** 2.714×10^{-9} m/s
Q.9 Streamline, turbulent	**Q.10** Turbulent	**Q.13** 60×10^8 Pa
Q.14 27.6 m	**Q.15** 9.56 N	**Q.16** 60 kg
Q.17 $72\pi \times 10^{-4}$ N	**Q.18** 0.0224	**Q.19** 3.5×10^{-2} Nm^{-1}
Q.20 1.785×10^{-4} m		

Exercise 2

Single Correct Choice Type

Q.1 C	**Q.2** B	**Q.3** D	**Q.4** D	**Q.5** A	**Q.6** A
Q.7 C	**Q.8** B	**Q.9** D	**Q.10** C	**Q.11** A	**Q.12** A
Q.13 C	**Q.14** C	**Q.15** D	**Q.16** C	**Q.17** B	**Q.18** D
Q.19 D					

Previous Years' Questions

Q.1 C   Q.2 A   Q.3 A   Q.4 B   Q.5 A   Q.6 C

Q.7 A   Q.8 D   Q.9 A   Q.10 B   Q.11 B   Q.12 D

Q.13 B   Q.14 C   Q.15 C   Q.16 D

JEE Advanced/Boards

Exercise 1

Q.1 $h = \dfrac{2m}{\pi}, H = \dfrac{11}{32\pi}m$ Q.2 19.6m, 4 sec.   Q.3 37.5 N

Q.4 (a) $9600\sqrt{2}$, (b) nH   Q.5 0.75   Q.6 $\sqrt{3} : \sqrt{2}$

Q.7 20N   Q.8 4.9 litre/min   Q.9 5 cm

Q.10 6.43×10^{-4} m³/s   Q.11 velocity 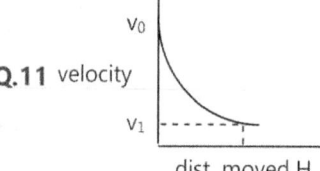

Q.12 20 m   Q.13 2.5 cm   Q.14 $\left(\dfrac{R}{24d}\right)^{\frac{1}{3}}$

Q.15 $r = \dfrac{r_1 r_2}{r_1 - r_2}$ Q.16 2.79 gm/cc

Q.17 $h_1 = \dfrac{2(3+\pi)}{15\pi} = 0.26$; $h_1 = \dfrac{3+\pi}{10\pi} = 0.195$ Q.18 2 sec., 1 sec

Q.19 $w = \sqrt{\dfrac{3g}{2L}\left(\dfrac{d_2 - d_1}{d_1}\right)}$ Q.20 1.9   Q.21 $R\sqrt{\dfrac{3}{2}}$

Q.22 $\dfrac{8h}{7}$

Q.23 $m_{min} = \pi r^2 \ell\left(\sqrt{\rho\sigma} - \rho\right)$; if tilted then it's axis should become vertical, C.M. should be lower than centre of buoyancy.

Q.24 2 : 1   Q.25 (a) $6\sqrt{2}$ m/s, (b) $9.6\sqrt{2} \times 10^{-3} M^3$ / sec, (c) 4.6×10^4

Q.26 $h_2 = 3h_1$ Q.27 $h = \dfrac{mg + 4Sa}{\rho_w a^2 g}$

Exercise 2

Single Correct Choice Type

Q.1 B   Q.2 D   Q.3 B   Q.4 D   Q.5 A   Q.6 A

Q.7 B   Q.8 B   Q.9 A   Q.10 D   Q.11 A   Q.12 C

Q.13 A   Q.14 C   Q.15 C   Q.16 C   Q.17 B   Q.18 D

Solutions

JEE Main/Boards

Exercise 1

Sol 1: Castor oil will come to rest first because its viscosity is greater than water

Sol 2: Acceleration is zero as velocity is constant

Sol 3: Flow rate is equal in any part of the body so

$A_1 V_1$ = constant

$\pi \left(\dfrac{D}{2}\right)^2 V$ = constant

Sol 4: Viscosity of gas increases with increase in temperature

Sol 5: For gas, viscosity of gases are independent of density and pressure but viscosity of gas increases with increase in temperature

For liquids:– Viscosity decreases with increase in temperature. Viscosity increase with increase in density viscosity of liquid is normally independent of pressure, but liquid under extreme pressure after experience an increase in viscosity

Sol 6: Volume remains same so

$2 \times 4/3\, \pi r^3 = 4/3\, \pi R^3$

$R = (2)^{1/3}\, r$

$V_T \propto r^2$

$\Rightarrow V'_T = k\, 2^{1/3}\, R^2 = (2)^{2/3}\, V_T$

Sol 7: r = 0.2 mm = 2×10^{-4} m

v = 4cm/s = 4×10^{-2} m/s

$F = 6\pi\, \eta\, rv$

$= 6\pi \times 1.8 \times 10^{-5} \times 2 \times 10^{-4} \times 4 \times 10^{-2}$

$= 6\pi \times 14.4 \times 10^{-11} = 2.714 \times 10^{-9}$ m/s

Sol 8: Refer Q-6 Exercise –I JEE Main

Sol 9: Critical velocity = $V_C = \dfrac{k\eta}{\rho r}$

k = Reynolds's number = 1000

$V_C = \dfrac{1000 \times 10^{-3}}{1000 \times \dfrac{1.25}{.200}} = \dfrac{1}{6.25} = 0.16$ m/s

Q = flow rate = 0.48L/min = $\dfrac{0.8}{100}$ L/sec = 8×10^{-6} m³/sec

Area = $\pi \left(\dfrac{1.25}{2}\right)^2 \times 10^{-4} = 1.227 \times 10^{-4}$

4.63

Velocity $= \dfrac{Q}{A} = \dfrac{8 \times 10^{-6}}{1.222 \times 10^{-4}}$

$= 6.5 \times 10^{-2}$ m/s

$V_1 < V_c \Rightarrow$ Streamline

When flow rate is 3L/min

$Q' = 3\text{L/min} = \dfrac{3}{60}$ L/sec $= \dfrac{1}{20} \times 10^{-3}$ m³/sec

$A = 1.227 \times 10^{-4}$

$V_2 = \dfrac{Q'}{A} = \dfrac{1 \times 10^{-3}}{20 \times 1.227 \times 10^{-4}}$

$= \dfrac{1}{2 \times 1.227} = 0.40$ m/s

$V_2 > V_c \Rightarrow$ turbulent flow

Sol 10: Refer Q – 9 Exercise–I JEE Main

Sol 11: Apparent weight of the floating block is zero.

Sol 12: Up thrust will be zero as body is not exerting any force on water during free fall and there is no buoyant force

Sol 13: Pressure $= \dfrac{F}{A} = \dfrac{60}{10^{-8}} = 60 \times 10^8$ Pa

Sol 14: $370 \times 10^3 = \rho gh + 10^5$

$\rho gh = (3.7-1) \times 10^5$

$h = \dfrac{2.7 \times 10^5}{9.8 \times 10^3} = 27.6$ m

Sol 15:

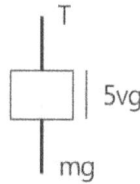

$T = mg - \rho_\omega vg$

$= 9000 \times 125 \times 10^{-6} \times 9.8 - 1200 \times 125 \times 10^{-6} \times 9.8$

$= 7800 \times 125 \times 10^{-6} \times 9.8$

$= 7.8 \times 125 \times 10^{-3} \times 9.8 = 9.56$ N

Sol 16: Change in depth corresponds to mass of man

$\rho \times 3 \times 2 \times \dfrac{1}{100} \times 10 = m \times 10$

$m = 1000 \times \dfrac{6}{100} = 60$ kg

Sol 17: Force $= 2\pi r\, S$

$= 2\pi \times \dfrac{5}{100} \times 72 \times 10^{-3}$

$= \pi \times 72 \times 10^{-4}$

Sol 18: $F = 2 \times$ perimeter $\times S$

$= 2 \times 4 \times \dfrac{1}{10} \times 28 \times 10^{-3}$

$= 8 \times 28 \times 10^{-4}$

$= 0.0224$

Sol 19: Pressure inside above atmospheric pressure

$\rho gh = \dfrac{4T}{r}$

$10^4 \times 8 \times 10^{-3} = \dfrac{4T \times 2}{3.5 \times 10^{-3}}$

$T = 3.5 \times 10^{-2}$ Nm⁻¹

Sol 20: $h = \dfrac{2T}{r\rho g}$

$r = \dfrac{2 \times 70 \times 10^{-3}}{8 \times 10^{-2} \times 10^4}$

$= \dfrac{70}{4} \times 10^{-5} = 1.785 \times 10^{-4}$ m

Exercise 2

Single Correct Choice Type

Sol 1: (C) Pressure due to difference in heights will be balanced by pressure due to 12 kg block

$\Rightarrow \rho gh = \dfrac{120}{800 \times 10^{-4}}$

$10^4 h = \dfrac{120}{800} \times 10^4$

$h = \dfrac{12}{80} = \dfrac{3}{20}$ m $= 15$ cm

Sol 2: (B)

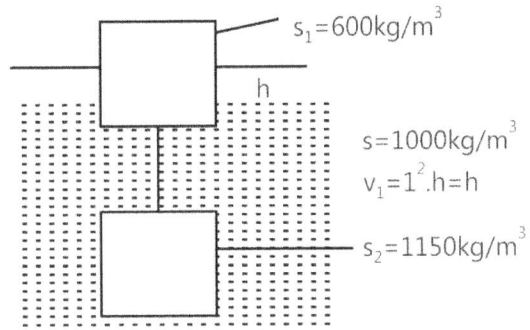

Downward force on the cubes = $(m_1 + m_2) g$

$= \rho_1 V g + \rho_2 V g$

$(1750) \times 10$

Upward force on the cubes = $\rho(V_1 + V) g$

$= 1000 (h + 1) \times 10$

Since cubes are in equilibrium

So $17500 = 10000 (h + 1)$

$1.75 - 1 = h$

$\Rightarrow h = 0.75$ m

Sol 3: (D)

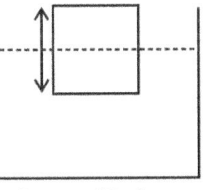 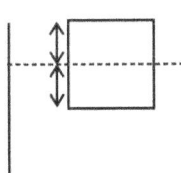

In equilibrium   Pushed down
                       by y distance

Initially in equilibrium

When pushed down by y distance, an extra upward force will act on the cube

$\rho (ybc) g = d \rho abc A$

[A = acceleration of the cube]

$y = \dfrac{da}{g} A \Rightarrow A = \dfrac{g}{da} y \Rightarrow \omega^2 = \dfrac{g}{da} \Rightarrow \omega = \sqrt{\dfrac{g}{da}} \Rightarrow T =$

$\dfrac{2\pi}{w} = 2\pi \sqrt{\dfrac{da}{g}}$

Sol 4: (D) $f \propto \sqrt{T}$, T = tension in the wire

in water frequency becomes $f/2$

$\Rightarrow$ Tension becomes ¼ of the initial

in liquid frequency becomes $f/3$

$\Rightarrow$ Tension become 1/9 of the initial

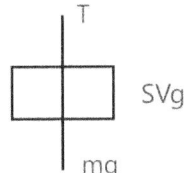

For water $\rho V g = \dfrac{3mg}{4}$

for liquid $d\rho V g = \dfrac{8}{9} mg$

$\Rightarrow d = \dfrac{8}{27} \times 4 = \dfrac{32}{27}$

Sol 5: (A) By work energy theorem

$W_{water} + W_{gravity} = \Delta KE = 0$

$W_{water} = (\rho V g) h$

$W_{gravity} = -(0.8 \rho V g) (h + 2)$

$\Rightarrow \rho V g h - 0.8 \rho V g (h + 2) = 0$

$h - 0.8 (h + 2) = 0 \Rightarrow \dfrac{5h}{4} = h + 2$

$\dfrac{h}{4} = 2 \Rightarrow h = 8m$

Sol 6: (A) The vertical component of tension balances out the net of weight & buoyancy.

Sol 7: (C) We know that time taken for the vessel to empty is $t_0 = \sqrt{\dfrac{2H}{g}}$, H = height of water

Time taken to empty vessel of height $\dfrac{H}{\eta}$ is t_2

$= \sqrt{\dfrac{2H}{g\eta}}$

$t_1 = t_0 - t_2$ and $t_1 = t_2$

$\Rightarrow \sqrt{\dfrac{2H}{g}} - \sqrt{\dfrac{2H}{g\eta}} = \sqrt{\dfrac{2H}{g\eta}} \Rightarrow \sqrt{\dfrac{2H}{g}} = 2\sqrt{\dfrac{2H}{g\eta}} \Rightarrow \eta = 4$

Sol 8: (B) Bernoulli's theorem is derived by the conservation of energy.

Sol 9: (D) Volume flow rate is same

So $\pi (1 \times 10^{-2})^2 \times 3$

$$= 100 \times \pi \left(\frac{0.05 \times 10^{-2}}{2}\right)^2 \times V$$

$$\pi \times 10^{-4} \times 3$$

$$= 100 \times \pi \times \tfrac{1}{4} \times 25 \times 10^{-8} \times V$$

$$V = \frac{4 \times 3}{25} \times 100 = 48$$

Sol 10: (C) We know that force exerted by fluid coming out on the container is $\rho A v^2$

v = velocity of fluid

$$v = \sqrt{2g \frac{H}{2}}$$

A = area of the hole

Acceleration of the tank $= \dfrac{\rho A v^2}{\rho (NAH)}$

$$= \frac{\rho (AgH)}{\rho NAH} = \frac{g}{N}$$

Sol 11: (A) $A_1 V_1 = A_2 V_2$

$$\pi (10^{-2})^2 V_P = \pi (2 \times 10^{-2})^2 V_Q$$

$$V_P = 4 V_Q$$

Sol 12: (A)

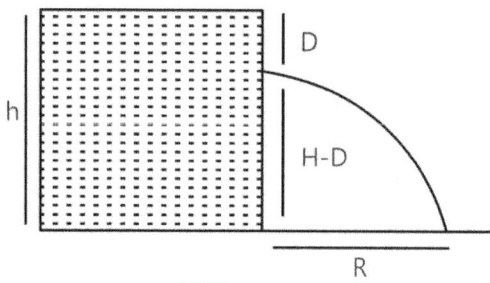

Velocity of water $= \sqrt{2Dg}$

Time taken by water to come to the ground

$$t = \sqrt{\frac{2(H-D)}{g}}$$

Distance where water hit the surface = vt

$$\sqrt{2Dg} \cdot \sqrt{\frac{2(H-D)}{g}} = 2\sqrt{D(H-d)}$$

Sol 13: (C) A = volume flow rate

$$= 10^{-2} \times 2 = 2 \times 10^{-2} \text{ m}^3/\text{s}$$

$$Q = A_1 v_1 = A_2 v_2$$

$$v_2 = \frac{Q}{A_2} = \frac{2 \times 10^{-2}}{1/2 \times 10^{-2}} = 4 \text{ m/s}$$

By Bernoulli equation

$$P_1 + \tfrac{1}{2} \rho v_1^2 = P_2 + \tfrac{1}{2} \rho v_2^2$$

$$8000 + \tfrac{1}{2} \times 1000 \times 4 = P_2 + \tfrac{1}{2} \times 1000 \times 16$$

$$10000 = P_2 + 8000$$

$$P_2 = 2000 \text{ Pa}$$

Sol 14: (C) Viscosity is not an assumption

Sol 15: (D) Frictional resistance $f \propto Av^2$

$$f = kAv^2 = k\pi r^2 v^2$$

k = constant

When ball acquires terminal velocity

$$f = mg$$

$$k\pi r^2 v^2 = mg$$

$$k \pi r^2 v^2 = (4/3 \pi r^3) \rho g$$

$$v^2 \propto r \Rightarrow v \propto r^{1/2}$$

Sol 16: (C)

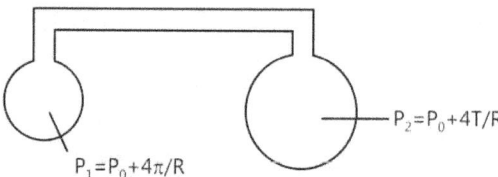

as $P_1 > P_2$ so air will flow out of the small bubble.

Sol 17: (B)

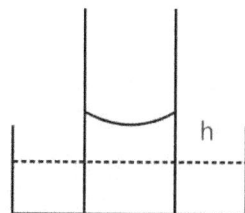

$$h = \frac{2T}{R\rho g}$$

R = radius of curvature

$$hR = \frac{2T}{\rho g} = \text{constant}$$

So graph of R vs. h will be hyperbola

Sol 18: (D) By Bernoulli equation

$P_0 + \rho g h_3 = P_0 + \tfrac{1}{2}\rho v^2$

$\tfrac{1}{2}\rho v^2 = \rho g h_3$

$P_0 = P_2 + \tfrac{1}{2}\rho v^2$

$P_2 = P_0 - \rho g h_3$

Sol 19: (D) By continuity

$A_x V_x = A_y V_y$

$A_x > A_y$

$\Rightarrow v_x < v_y$

By Bernoulli equation

$P_x + \tfrac{1}{2}\rho v_x^2 = P_y + \tfrac{1}{2}\rho v_y^2$

$v_x < v_y$

$\Rightarrow P_x > P_y$

KE per m^3 of water $= \tfrac{1}{2}\rho v^2$

$KE_x = \tfrac{1}{2}\rho v_x^2$

$KE_y = \tfrac{1}{2}\rho v_y^2$

$KE_x < KE_y$

Previous Years' Questions

Sol 1: (C) $W_{app} = W_{actual} -$ Upthrust

Upthrust $F = V_s \rho_L g$

Here, V_s = volume of solid,

r_L = density of liquid.

At higher temperature $F' = V'_s \rho'_L g$

$\therefore \dfrac{F'}{F} = \dfrac{V'_s}{V_s} \cdot \dfrac{\rho'_L}{\rho_L} = \dfrac{(1 + \gamma_s \Delta\theta)}{(1 + \gamma_L \Delta\theta)}$

Since $\gamma_s < \gamma_L$ (given)

$\therefore F' < F$ or $W'_{app} > W_{app}$

Sol 2: (A) Net force on the free surface of the liquid in equilibrium (from accelerate frame) should be perpendicular to it.

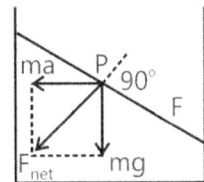

Force on a water particle P on the free surfaces have been shown in the figure. In the figure ma is the pseudo force.

Sol 3: (A) In a freely falling system $g_{eff} = 0$ and since,

Upthrust $= V_i \rho_L g_{eff}$

(V_i = immersed volume, ρ_L = density of liquid)

Upthrust $= 0$.

Sol 4: (B)

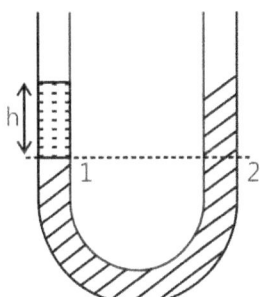

$p_1 = p_2 \Rightarrow p_0 + \rho_I g h = p_0 + \rho_{II} g h$

$\therefore \rho_I = \rho_{II}$

Sol 5: (A)

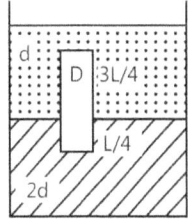

Considering vertical equilibrium of cylinder

Weight of cylinder = Upthrust due to upper

liquid+ upthrust due

to lower liquid.

$\therefore (A/5)(L) D_g = (A/5)(3L/4)(d)g$

$\qquad\qquad\qquad + (A/5)(L/4)(2d)(g)$

$\therefore D = \left(\dfrac{3}{4}\right)d + \left(\dfrac{1}{4}\right)(2d)$

$D = \dfrac{5}{4}d$

Sol 6: (C) From conservation of energy

$v_2^2 = v_1^2 + 2gh$ ......(i)

[can also be found by applying Bernoulli's theorem]

From continuity equation

$$A_1 v_1 = A_2 v_2$$

$$v_2 = \left(\frac{A_1}{A_2}\right) v_1 \qquad\qquad .....(ii)$$

Substituting value of v_2 from Eq. (ii) in Eq. (i)

$$\frac{A_1^2}{A_2^2} . v_1^2 = v_1^2 + 2gh$$

$$\text{or } A_2^2 = \frac{A_1^2 \, v_1^2}{v_1^2 + 2gh} .$$

$$\therefore A_2 = \frac{A_1 v_1}{\sqrt{v_1^2 + 2gh}}$$

Substituting the given value

$$A_2 = \frac{(10^{-4})(1.0)}{\sqrt{(1.0)^2 + 2(10)(0.15)}}$$

$$A_2 = 5.0 \times 10^{-5} \text{ m}^2$$

Sol 7: (A) Velocity of efflux at a depth h is given by v $= \sqrt{2gh}$. Volume of water flowing out per second from both the holes are equal.

$$\therefore \quad a_1 v_1 = a_2 v_2$$

$$\text{or} \quad (L^2)\sqrt{2g(y)} = \pi R^2 \sqrt{2g(4y)}$$

$$\text{or} \quad R = \frac{L}{\sqrt{2\pi}}$$

Sol 8: (D) l will decreases because the block moves up and h will decrease because the coin will displace the volume of water (V_1) equal to its own volume when it is in the water whereas when it is on the block it will displace the volume of water (V_2) whose weight is equal to weight of coin and since density of coin is greater than the density of water, $V_1 < V_2$.

Sol 9: (A) Applying continuity equation at 1 and 2, we have

$$A_1 v_1 = A_2 v_2 \qquad\qquad .....(i)$$

Further applying Bernoulli's equation at these two points, we have

$$p_0 + \rho gh + \frac{1}{2}\rho v_1^2 = p_0 + 0 + \frac{1}{2}\rho v_2^2 \qquad ....(ii)$$

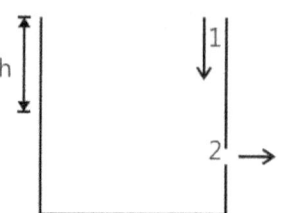

Solving eq. (i) and (ii), we have

$$v_2^2 = \frac{2gh}{1 - \dfrac{A_2^2}{A_1^2}}$$

Substituting the values, we have

$$v_2^2 = \frac{2 \times 10 \times 2.475}{1 - (0.1)^2} = 50 \text{m}^2 / s^2$$

Sol 10: (B) $\Delta p_1 = \dfrac{4T}{r_1}$ and $\Delta p_2 = \dfrac{4T}{r_2}$

$$r_1 < r_2$$

$$\therefore \quad \Delta p_1 > \Delta p_2$$

$\therefore$ Air will flow from 1 to 2 and volumes of bubble at end 1 will decrease.

Therefore, correct option is (B).

Sol 11: (B) When cylinder is displaced by an amount x from its mean position, spring force and upthrust both will increase. Hence, Net restoring fore = extra spring force + extra upthrust

$$\text{or } F = - (kx + Ax \rho g) \text{ or } a = - \left(\frac{k + \rho Ag}{M}\right) x$$

Now, f = $\dfrac{1}{2\pi} \sqrt{\left|\dfrac{a}{x}\right|} = \dfrac{1}{2\pi} \sqrt{\dfrac{k + \rho Ag}{M}}$

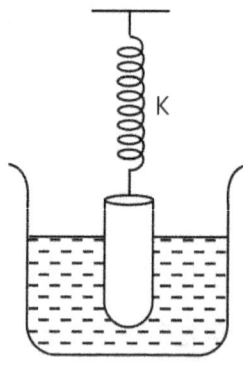

Sol 12: (D) The force of surface tension acting on the slider balances the force due to the weight.

$$\Rightarrow F = 2T \, \ell = w$$

$$\Rightarrow 2T(0.3) = 1.5 \times 10^{-2}$$

$$\Rightarrow T = 2.5 \times 10^{-2} \text{ N/m}$$

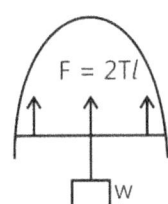
F = 2Tl

w

Sol 13: (B)

At equilibrium $\sum F = 0$

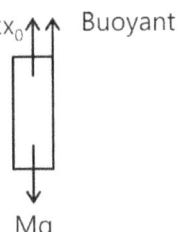
kx_0 Buoyant force

Mg

$$kx_0 + \left(\frac{AL}{2}\sigma g\right) - Mg = 0$$

$$x_0 = Mg\left[1 - \frac{LA\sigma}{2M}\right]$$

Sol 14: (C)

$$\rho 4\pi R^2 \Delta RL = T4\pi\left[R^2 - (R - \Delta R)^2\right]$$

$$\rho R^2 \Delta RL = T\left[R^2 - R^2 + 2R\Delta R - \Delta R^2\right]$$

$$\rho R^2 \Delta RL = T2R\Delta R \quad (\Delta R \text{ is very small})$$

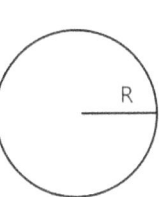
R

$$R = \frac{2T}{\rho L}$$

Sol 15: (C)

$$(76)(8) = (54 - x)(76 - x)$$

$$x = 38 \text{ cm}$$

Length of air column = 54 − 38 = 16 cm.

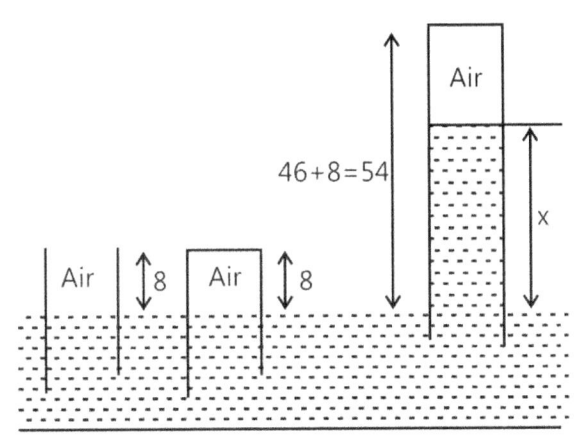
Air

46+8=54

x

Air 8 Air 8

Sol 16: (D)

$$(2\pi r T)\sin\theta = \frac{4}{3}\pi R^3 \rho_w \cdot g$$

$$T \times \frac{r}{R} \times 2\pi r = \frac{4}{3}\pi R^3 \rho_w g$$

$$r^2 = \frac{2}{3}\frac{R^4 \rho_w g}{T}$$

$$r = R^2\sqrt{\frac{2\rho_w g}{3T}}$$

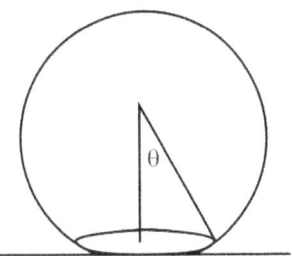
θ

JEE Advanced/Boards

Exercise 1

Sol 1: Pressure at A = $P_0 + \dfrac{Mg}{A} = P_0 + \rho g h$

$A = \pi\left[(0.04)^2 - (0.01)^2\right] = \pi \times 15 \times 10^{-4}$

$\dfrac{M}{A} = \rho h$

$h = \dfrac{M}{A\rho} = \dfrac{3}{\pi \times 15 \times 10^{-4} \times 1000} = \dfrac{2}{\pi}$ m

mass of water = 750 gm = 0.75 kg

mass of water below piston

= 0.75 − (1000)($\pi \times (0.01)^2$) × h

= (1000) × π × (0.04)² × H

$0.75 = \dfrac{\pi}{10} \times \dfrac{2}{\pi} + \dfrac{16\pi}{10} \times H$

$0.55 = \dfrac{16\pi}{10} H \Rightarrow H = \dfrac{5.5}{16\pi} = \dfrac{11}{32\pi}$ m

Sol 2: Net force on the ball in water

$F_B = \rho_w Vg - \dfrac{\rho_w Vg}{2} = \dfrac{\rho_w Vg}{2} = mg$

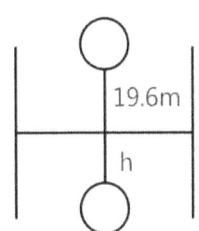
19.6m

h

Let us assume that ball will go up to depth h in water. By work energy theorem

−mg (19.6) + mg h = 0 ⇒ h = 19.6 m

Upward force F = mg in water

Acceleration = g

Time required to come on surface

$$= \sqrt{\frac{2s}{g}} = \sqrt{\frac{2 \times 19.6}{9.8}} = 2 \text{ sec}$$

Time required to go inside surface is also 2 sec

So total time required = 4 sec

Sol 3:

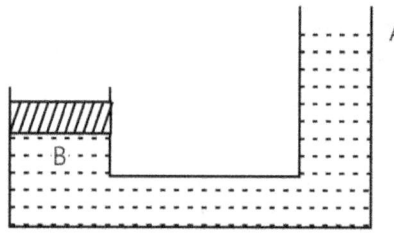

Pressure at point A = $P_0 + \dfrac{F}{A_1} = P_A$

Pressure at point B = $P_0 + \dfrac{mg}{A_2} = P_B$

Difference in pressure = $\rho g \times 8 = P_B - P_A$

$$\Rightarrow P_0 + \frac{F}{A_1} - P_0 - \frac{mg}{A_2} = -\rho g \times 8$$

$$\frac{F}{A_1} = \frac{6000}{800 \times 10^{-4}} - 750 \times 10 \times 8 = \frac{15}{2} \times 10^4 - 6 \times 10^4$$

$$\frac{F}{A_1} = \frac{3}{2} \times 10^{-4}$$

$$F = 1.5 \times 10^4 \times 25 \times 10^{-4}$$

$$F = 37.5 \text{ N}$$

Sol 4:

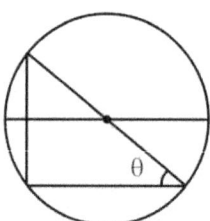

(a) $\tan \theta = \dfrac{a}{g} = 1$

$\theta = 45°$

Maximum gang pressure = $\rho \sqrt{a^2 + g^2}\ r$

$= 800 \times 10\sqrt{2} \times 1.2 = 9600\sqrt{2} \text{ N/m}^2$

(b) h = depth of lake

When bubble is at bottom pressure inside

$= P_0 + \rho g h$

When bubble is at surface pressure is $= P_0$

T_0 = surface tension

$$P_0 = \frac{2T}{2r} = \frac{T}{r}$$

$$P_0 + \rho g h = \frac{2T}{r}$$

$$\frac{T}{r} + \rho g h = \frac{2T}{r}$$

$$\rho g h = \frac{T}{r} = P_0 = \rho_m g h_m$$

$$\rho_m = n\rho$$

$$h = nH$$

Sol 5: Upward force on test tube initially $= \rho_s A \times 0.1\, g$

Upward force after adding liquid $= \rho_w A \times 0.4 g$

Weight of the fluid $= \rho_w A \times 0.4 g - \rho_w A \times 0.1 g = \rho' A \times 0.4\, g$

$$\Rightarrow \rho' = \frac{3\rho_w}{4}$$

Sol 6: At point A by Bernoulli equation

$$P_0 + \sigma g \frac{h}{4} = \frac{1}{2} \sigma v^2 + P_0 \Rightarrow v = \sqrt{\frac{gh}{2}}$$

Time take $= \dfrac{\sqrt{2(3h/4)}}{g} = \sqrt{\dfrac{3h}{2g}}$

Distance travelled = vt

$$= \sqrt{\frac{gh}{2}} \times \sqrt{\frac{3g}{2g}}$$

Distance $= \dfrac{h}{2}\sqrt{3}$

At point B

$$P_0 = \sigma g \frac{h}{2} = P_0 + (+2\sigma g\, (-h/4)) + \tfrac{1}{2}\, 2\sigma v'^2$$

$$\frac{gh}{2} = -\frac{gh}{2} + v'^2$$

$$V' = \sqrt{gh}$$

Time $t' = \sqrt{\dfrac{2(h/4)}{g}} = \sqrt{\dfrac{h}{2g}}$

Distance travelled $= vt' = \sqrt{gh}\,\sqrt{\dfrac{h}{2g}} = \dfrac{h}{\sqrt{2}}$

Ratio of distance travelled $= \dfrac{h\sqrt{3}}{2\dfrac{h}{\sqrt{2}}} = \dfrac{\sqrt{3}}{\sqrt{2}}$

Sol 7: Force exerted is change in momentum per sec

$= \dfrac{d(mv)}{dt} = v\dfrac{dm}{dt}$

$= v\rho Av$

$= \rho Av^2 = 1000 \times 2 \times 10^{-4} \times 100$

$= 20$ N

Sol 8: By Bernoulli's equation

$P_0 + \rho gh + \tfrac{1}{2}\rho v_1^2 = P_0 + \tfrac{1}{2}\rho v_2^2$

By continuity equation

$A_1 v_1 = A_2 v_2$

$v_1 = \dfrac{v_2}{2}$

$gh + \tfrac{1}{2}\left(\dfrac{v_2}{2}\right)^2 = \dfrac{1}{2}v_2^2$

$gh = \dfrac{v_2^2}{2}[1 - 1/4] = 3/8\, v_2^2$

$v_2 = \sqrt{\dfrac{8gh}{3}} \Rightarrow v_1 = \sqrt{\dfrac{2gh}{3}} = \sqrt{\dfrac{2}{3}}$

Volume flow rate $= \sqrt{\dfrac{2}{3}} \times 10^{-4}$ m^3/s

$= 60\sqrt{\dfrac{2}{3}} \times 10^{-4}$ m^3/min

$= 60\sqrt{\dfrac{2}{3}} \times \dfrac{1}{10}$ litre/min

$= 4.9$ litre/min

Sol 9:

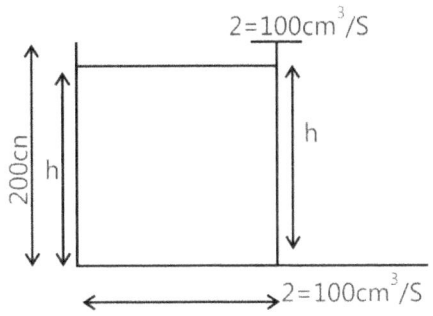

By Bernoulli's equation

$P_0 + \rho gh = P_0 + \tfrac{1}{2}\rho v^2$

$v = \sqrt{2gh}$

$Q = 100$ cm^3/s

$A = 1$ cm^2

$v = 100$ cm/s $= 1$ m/s

$v = \sqrt{2 \times 10 \times h} = 1$

$h = \dfrac{1}{20}$ m $= 5$ cm

Sol 10:

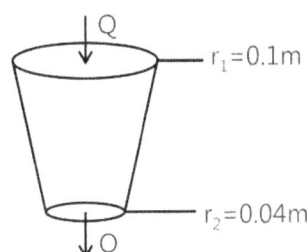

By Bernoulli's equation

$P_1 + \tfrac{1}{2}\rho V_1^2 = P_2 + \tfrac{1}{2}\rho V_2^2$

$-P_2 + P_1 = 10$ N/m^2

$10 + \tfrac{1}{2} \times 1250\, v_1^2 = \tfrac{1}{2} \times 1250\, v_2^2$ ....(i)

Continuity equation

$A_1 v_1 = A_2 v_2$

$\pi (0.1)^2\, v_1 = \pi (0.04)^2\, v_2$

$\dfrac{1}{100}\, v_1 = \dfrac{4 \times 4}{10000}\, v_2$

$v_1 = 0.16\, v_2$ .....(ii)

By (i) and (ii)

$10 + 625\,(0.16\, v_2)^2 = 625\,(v_2)^2$

$625\,(0.9744)\, v_2^2 = 10$

$v_2 = 0.128$ m/s

$Q = A_2v_2$

$= \pi (0.04)^2 \times 0.128$

$= \pi \times 16 \times 10^{-4} \times 0.128$

$= 6.44 \times 10^{-4}$ m³/s

Sol 11: Since velocity is greater than terminal velocity, so it will decrease until velocity reaches terminal velocity

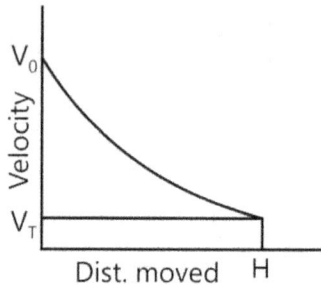

Sol 12: Terminal velocity $= V = \dfrac{2}{9} r^2 \dfrac{(\rho - \sigma)}{\eta} g$

$= \dfrac{2}{9} 10^{-8} \dfrac{(10^4 - 10^3) \times 9.8}{9.8 \times 10^{-6}}$

$v = 20$ m/s

Distance required to reach terminal velocity is

$h = \dfrac{v^2}{2g} = \dfrac{(20)^2}{2 \times 10} = \dfrac{400}{20} = 20$ m

Sol 13:

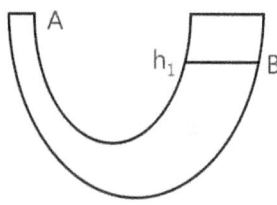

$P_A = P_0 - \dfrac{2T}{r_1}$

$P_B = P_0 - \dfrac{2T}{r_2}$

$P_B - P_A = \rho g h = 2T \left[\dfrac{1}{r_1} - \dfrac{1}{r_2} \right] = 2T [2000 - 200]$

$\rho g h = 2 \times 2 \times 7 \times 9 = 252$

$h = \dfrac{252}{10 \times 10^3} = 2.5$ cm

Sol 14: Surface energy of bubble $= 4\pi r^2 T_1$

Surface energy of drop $= 4\pi r_2^2 T_2$

Volume of bubble and drop is same so

$4\pi R^2 d = (4/3) \pi r_2^3$

$r_2^3 = 3R^2 d$

$r_2^2 = (3R^2 d)^{2/3}$

$\dfrac{P_1}{P_2} = \dfrac{4T}{R2T} \quad r_2 = \dfrac{2r_2}{R} = \sqrt[3]{\dfrac{24d}{R}}$

$\dfrac{P_2}{P_1} = \sqrt[3]{\dfrac{R}{24d}}$

Sol 15: $P_0 + \dfrac{4T}{r_1} + \dfrac{4T}{r} = P_0 + \dfrac{4T}{r_2}$

$\Rightarrow r = \dfrac{r_1 r_2}{r_1 - r_2}$

Sol 16: Let the volume of bearer be V

Then balancing force on beaker

$\rho_w V.g = (0.39)g + \rho_w \times 250\, g$

$10^3 \times V = 0.39 + 0.25$

$V = 640 \times 10^{-6}$ m³

$V = 60$ cm³

Volume of glass $= 640 - 500 = 140$ cm³

Density $= \dfrac{m}{v} = \dfrac{390}{140} = 2.785$ gm/cc

Sol 17: Plug will open when float is lifted upwards due to buoyant force

Balancing force we get

$\rho\, h' \times \pi (0.1)^2 \times g = 3g$

$h' = \dfrac{3}{1000 \times \pi \times \dfrac{1}{100}} = \dfrac{3}{10\pi}$

height $h = h' + 10$ cm $= \dfrac{3}{10\pi} + \dfrac{1}{10} = \dfrac{3 + \pi}{10\pi}$

Sol 18:

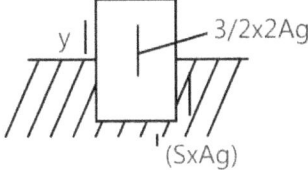

(a) By Newton's second law

a = Upward acceleration

$$\frac{-\rho}{2} \times 2Ag + \rho(1-y)Ag = (\rho/2)2Aa$$

$$-\rho Ag + \rho(1-y)Ag = \rho Aa$$

$$-g + g - gy = a$$

$$a = -gy$$

acceleration is density proportional to the displacement so it will perform SHM

$$a = -\omega^2 y$$

$$\omega^2 = g$$

$$\omega = \sqrt{g} = \sqrt{\pi^2} = \pi$$

Time period = $\dfrac{2\pi}{\omega}$ = 2 sec

(b) Time taken for rod to go from 1 extreme position to other is half of the time period

So time taken = 2/2 = 1 sec

Sol 19:

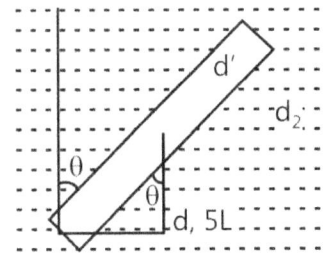

Torque on rod = $(d_1 \ell Sg - d_2 \ell sg) l/2 \sin\theta$

for small θ

$$\tau = (d - d_2)(\ell^2/2) s g \theta$$

Net torque = $I\alpha$

$$I\alpha = (d_1 - d_2)\ell^2 s g \theta$$

$$I = \frac{(d_1 s\ell)\ell^2}{3}$$

$$\frac{d_1 s \ell^3}{3}\alpha = (d_1 - d_2)\frac{\ell^2}{2}\rho g \theta$$

$$\frac{d_1 \ell \alpha}{3} = \frac{(d_1 - d_2)}{2}g\theta$$

$$\alpha = \frac{3(d_1 - d_2)g}{2d_1 \ell}\theta$$

$$\omega = \sqrt{\frac{3(d_1 - d_2)g}{2d_1 \ell}}$$

Sol 20:

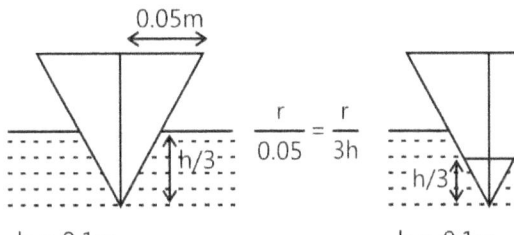

h = 0.1m   h = 0.1m

$$\frac{r}{0.05} = \frac{r}{3h}$$

h = 0.1m

By force balance

$$mg = 0.8\left(\frac{1}{3}\pi\left(\frac{0.05}{3}\right)^2\left(\frac{h}{3}\right)\right) \qquad \ldots\text{(i)}$$

When liquid is added

$$Mg + \rho\left(\frac{1}{3}\pi\left(\frac{0.05}{3}\right)^2\left(\frac{h}{3}\right)\right)$$

$$= 0.8\left(\frac{1}{3}\pi\left(\frac{0.05}{2}\right)^2\left(\frac{h}{2}\right)\right) \qquad \ldots\text{(ii)}$$

By (i) and (ii)

$$(0.8)\left(\frac{1}{3}\pi\frac{(0.05)^2 h}{3^3}\right) + \rho\left(\frac{1}{3}\pi\frac{(0.05)^2 h}{3^3}\right)$$

$$= 0.8\left(\frac{1}{3}\pi\frac{(0.05)^2 h}{2^3}\right)$$

$$\frac{0.8 + \rho}{3^3} = \frac{0.8}{8} = 0.1 \Rightarrow 0.8 + \rho = 2.7$$

$$\rho = 1.9$$

Sol 21:

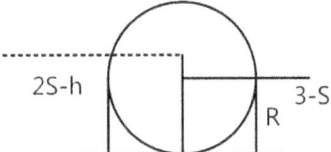

Balancing force on both sides

Horizontal force acting on the cylinder can be assumed to be acting on the cross–sectional area in the vertical direction

$$2\rho gh. \frac{h}{2} = \frac{3\rho gR.R}{2}$$

$$h^2 = \frac{3}{2} R^2$$

$$h = \sqrt{\frac{3}{2}} R$$

Sol 22:

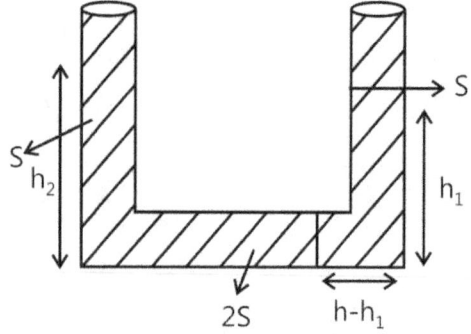

$h = h_2 - (h - h_1)$

$h = h_2 - h + h_1$

$h = \dfrac{h_1 + h_2}{2}$

$P_0 + \rho gh_1 + \rho a (h - h_1) + 2\rho a (h + h_1) - 2\rho g (h - h_1) - \rho gh = P_0$

$gh_1 + ah - ah_1 + 2ah + 2ah_1 - 2gh + 2gh_1 - gh = 0$

$h_1 (g - a + 2a + 2g) + h (a + 2a - 2g - g) = 0$

$h_1 = \dfrac{3(g - a)h}{3g + a}$

$a = g/2$

$$\Rightarrow h_1 = \frac{\frac{3h}{2}}{3 + \frac{1}{2}} = \frac{3h}{7}$$

$h_2 = h + h - h_1$

$h_2 = 2h - h_1$

Difference in height = $h_2 - h_1$

$= 2h - h_1 - h_1 = 2(h - h_1) = \dfrac{8h}{7}$

Sol 23:

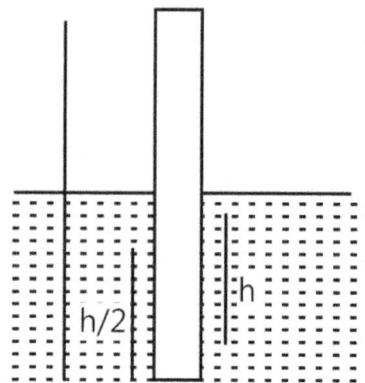

For the rod to be in equilibrium centre of mass of (rod + mass m) system should be below centre of gravity of the volume displaced by the rod.

For minimum m should coincide so.

Suppose h length of rod is below water then, by force balance

$\sigma(\pi R^2 h) g - \rho l \pi R^2 g - mg = 0$

$(\pi R^2) (\sigma h - \rho l) = m$ .....(i)

Reaction of centre of mass should be at h/2 distance from bottom

$$\Rightarrow h/2 = \frac{\frac{\rho l \pi R^2 \times l}{2}}{m + \rho l \pi R^2}$$ .....(ii)

$$\Rightarrow h^2 = \frac{\rho l^2}{\sigma} \Rightarrow h = l \sqrt{\frac{\rho}{6}}$$ .... (iii)

By (i) and (iii)

$m = (\sqrt{\sigma l} - \rho) \pi R^2 l$

Sol 24: Volume of water collected = A.V

A = cross sectional area perpendicular to the rain.

v = velocity of rain

in 1st beaker $A_2 = A_1 \cos 30°$

in 2nd beaker $A'_3 = A \cos 60°$

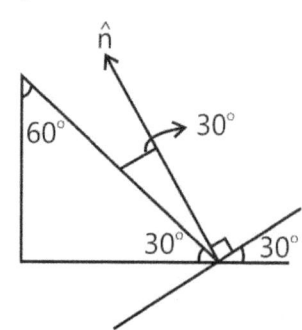

$A_1 = \dfrac{A}{\cos 30°}$

$\Rightarrow A_2 = \dfrac{A}{\cos 30°} \cos 30° = A$

$\Rightarrow A_3 = \dfrac{A}{2}$

So $\dfrac{Q_2}{Q_3} = \dfrac{A_2 V}{A_3 V} = 2 : 1$

Sol 25: (a) By Bernoulli's equation

$P_0 + \rho g(3.6) = P_0 + \frac{1}{2}\rho v^2$

$v = \sqrt{2g(3.6)} = \sqrt{72}$

$v = 6\sqrt{2}$ m/s

(b) Discharge rate $= \pi r^2 v$

$= \pi \times \dfrac{16}{\pi} \times 10^{-4} \times 6\sqrt{2}$ m^3/s

$= 9.6\sqrt{2} \times 10^{-3}$ m^3/s

(c) By Bernoulli equation

$P_A + \frac{1}{2}\rho v^2 + \rho g(5.4) = P_0 + \frac{1}{2}\rho V^2$

$P_A = P_0 - \rho g \times 5.4$

$= 10 \times 10^4 - 5.4 \times 10^4$

$= 4.6 \times 10^{-4}$ Pa

Sol 26: Pressure at C $= P_c$

By Bernoulli's equation

$P_0 + \rho g h_1 = P_0 + \frac{1}{2}\rho V_D^2$

$\frac{1}{2}\rho V_D^2 = \rho g h_1$

$P_0 + \rho g h_1 = P_c + \frac{1}{2}\rho V_C^2$

$A_C V_C = A_D V_D$

$A_C = \dfrac{A_D}{2}$

$\Rightarrow v_C = 2v_D$

$P_0 + \rho g h_1 = P_c + 2\rho V_D^2$

$\Rightarrow P_c = P_0 + \rho g h_1 - 4\rho g h_1 = P_0 - 3\rho g h_1$ ....(i)

Pressure at C can also be written as

$P_C + \rho g h_2 = P_0$

$P_C = P_0 - \rho g h_2$ ...... (ii)

By (i) and (ii)

$\rho g h_2 = 3\rho g h_1$

$h_2 = 3h_1$

Sol 27: By force equilibrium on the cube

$F_{gravity} + F_{buoyant} + F_{surface\ tension} = 0$

$-mg + \rho_w a^2 hg - S \times 4a = 0$

$h = \dfrac{mg + 4Sa}{\rho_w a^2 g}$

Exercise 2

Sol 1: (B) By Newton's second law

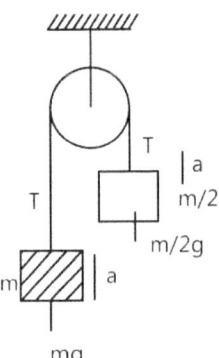

$T - \dfrac{mg}{2} = \dfrac{ma}{2}$

$T = \dfrac{m}{2}(g + a)$ ..... (i)

$-T + mg = me$ ..... (ii)

By (i) and (ii)

$\dfrac{-m}{2}(g + a) + mg = ma$

$\dfrac{-g}{2} \ \dfrac{-a}{2} + g = a$

$\dfrac{g}{2} = \dfrac{3a}{2}$; $a = \dfrac{g}{3}$

Effective acceleration of the bucket is $\left(g - \dfrac{g}{3}\right)$ downwards water pressure at the bottom above atmospheric pressure is

$P = \dfrac{2g}{3}h\rho = 1000 \times \dfrac{2}{3} \times 10 \times \dfrac{15}{100} = 1$ kPa

Sol 2: (D) Buoyant force = sum of all forces acting on the body

= force acting on the slant surface

+ force acting on the bottom surface

$F_B = F_s + F_b$

$F_B = (1/3) \pi R^2 H\rho g$

$F_b = \pi R^2 \rho g H$

$\Rightarrow F_s = (-2/3) \pi R^2 \rho g H$

Sol 3: (B)

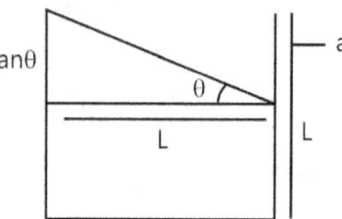

A = area of the base

$\tan \theta = a/g$

Finally 1/3 rd of the water spilled out

So volume of water spilled out finally

$= V_f = \dfrac{2\tan\theta \times A}{2} = \dfrac{L^3 \tan\theta}{2}$

this is 1/3 volume of L^3

$\Rightarrow \dfrac{\tan\theta}{2} = \dfrac{1}{3} \Rightarrow \tan\theta = 2/3 = a/g$

$a = 2g/3$

Sol 4: (D) Force applied by the liquid will be same on both the vessels as the mass of liquid is same in both the vessels

Sol 5: (A) Total force exerted on the base by water and cane's slant surface = mg

$= 1/3 \pi R^2 H\rho g$ downwards

Force exerted by the water =

$(\rho g H)(\pi R^2)$ downwards

So force exerted by the slant surface =

$2/3 \rho g H \pi R^2$ upwards

So force exerted by water on slant surface = $2/3 \rho g H \pi R^2$

Sol 6: (A)

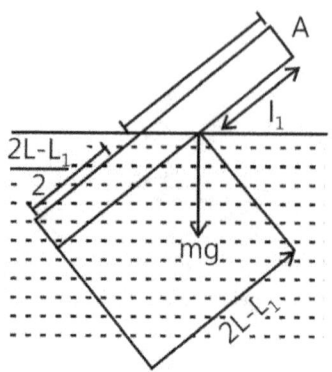

Let the length of rod that extends out of water is λ_1, since the rod is in equilibrium

So balancing net torque about point A

we get $(\rho A(2L-L_1)g)\left(\dfrac{2L+L_1}{2}\right) \cos\theta$

$= 0.75 \rho AL \, g \, L \cos\theta$

$\dfrac{4L^2 - L_1^2}{2} = \dfrac{3}{4}L^2$

Sol 7: (B)

By force equilibrium we get

$-Mg - 2Mg - mg + \dfrac{\rho Vg}{2} + \dfrac{\rho Vg}{2} = 0$

$\Rightarrow m = \rho v - 3M$ .....(i)

By torque equilibrium about mass M we get

$-mg(d - \ell) - 2\,Mgd + \dfrac{d\rho Vg}{2} = 0$

$m\ell - d\left(m + 2M - \dfrac{\rho V}{2}\right) = 0$

$\ell = \dfrac{d\left(2M + m - \dfrac{\rho v}{2}\right)}{m}$ ...(ii)

By (i) and (ii) we get $\ell = \dfrac{d(\rho V - 2M)}{2(\rho V - 3M)}$

Sol 8: (B) By work energy theorem

$W_{water} + W_{gravity} = \Delta KE = 0$

$(\sigma v g h) - \rho v g (h + H) = 0$

$\sigma h = \rho (h + H)$

$H = \dfrac{(\sigma - \rho)h}{\rho} = \left(\dfrac{\sigma}{\rho} - 1\right) h$

Sol 9: (A) Buoyant force = $\rho_w \times 4/3\ \pi R^3\ g$

Gravitational force = $(\sigma \rho_w)\ (4/3\ \pi(R^3 - r^3))\ g$

Sphere is in equilibrium so

$\rho_w\ 4/3\ \pi\ R^3\ g = (\sigma \rho_w)\ (4/3\ \pi(R^3 - r^3)\ g)$

$R^3 = \sigma(R^3 - r^3)$

$\dfrac{1}{\sigma} = 1 - \dfrac{r^3}{R^3}$

$\dfrac{r^3}{R^3} = 1 - \dfrac{1}{\sigma} = \dfrac{\sigma - 1}{\sigma}$

$\dfrac{R}{r} = \left(\dfrac{\sigma}{\sigma - 1}\right)^{1/3}$

Sol 10: (D) Force exerted = change in momentum per sec

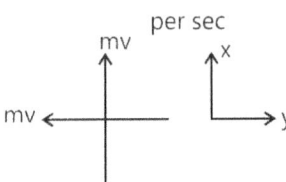

$= \dfrac{mv\hat{j} - mv\hat{i}}{t} = \rho L v\hat{j} - \rho L v\hat{i} = \sqrt{2}\ \rho\, v\, L$

Sol 11: (A) Force exerted by water = $\rho A V^2$

A = area of hole

V = velocity of water through hole

Friction force = $\mu\, Mg$

for the vessel to just move

$\rho A V^2 = \mu\, Mg$

$\rho \times \dfrac{\pi D^2}{4} \times 2\, g\, H = \mu Mg \Rightarrow D = \sqrt{\dfrac{2\mu M}{\pi\rho H}}$

Sol 12: (C) We know that force applied is proportional to velocity of shaft. So if the force is increased three times, velocity will also increase three times.

Sol 13: (A) Viscous force $F = -\eta\, A\, \dfrac{dv}{dx}$

$F = -\eta\, A\, \dfrac{v}{t}$

$F = mg \sin 37° = \dfrac{3mg}{5}$

$\eta = \dfrac{3mgt}{5AV} = \dfrac{3\rho a^3 gt}{5a^2 V} = \dfrac{3\rho agt}{5V}$

Sol 14: (C) Graph (c) best represents the motion of raindrop because velocity of rain approaches the terminal velocity.

Sol 15: (C) Graph (D) incorrect because at t = 0; x = 0 and graph will not be straight time

Sol 16: (C)

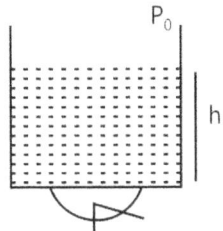

$P = P_0 + 2T/r$

$P_0 + 2T/r - \rho gh = P_0$

$\dfrac{2T}{r} = \rho gh$

$h = \dfrac{2T}{r\rho g} = \dfrac{2 \times 75 \times 10^{-3}}{\dfrac{10^{-4}}{2} \times 1000 \times 10} = 0.30\ m$

h = 30 cm

Sol 17: (B)

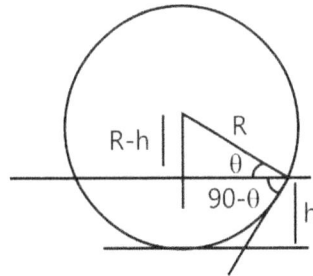

$\cos(90 - \theta) = \sin\theta = \dfrac{R - h}{R}$

Angle of contact = $90 - \theta = \cos^{-1}\left(\dfrac{R - h}{R}\right)$

Sol 18: (D) Force exerted by liquid = $\rho g\,(2h).A = F$

weight of liquid is W

Force exerted by liquid on walls = F – W (upwards)

So force exerted by the walls on the liquid

= (F – W) downwards

4.77

Multiple Correct Choice Type

Sol 19: (C, D)

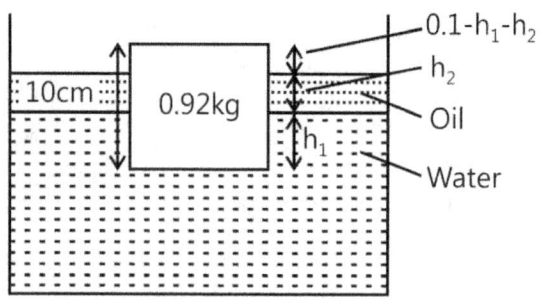

Balancing net force on the block we get

$-0.92 \times 10 + (1000) \times h_1 \times (0.01) \times 10$

$\qquad + (600) h_2 \times (0.01) \times 10 = 0$

$10\,h_1 + 6h_2 = 0.92$

if $h_2 = 4$ cm

then $10 h_1 + 6 \times 0.04 = 0.92$

$10 h_1 = 0.68$

$h_1 = \dfrac{0.68}{10} = 0.068$ m $= 6.8$ cm

$h_1 + h_2$ should be less than 10 cm so

$h_2 < 4$ cm

and $h_1 + h_2 = 10$ cm

$\Rightarrow 10\,h_1 + 6\,(0.1 - h_1) = 0.92$

$4h_1 + 0.6 = 0.92$

$4h_1 - 0.32$

$h_1 = 0.08$ m

$\Rightarrow h_1 = 8$ cm

$h_2 = 2$ cm

Sol 20: (B, D) Thrust exerted by the water is $\rho A V^2$ if velocity is doubled then thrust will increase 4 times.

Energy lost per second $= \frac{1}{2} \dfrac{dm}{dt} v^2$

$= \frac{1}{2}\, \rho A v \cdot v^2 = \frac{1}{2}\, \rho A v^3$

If velocity is doubled then energy lost per second will be 8 times

Sol 21: (A, C)

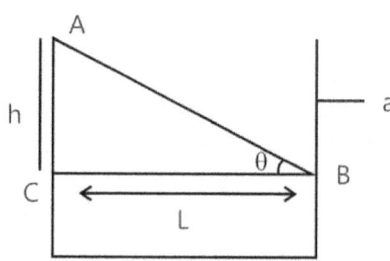

$P_A = P_0$

$P_B = P_0$

$P_B = P_A + \rho g h - \rho a \ell = P_0$

$gh = a\ell$

$\tan \theta = \dfrac{h}{L} = \dfrac{a}{g}$

Sol 22: (B, C)

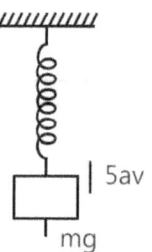

Balance A will read less than 2 kg as an upward buoyant force is acting on the block. Balance B will read more than 5 kg as downward reaction of the block due to buoyant force is acting on beaker.

Sol 23: (A, D)

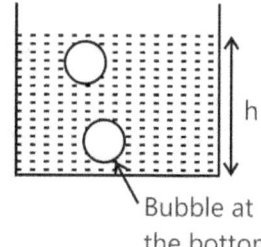

Pressure inside the bubble at the bottom is

$P_1 = P_0 + \rho g h + \dfrac{2T}{r_1}$

Pressure inside the bubble near the surface is

$P_2 = P_0 + \dfrac{2T}{r_2}$

Pressure inside the bubble near the surface is

$P_2 = P_0 + \dfrac{2T}{r_2}$

So pressure will decrease as we move upwards.

Sol 24: (A, C)

Velocity of fluid coming out of the hole =

$v = \sqrt{2gy}$

time taken by the fluid to collide with surface =

$t = \sqrt{\dfrac{2(h+h-y)}{g}}$

range = vt

$= \sqrt{2gy} \cdot \sqrt{\dfrac{2(h+h-y)}{g}}$

$R = \sqrt{4y(2h-y)}$. For maximum R, $\dfrac{dR}{dy} = 0$

$\Rightarrow \dfrac{1}{\sqrt{4y(2h-y)}} (2h - 2y) = 0$

$\Rightarrow y = h$

$R_{max} = \sqrt{4h^2} = 2h$

Assertion Reasoning Type

Sol 25: (B) Pressure of air decreases with increase in height so when pressure outside the balloon is equal to balloon pressure, it will not size up.

Sol 26: (D)

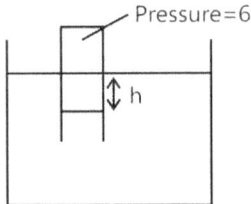

Pressure inside the tube is $P = P_0 + \rho gh$

When pressure changes height will also change.

So Statement-I is true.

Buoyant force is independent of atmospheric pressure.

Sol 27: (A) Suppose submarine is resting on the floor, then water is exerting only net downward force on the submarine as lower surface is not available for the upward force.

Sol 28: (D)

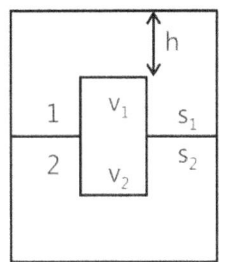

Force exerted by liquid – 1 = $(\rho gH + P_0)$ A downwards

So statement-I is false

Sol 29: (D) Coefficient of viscosity of gases increase with increasing temperature

Sol 30: (A) Free surface is always perpendicular to the g_{eff}. Liquids at rest can have only normal forces.

Previous Years' Questions

Sol 1: (D) $F_2 - F_1$ = upthrust

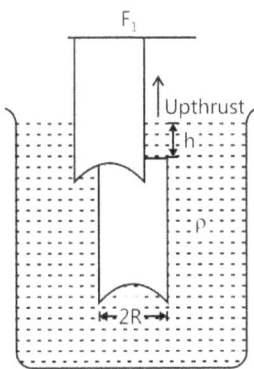

$\therefore F_2 = F_1$ + upthrust

$F_2 = (p_0 + \rho gh) \pi R^2 + V\rho g$

$= p_0 \pi R^2 + \rho g(\pi R^2 h + V)$

$\therefore$ Most appropriate option is (D).

Sol 2: (A) $k_1 = \left(\dfrac{\rho_{Fe}}{\rho_{Hg}}\right)_{0°C}$ and $k_2 = \left(\dfrac{\rho_{Fe}}{\rho_{Hg}}\right)_{60°C}$.

Here, ρ = Density

$\therefore \dfrac{k_1}{k_2} = \dfrac{(\rho_{Fe})_{0°C}}{(\rho_{Hg})_{0°C}} \times \left(\dfrac{\rho_{Hg}}{\rho_{Fe}}\right)_{60°C} = \dfrac{(1+60\gamma_{Fe})}{(1+60\gamma_{Hg})}$

Note: In this problem two concepts are used:

(i) When a solid floats in a liquid, then

Fraction of volume submerged $(k) = \dfrac{\rho_{solid}}{\rho_{liquid}}$

This result comes from the fact that

Weight = Upthrust

$$V\rho_{solid}g = V_{submerged}\,\rho_{liquid}g$$

$\therefore$ $\quad \dfrac{V_{submerged}}{V} = \dfrac{\rho_{solid}}{\rho_{liquid}}$

(ii) $\dfrac{\rho_{\theta°C}}{\rho_{0°C}} = \dfrac{1}{1+\gamma\cdot\theta}$

This is because $\rho \propto \dfrac{1}{\text{Volume}}$ (mass remaining constant)

$\therefore$ $\quad \dfrac{\rho_{\theta°C}}{\rho_{0°C}} = \dfrac{V_{0°C}}{V_{\theta°C}} = \dfrac{V_{0°C}}{V_{0°C}+\Delta V}$

$\quad = \dfrac{V_{0°C}}{V_{0°C}+V_{0°C}\gamma\theta} = \dfrac{1}{1+\gamma\theta}$

Sol 3: (B) Force from right hand side liquid on left hand side liquid.

(i) Due to surface tension force

$\quad\quad\quad = 2RT$ (towards right)

(ii) Due to liquid pressure force

$$= \int_{x=0}^{x=h}(p_0 + \rho gh)(2R.x)dx$$

$$= (2p_0 Rh + R\rho gh^2)\text{(towards left)}$$

$\therefore$ Net force is $|2p_0 Rh + R\rho gh^2 - 2RT|$

Sol 4: (C) Let

A_1 = Area of cross-section of cylinder = $4\pi r^2$

A_2 = Area of base of cylinder in air = πr^2

and A_3 = Area of base of cylinder in water

$\quad = A_1 - A_2 = 3\pi r^2$

Drawing free body diagram of cylinder

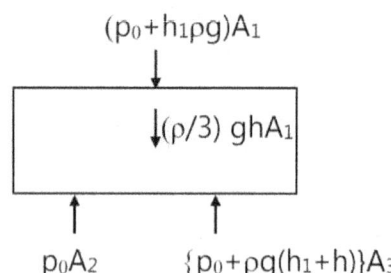

Equating the net downward forces and net upward forces, we get, $h_1 = \dfrac{5}{3}h$.

Sol 5: (A) Again equating the forces, we get

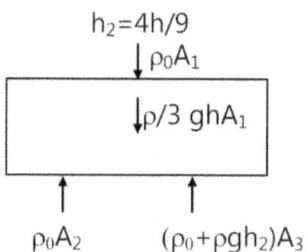

Sol 6: (A) For $h_2 < \dfrac{4h}{9}$, buoyant force will further decrease. Hence, the cylinder remains at its original position.

Sol 7: (C) Vertical force due to surface tension.

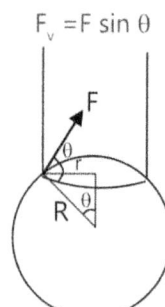

$= (T2\pi r)(r/R)$

$= \dfrac{2\pi r^2 T}{R}$

Sol 8: (A) $\dfrac{2\pi r^2 T}{R} = mg = \dfrac{4}{3}\pi R^3 . \rho.g$

$\therefore R^4 = \dfrac{3r^2 T}{2\rho g} = \dfrac{3\times(5\times10^{-4})^2(0.11)}{2\times10^3\times10}$

$\quad = 4.125 \times 10^{-12}\ m^4$

$\therefore R = 1.425 \times 10^{-3}\ m$

$\quad \approx 1.4\times10^{-3}m$

Sol 9: (B) Surface energy,

$\quad E = (4\pi R^2)T$

$\quad = (4\pi)(1.4\times10^{-3})^2(0.11)$

$\quad = 2.7\times10^{-6}\ J$

Sol 10: (B, C) Liquid will apply an upthrust on m. An equal force will be exerted (from Newton's third law) on the liquid. Hence, A will read less than 2 kg and B more than 5 kg. Therefore, the correct options are (B) and (C).

Sol 11: (A, B, D)

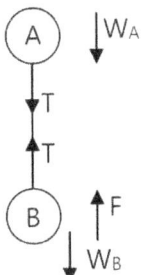

F = upthrust = Vd_Fg

Equilibrium of A

$$Vd_Fg = T + W_A$$

$$= T + Vd_Ag \qquad ....(i)$$

Equilibrium of B

$$T + Vd_Fg = Vd_Bg \qquad ....(ii)$$

Adding eqns. (i) and (ii), we get

$$2d_f = d_A + d_B$$

∴ Option (D) is correct.

From Eq. (i) we can see that

$$d_F > d_A \qquad (as\ T > 0)$$

∴ Option (A) is correct.

From equation (ii) we can see that,

$$d_B > d_F$$

∴ Option (B) is correct.

∴ Correct options are (A), (B) and (D).

Sol 12: (A)

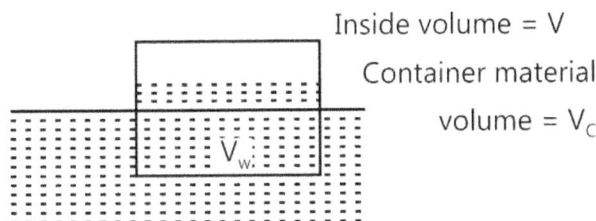

Inside volume = V
Container material volume = V_C

$$m_cg + m_wh = F_B$$

$$\rho_c V_c g + 1 V_w g = 1 \left[\frac{V}{2} + \frac{V_c}{2} \right] g$$

$$V_w = \frac{V}{2} + V_c \left[\frac{1}{2} - \rho_c \right]$$

If $\rho_c < \frac{1}{2}$; $V_w > \frac{V}{2}$

Sol 13: (A, D) At equilibrium,

$$\frac{4}{3} \pi R^3 2\rho g = \frac{4}{3} \pi R^3 \rho g + T$$

$$T = \frac{4}{3} \pi R^3 \rho g$$

$$\therefore \Delta \ell = \frac{4}{3k} \pi R^3 \rho g$$

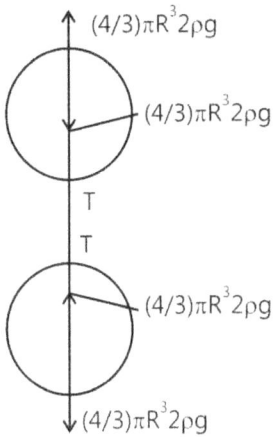

For equilibrium of the complete system, net force of buoyancy must be equal to the total weight of the sphere which holds true in the given problem. So both the spheres are completely submerged.

Sol 14: (C) By $A_1 V_1 = A_2 V_2$

$$\Rightarrow \pi (20)^2 \times 5 = \pi (1)^2 V_2 \Rightarrow V_2 = 2\ m/s^2$$

Sol 15: (A) $\frac{1}{2} \rho_a V_a^2 = \frac{1}{2} \rho_\ell V_\ell^2$

For given V_a

$$V_\ell \propto \sqrt{\frac{\rho_a}{\rho_\ell}}$$

Sol 16: (C) In P, Q, R no horizontal velocity is imparted to falling water, so d remains same.

In S, since its free fall, $a_{eff} = 0$

∴ Liquid won't fall with respect to lift.

Sol 17: (A, D) From the given conditions,

$\rho_1 < \sigma_1 < \sigma_2 < \rho_2$

$$V_P = \frac{2}{9}\left(\frac{\rho_1 - \sigma_2}{\eta_2}\right)g \text{ and } V_Q = \frac{2}{9}\left(\frac{\rho_2 - \sigma_1}{\eta_1}\right)g$$

So, $\dfrac{\left|\vec{V}_P\right|}{\left|\vec{V}_Q\right|} = \dfrac{\eta_1}{\eta_2}$ and $\vec{V}_P \cdot \vec{V}_Q < 0$

Sol 18: (B, C) $P(r) = K\left(1 - \dfrac{r^2}{R^2}\right)$

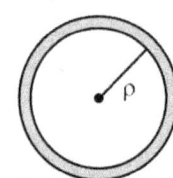

Sol 19: Terminal velocity $v_T = \dfrac{2}{9}\dfrac{r^2}{\eta}(\rho - \sigma)g$, where ρ is the density of the solid sphere and σ is the density of the liquid

$$\therefore \frac{v_P}{v_Q} = \frac{(8 - 0.8) \times \left(\frac{1}{2}\right)^2 \times 2}{(8 - 1.6) \times \left(\frac{1}{4}\right)^2 \times 3} = 3$$